Sex Isn't Real

Sex Isn't Real

The Invention of an Incoherent Binary

Beans Velocci

DUKE UNIVERSITY PRESS *Durham and London* 2026

Project Editor: Bird Williams
Designed by David Rainey
Typeset in Portrait Text Regular and Meshed Display by
Westchester Publishing Services

Library of Congress Cataloging-in-Publication Data
Names: Velocci, Beans, [date] author.
Title: Sex isn't real : the invention of an incoherent binary / Beans Velocci.
Description: Durham : Duke University Press, 2026. | Includes
bibliographical references and index.
Identifiers: LCCN 2025021625 (print)
LCCN 2025021626 (ebook)
ISBN 9781478033028 (paperback)
ISBN 9781478029595 (hardcover)
ISBN 9781478061779 (ebook)
Subjects: LCSH: Sexology—Research. | Science—Social aspects. | Transgender
people. | Gender nonconformity. | Gender identity. | Cisgender people. |
Feminist theory. | Queer theory.
Classification: LCC HQ60 .V45 2026 (print) | LCC HQ60 (ebook)
DDC 305.3—dc23/eng/20250616
LC record available at https://lccn.loc.gov/2025021625
LC ebook record available at https://lccn.loc.gov/2025021626

Cover art: Vallabh Soni, stock.adobe.com

for every trans person who is sick of reading the news

I offer you this warning: the Nature you bedevil me with is a lie. Do not trust it to protect you from what I represent, for it is a fabrication that cloaks the groundlessness of the privilege you seek to maintain for yourself at my expense. You are as constructed as me; the same anarchic womb has birthed us both. I call upon you to investigate your nature as I have been compelled to confront mine. I challenge you to risk abjection and flourish as well as have I. Heed my words, and you may well discover the seams and sutures in yourself.

—SUSAN STRYKER, "My Words to Victor Frankenstein Above the Village of Chamounix: Performing Transgender Rage" (1994)

There is no such biological entity as sex.

—FRANK LILLIE, introduction to *Sex and Internal Secretions* (1939)

Contents

A Note on Notes

This book contains both footnotes and endnotes. You'll find immediately relevant commentary at the bottom of the page, indicated by Roman numerals. The endnotes, indicated by Arabic numerals, contain suggestions for further reading of the "see also" variety, basic citations, and occasional notes that, despite their discursive nature, wound up at the back of the book because they mixed several of the above elements and made everything look too cluttered. It is only fitting that a book about a system that remains in use despite constant exceptions would employ just that kind of system for its notes.

Acknowledgments

———————

Writing this book was hard. Writing any book is hard, I think, but I wrote this one in the midst of an ever-worsening surge of anti-transness (see also climate change, fascism, genocide, pandemic, police violence, among others) and there were days and weeks and sometimes even months when the thought of confronting the source material and the manuscript was somewhere between nausea-inducing and impossible. I would not have been able to do it alone.

I'd like to begin by thanking every newspaper columnist, congressperson, evolutionary biology troll, children's author, and general malfeasant hiding behind and weaponizing incoherent sex—this book was largely fueled by spite, and you have collectively given this work of history tremendous contemporary relevance. Your contributions to the deconstruction of sex are duly noted.

More important than the aforementioned, though, are all of the people who have nurtured and cared for and thought (and laughed and cried) with me over the last many years. Liz Ault implied that twenty pages of acknowledgments would be excessive, but I hope you will understand this is the short version and it would be as voluminous as my discursive notes if possible.

There are too many people who have contributed to my development as a scholar to even begin to name—I'm grateful to each and every one of you. Thank you to Darcy Buerkle, Beth Clement, and Nadja Durbach, who first taught me how to be a historian. Joanne Meyerowitz, Joanna

Radin, and Greta LaFleur helped me cultivate the earliest iterations of this project and have continued to be steadfast mentors and true friends. I arrived at Penn still shocked that despite the carnage of the early COVID job market, I had wound up with stable employment after my March 13, 2020, campus visit went on indefinite hiatus. I'm grateful to all of my colleagues in History and Sociology and Gender, Sexuality, and Women's Studies who welcomed me to a new stage of life, and who continue to put up with me having opinions in faculty meetings. Particular thanks to Gwendolyn Beetham, Susan Lindee, Beth Linker, Melissa Sanchez, Elly Truitt, and Heidi Voskuhl for their mentorship and support, and to Kristen Ramsey and Riley Fortier for their administrative prowess and institutional knowledge. I'd like to also thank the Cohen Hall staff members whose labor makes this entire enterprise possible: in the business office, Chip Bagnall, Susan Cerrone, Hyemi Ghilardi, and Roxanne Ortiz; in IT, Rich King and Adam Podlaski; and in housekeeping, Maurica Blaylock, Glenn Butler, Seidy Ibrahim, and Madelin Miranda.

This project is indebted to archivists and librarians, whose expertise, guidance, and careful work undergird this entire book. Endless thanks to Liana Zhou, Shawn Wilson, and Bri Watson at the Kinsey Institute; Joe Di-Lullo, Cynthia Heider, Susan Laquer, Adrianna Link, Valerie-Ann Lutz, Estelle Markel-Joyet, Melanie Rinehart, and Paul Sutherland at the American Philosophical Society; Dominic Hall and Jessica Murphy at the Countway Library of Medicine at Harvard Medical School; Jennifer Walton at the Marine Biological Laboratory/Woods Hole Oceanographic Institution Archives; Clare Clark and Stephanie Satalino (who truly came to the rescue with a last-minute scan of what is now figure 2.2) at the Cold Spring Harbor Laboratory Archives; and Melissa Grafe at the Yale Medical Historical Library. Some of you have moved on from these posts, but you were there when I needed you! Thank you also to David Azzolina and John Pollack, who introduced me to the collections at Penn with great warmth. Generous funding from multiple sources enabled the travel necessary to do the aforementioned archival research: the Yale University Fund for Lesbian and Gay Studies, the John Money Fellowship for Scholars of Sexology at the Kinsey Institute, and the William T. Golden Fellowship at the American Philosophical Society. My thanks also to the University of Pennsylvania for grant funding in support of a manuscript workshop and publication subvention, and for a research account that covered access to published materials, image use fees, and costs of travel to conferences and workshops where I received feedback on this work.

Thanks to everyone who has engaged with the manuscript and helped me make it better at various stages. Attendees at workshops at Princeton University and Johns Hopkins University and panel audiences at History of Science Society, 4S, and American Association for the History of Medicine conferences asked the right questions at the right time. Dan Bouk read a draft of chapter 4 at a crucial moment and provided some of the most thoughtful and insightful comments and questions anyone could hope for; thank you, especially, for reminding me that queerness exceeds even the most regulatory of mathematical methods. Enormous gratitude to those who participated in my book manuscript workshop for your commitment to the text and for your expertise: Pearl Brilmyer, Kathy Brown, Colby Gordon, Beth Linker, Donovan Schaefer, and especially Jules Gill-Peterson and Sarah Richardson. Your feedback unquestionably made this book stronger and more cogent.

My work and intellectual well-being have been enriched by so many kind and brilliant scholars and thinkers. I've had powerful comrades in Colby Gordon, Emmett Harsin-Drager, Os Keyes, Scott Larson, Zavier Nunn, Nikita Shepard, Ketil Slagstad, and the growing numbers of scholars who are making trans history impossible to ignore. Kelsey Henry, Tess Lanzarotta, Ayah Nuriddin, and Miriam Rich have made conference attendance worth it, answered obscure historical questions at all hours, and given me a plethora of things to look forward to as we work to make historical spaces and scholarship more inclusive and equitable. Ambika Kamath has taught me so much about science and even more about demanding better from it. Perrin Ireland reminded me when I needed it most that this stuff can be fun. Emma Heaney gave me the platform and the impetus to turn a rant into a method. Ahmed Ragab has on more than one occasion seen the forest while I was counting leaves. Thank you to Henry Cowles for building an infrastructure with which to embrace the Unknown, and to everyone who constituted the Science Studies Beyond History Epistemology workshop (sanctuary? support group? séance? revival?) and its afterlives: Karen Darling, Steph Dick, Cathy Gere, Pablo Gómez, Alex Hui, Edward Jones-Imhotep, Julia Menzel, Taylor Moore, Joanna Radin, Marco Ramos, Myrna Perez, and Dora Zhang. That moment of otherwise has sustained me; more birdsong soon, I hope.

The publication process with Duke University Press—the dream all along—has gone absurdly smoothly. Much gratitude to Joshua Gutterman Tranen, who, back when he was at Duke, took notice of my work and emailed out of the blue to ask if I'd be interested in talking. Liz Ault has been a tremendous advocate for the book (and its footnotes!) and an absolute pleasure to work with. Ben Kossak has ensured that I don't lose track of the details and was especially integral in sorting out images. Two anonymous readers of the manuscript gave detailed, constructive feedback that enabled the book to attain its best possible form—I'm particularly grateful for their notes on weaving the chapters into a more cohesive narrative and for their palpable enthusiasm, which made the revision process infinitely more doable.

It turns out, contrary to my deepest hopes, that I'm neither a robot nor a brain in a jar, and I have needed a few people to keep things functioning. Thank you to Lane DiFlavis, who is rude and annoying and almost always right; to Rachel Brandstadter, who takes quality of life seriously; and to Daire O'Boyle, who gets me unstuck.

Gratitude as always to Mither and Mark—you've always accepted my tendency to do things the hard way, and you've got the car decals/T-shirts/hats/student loan help/Penn pens to prove it. I'm honored to be the subject of bragging to your friends (that is, thanks for always being proud).

This wouldn't be a queer acknowledgments section without a segment on chosen family, so here's to you, Kath Weston. Academic life has meant I'm in more long-distance relationships than I'd like, but they're made no less important by geographic circumstances. Thank you to Tam Bumpas, who has proven over a decade and a half that romantic friendship really does flourish at and beyond (historically) women's colleges; Nat Cohen, who's been around longest; Cat Dawson, the best brodent I've got; and Maria Murphy, blessed by nine priests. This is now Joanna Radin's third acknowledgment mention, which tracks given how frequently I ask myself what Joanna would do—thank you for helping me see my own wisdom and for the joy of being co-conspirators. Salonee Bhaman and Monique Flores Ulysses are a constant presence in my life via the little computer in my pocket, and I'm so grateful that we get to spend every day together from "good morning group chat" onward. I'm not sure what I'd do without your guidance, whether "yes, you should get takeout" or "you've done SO much work today, time to relax!" or "just because it's an actors' category doesn't mean you have to use it." I look forward to the day one of us turns out to be the heir(ess) to that mayonnaise fortune and we can afford a luxury retreat of our own. I'm also nourished by the love and care of dear friends in Philadelphia. Nat DiFrank, Moss Graves, and Sam Nasstrom are the brightest blossoms in the trans garden. Kristen Rosa-Houlihan, Molly Rosa-Houlihan, Ayla Taffel, and Andrew Watring became immediately and irrevocably kin. Angélica Clayton, Chris Chambers, and Anna Duensing, I'm so glad you're here with me. Thank you, all of you, for making this a place I feel rooted in. And thanks to the smallest ones, near and far—Rafa, Pili, Layla, and Lucy—whose incisive questions, fashion sense, and tiny-lunged-yet-righteous screams are an inspiration.

The traditional end-of-acknowledgments place of honor is shared. Molly Norris has stuck with me for a full twenty years, through teenage hair choices, various iterations of mental health and illness, multiple instances when we were "not dating," and too many cross-country flights to count. I couldn't have written this book without the deep sense of safety they emanate or their literal and metaphorical habit of taking roads marked "primitive—no warning signs" just to see where they lead. Caz Batten came into our lives when we least expected it, both conceptually and because it was at a Welcome New Faculty Zoom. They've given me the bravery to believe that sometimes good things really do happen, not to mention a number of extremely smart ideas about this book. When on several occasions finishing this project got to me with far more intensity than I thought possible (at one point I asked them, distraught, "what if sex is actually real?"), Caz offered patient reminders that trans embodiment is the realest thing of all. I'm perpetually in awe of our little communism of three and the life that we've decided to build together.

We had to say goodbye to Tater Tot, our sweetest goober, just as I was making final edits to the manuscript. This book is imbued with his influence. The Potato protected me from century-old ghosts and contemporary mail carriers through the entire writing process; he taught me about asking for help (he did not mind seeking assistance from people with thumbs), sharing my feelings (he let us know his emotional state by barking excitement and/or concern), and showing affection (he clearly believed that friends don't mind when you drool on them); and he made me stop typing and take breaks by nudging my arm away from my keyboard with his giant fuzzy head. I considered dedicating this book to him, but Tater Tot didn't remotely care or even know about sex/gender, and I think his twelve and a half years were better for it. May we all one day pay more attention to the squeaky tennis ball of our dreams than to any of the nonsense contained in this book.

Introduction

A Trans History of Classification

After nearly two decades of effort to understand and clinically manage transsexuality, Harry Benjamin opened his 1966 book *The Transsexual Phenomenon* with a reflection on the continued mystery of sex.[i] "There is hardly a word in the English language comparable to the word 'sex' in its vagueness and in its emotional content," he began the first chapter, titled "The Symphony of Sexes." The symphony, alas, was rather discordant. "It seems definite (male or female) and yet it is indefinite (as we will see)," Benjamin continued. "The more sex is studied in its nature and implications, the more it loses an exact scientific meaning."[1]

Benjamin was, as will become clear by the end of this book, wrong about a lot of things. He was a consummate medical gatekeeper, disgusted and annoyed by his trans patients, and he and his collaborators laid the groundwork for medical approaches that still make life harder for trans people more than fifty years later. For one shining moment, though, Benjamin got something right. He was fighting to make sense of sex in the wake of a century of scientific thought and practice that had produced multiple incoherent versions of it: sometimes a binary of male and female, sometimes a more expansive array; sometimes flexible, sometimes immovably

---

i   A discussion of what I mean by *sex* will follow. For now, think of something expansive.

static; defined by anatomy, by gametes, by gonads, by chromosomes.[ii] Benjamin could see the failure of science to develop a robust explanation of sex in the transsexual bodies before his eyes. To attempt to classify the sex of his patients who had modified their genitals and endocrine profile, who lived in the world as women even though legal documents insisted they were men, whose distribution of body hair or breast development was not what was expected for their karyotype, Benjamin had to contend with a system of sorting that presumed all of these traits went together, even as scientific knowledge increasingly demonstrated that they didn't. Benjamin would puzzle through this conundrum in the context of transsexuality, but transsexuals weren't the only people for whom sex consisted of multiple, often conflicting parts. He was ensnared in not just a transsexual phenomenon but a fundamental condition of sex science.

This book historicizes the mess of sex and how that mess was and remains central to the tenacity of sex as a classification system. Between the mid-nineteenth and mid-twentieth centuries, American researchers failed to get sex categories to cohere, and sex nonetheless continued to serve as the bedrock of both social arrangements and theories of life itself. The incoherence of sex—its multiple, coexisting, conflicting meanings—made it infinitely flexible in the face of evidence that the living world cannot be split so easily into male and female categories.[2] Scientists peered at and into the bodies of all kinds of organisms and, with their specialized powers of observation, made determinations about what was male and what was female with escalating specificity. New fields and institutions of study organized themselves into existence, and they accrued funding and institutional stature to match their increasing importance.[3] Scientists and clinicians alike staked their own claims to expertise on the rarified knowledges and techniques they possessed, which granted them insight into what they valued as one of the driving forces of life.[iii] Sexuality and its accompanying taxonomies—heterosexual, homosexual, invert, nymphomaniac, fetishist, fairy—became a defining feature of the self.[4]

ii   Throughout, I use *binary* and *male and female* roughly interchangeably. A discussion of *binary* as a useful analytic but ahistorical term appears later in this introduction.

iii   By the construction of expertise, I don't mean some kind of nefarious campaign. Rather, all science requires negotiation around what facts will be true, and which field or discipline will be the privileged site for knowing things about a given topic. See Shapin and Schaffer, *Leviathan*; Gieryn, "Boundary-Work"; Latour, "Give Me a Laboratory"; and Longino, *Science as Social Knowledge*.

This outward crystallization of categories and precision of definitions of sex, however, belies the persistent incoherence that scientists confronted and created in their efforts to stabilize the meanings of male, female, and sex itself. As scientists poked, prodded, and gazed, they encountered a vast quantity of evidence that did not show an obvious or stable division between male and female bodies, or a singular thing called sex at all. Exceptions and anomalies piled up, conflicting accounts of how sex functioned and what it was filled the leading journals, and researchers deployed wildly different definitions and practices for their investigations of sex. These contradictions did not undermine the idea that sex was knowable or the primacy of science as a way to know it. On the contrary, this book argues, the incoherent multiplicity of sex was a feature, not a bug. It was both opportunity and release valve, a means of ensuring near limitless claims to scientific innovation and a built-in dexterity that could cope with the range of its own otherwise threatening discoveries.

This book follows an interconnected cast of researchers across five sites that make the management of sexual incoherence particularly visible. It examines how zoologists navigated their encounters with animals that did not neatly fit into male and female categories even as they used the so-called natural world as fodder to theorize human racial hierarchy; how eugenic scientists simultaneously harnessed the malleability of sex while organizing their research according to a static binary; how a gynecologist's theory of sex as a matter of degree, not kind, clashed with his reluctance to deem any of his patients not women; how the statistical Kinsey studies calcified binary sex as a variable with which to understand the diversity of sexual behavior; and how the early years of American trans medicine's straightforward definition of the transsexual crashed into the anxiety of medical doctors convinced they might allow the wrong people to transition. In these spaces, sex amassed its power to sort bodies not from fixed, agreed-upon parameters, but from a tacit agreement that it could be multiple, often contradictory things at once. Sex has worked as a way of socially ordering the world with the cultural weight of science behind it because researchers could and did reclassify bodies, redefine categorical criteria, and reconstitute what they considered sex. As a result, unruly bodies could be recaptured back into normative male and female categories, with no need to question a binary sex system, the right of science to serve as a privileged site of sexual knowledge, or a racial hierarchy defined in part by sexual difference. Sex, it turns out, has very little to do with bodies. It's about the categories and who controls them.

Ripping the Seams and Sutures

This is a work of trans history.[iv] It interrogates the assumption that most people happen to fit into the binary sex categories they were assigned at birth, in contrast to a trans minority that does not. While one of its chapters does focus on trans medicine specifically, the forms of troublesome sex that are the book's primary target are neither referred to by historical actors as transgender (or transsexual or transvestites or inverts, or any of the many other historical terms for people whose sense of self or social role does not match their assigned sex) nor recognizable as related to contemporary definitions of transness.[5] I focus instead on human and non-human subjects whose bodily deviance was ultimately drawn back *into* normative categories rather than excluded from them. Doing so shows the tremendous amount of work that has gone into making it appear that "non-trans" forms of sex and gender are not just as constructed as trans ones. I aim to make that work visible and demonstrate that even if you don't think trans history is relevant to your work on gender, sex, or sexuality, it almost certainly is.[v]

Trans people and the concept of transness as a whole have, until quite recently, largely been left out of historical narratives.[vi] More urgently, contemporary denials of trans legitimacy and attempts to eject us out of

iv   This section is adapted from Velocci, "Denaturing Cisness."
v   My hope is this book will reach a range of audiences, including some who are well versed in science and technology studies (STS) or history of science but not trans studies, and vice versa. This section and the next go into the detail and context they do to provide an overview of the relevant historiographical issues from each of these fields and some conceptual and methodological translation.
vi   Trans people, of course, have been thinking about trans history for quite some time, with frequent references to historical precedent in transvestite and transsexual publications like *The Femme Forum* in the late 1960s, *Renaissance* in the 1970s, and *Our Sorority* in the 1980s. Even before that, sexological treatments of what would come to be known as transness, like Magnus Hirschfeld's *Transvestites* (1910) and Harry Benjamin's *The Transsexual Phenomenon* (1966)—both widely read by trans people—contained extensive reference to historical precedent. It took, however, quite a bit longer for trans history as such to coalesce. Leslie Feinberg's *Transgender Warriors* (1996), written for a popular press, is generally regarded as the first book-length study of trans history. Academic trans history would not get going until 2002 with the publication of Joanne Meyerowitz's *How Sex Changed*, and it did not take off in a sustained manner until the final years of the 2010s with the rapid-fire publication of Emma Heaney's *The New Woman*, Emily Skidmore's *True Sex*, and C. Riley Snorton's *Black on Both Sides*

public and epistemological existence have coalesced around the argument that transness is some kind of dangerous, newfangled trend.[vii] To counter this, trans people and academic historians have worked to show that even if named categories have shifted over time, people who might be considered trans today have a long history. This recovery of stories of pre-1950s figures stands to solve one problem by refuting a version of history in which trans people did not exist in the past. As someone who came to transness because of a history of sexuality class, I fully appreciate that this recovery work makes clear how trans people have been rendered invisible and that people have existed outside the bounds of normative gender, sex, and sexuality across time and place. This crucial work, however, is not enough on its own.

These histories hold fast to the idea that trans people have always existed as a small minority, while most people's sex and gender just happen to coincide. They project what we now call cisness—a match between sex assigned at birth and gender identity, the opposite of transness—onto the majority of people who have ever lived. The methods deployed to write about trans people before explicit trans categories, especially, imply that nearly everyone fits neatly into the sex and gender category they were assigned at birth. Most people, in this model, are presumed cis.[viii] At the

———————

in 2017, and Jules Gill-Peterson's *Histories of the Transgender Child* and Howard Chiang's *After Eunuchs* in 2018.

vii   The *New York Times* loves this vibe. But it's everywhere, from Ab*gail Shr*er, whom I'm not going to cite (elided like the torso of the trans person on her book cover!), to Samuel Alito's dissenting opinion in *Bostock v. Clayton County*, where he argued that trans people cannot be protected from employment discrimination because when Title VII was written, no one would have been understood as trans. Alito cited Joanne Meyerowitz, on whose behalf I remain outraged, to claim that while some people living in 1964 may have experienced gender dysphoria, "terms like [*transgender status* and *gender identity*] would have left people at the time scratching their heads"—never mind that *How Sex Changed* literally opens with the 1952 publicity surrounding Christine Jorgensen's transition. Like, it's ultimately way more egregious that he's a shitty transphobe, not to mention his role in trading *Roe v. Wade* for the ongoing catastrophe of restrictions on abortion, but come on, dude, 1964 is twelve years later than 1952! It's the first fucking sentence of the book!

viii   It's important to note that I don't use *cisgender* as an ontological state, but as an approximation using the most concise language presently available. I mean something more akin to the rather wordy "people who experience life as not-trans, whether materially or in terms of identity." Historically, something like "normal men and women" might be more accurate—which is so vague as to

same time that historians have fruitfully told trans stories, we have unintentionally naturalized a cis/trans binary that persists on its own over time. This book argues that cisness is not a natural state—instead, the idea that most people fit into their assigned sex category took a tremendous amount of work to construct and takes a tremendous amount of work to maintain.

Because trans history emerged as a subset of the history of sexuality, those studying transness in the past have largely followed its norms regarding the use of contemporary categories to describe historical actors, namely, don't refer to people using a category before that category existed.[6] In the case of trans history, that means (ostensibly) no "transvestites" before the 1910s, "transsexuals" before the 1950s and 1960s, or "transgender" people before the 1960s, or if you're being extra careful, before the word came into widespread use in the 1990s.[7] While there's debate about the precise use of these terms, and even more about whether pre-1950s figures like the "invert," "fairy," and "passing woman" should be considered trans, the idea that categories need to be narrowly historicized remains.[ix] Hewing closely to this disciplinary norm is one way that trans history has legitimized itself while having to insist on its objects of analysis as worth spending time and tenure lines on, particularly for historians

---

be useless. Part of the problem is that we don't yet have a robust conceptualization of how to talk about the many different ways of failing to fit into normative gender categories. A cis/trans binary is a wild oversimplification, but I still run into a different set of bureaucratic, medical, and social roadblocks than, say, my decidedly not trans mother, which makes cisness, though imperfect, a useful shorthand. We also have not yet sufficiently theorized the relationship between *binary* (i.e., male or female) and *static* (i.e., not changing categories); so far, that has resulted in more lateral violence about whether nonbinary people are really trans than a nuanced unpacking of how there are multiple ways to do gender wrong. Regulatory and knowledge systems in the United States and elsewhere demand both categorization as male or female *and* fixity within that category, and they punish or refuse to recognize various combinations of wrongness—this can happen for different reasons for violating different norms, all of it sucks, and I'm pretty confident the problem is actually cis people. *Cisgender*, as I use it, refers to a male or female sex classification assumed to be static in a system that penalizes anyone who doesn't do both (including people who don't identify as trans). Cf. Amin, "We Are All Nonbinary."

ix   This means that books like *Transgender Warriors* are not considered rigorous history.

who are themselves trans while navigating trans-exclusive or at the very least trans-disinterested universities.

In the process of displaying acceptable sorting processes, trans history runs up against the problem of who counts as a subject of trans history. This is particularly fraught when writing about people who lived before the invention of the aforementioned trans categories.[x] One can't simply run a keyword search for "transgender" in a nineteenth-century newspaper database; decisions must be made. As a result, historians often supply a performance of uncertainty regarding what transness might mean before the mid-twentieth century. Trans histories generally contain a paragraph, sometimes pages, explaining their use of terminology and how the author decided who to count as trans. If you read through these paragraphs in quick succession, it becomes clear that for all of the expressed anxiety about who to identify (or not) as trans, they're pretty much in agreement.

Processes of crossing and movement are key. The subject of trans history, especially in the pre-twentieth-century context, is someone who engages in "various forms and degrees of cross-gender practices and identifications," or who persists in a gender presentation other than the one they were assigned at birth, one that is, again, "cross-gender."[8] Emily Skidmore has argued that her turn-of-the-twentieth-century subjects can rightfully be called "trans men," because they "transitioned from the gender assigned to them at birth to the one with which they identified."[9] Jen Manion

---

x   Authors writing about the twentieth century seem less concerned with bounding the category "trans." An irony: Although much proverbial ink has been spilled over this, most trans histories have been written about periods before the mid-twentieth century. See, e.g., Boag, *Re-Dressing*; Sears, *Arresting Dress*; Skidmore, *True Sex*; and Manion, *Female Husbands*, and the recent uptick in pre- and early modern trans history (e.g., Gordon, *Glorious Bodies*, and LaFleur, Raskolinikov, and Kłosowska, *Trans Historical*). Meyerowitz's *How Sex Changed* and Gill-Peterson's *Histories of the Transgender Child* were, throughout most of this writing, the only monographs on twentieth-century trans history in the United States, joined in international perspective by Chiang's *After Eunuchs*. There have been a few additions in the latter stages of preparing this manuscript, like Avery Dame-Griff's *The Two Revolutions: A History of the Transgender Internet* and Gill-Peterson's *A Short History of Trans Misogyny* in 2024. Recent doctoral work by Emmett Harsin Drager and Os Keyes will hopefully soon add two more monographs to this count. Still, given the prolonged paucity of post-1950s trans history when so many sources are easily accessible and quite obvious, one wonders if the fuss over pre-1950s trans history is less a "methodology" problem and more a "lack of institutional support for trans history" problem.

proposes that the pre-twentieth-century "female husband" was "effectively a trans position," citing Susan Stryker's definition of *transgender* as referring to "people who move away from the gender they were assigned at birth."[10]

The practices that signify this movement also align across texts. Historians often search for identifying features and behaviors that correspond to something that looks like contemporary transness to locate transness in the past. Trans history thus tends to be populated by people who were arrested for cross-dressing, who appear in sensational accounts of "women masquerading as men," or who sought to make inhabiting their bodies more comfortable with physical interventions.[11] They may have changed their names and pronouns, worked a job that matched the gender they transitioned to within strictly gender-segmented labor markets, or had what at least looked like male/female romantic and sexual relationships (i.e., they were "masquerading as a man" while married to a woman).[xi] Frequently they make it into the historical record because they were outed, often through some kind of interaction with carceral regimes and medical or psychiatric institutions, or while being prepared for burial.[12] While it's the movement from one gender to another that constitutes transness, these other traits are evidence that movement has taken place.

The problem is this: Searching for a type of person who engages in certain behaviors unexpected for their assigned sex requires the formulation of and adherence to a classification system for transness. This book takes seriously foundational work in the study of classification that offers numerous cautionary tales about the consequences of cutting up the world. As Geoffrey Bowker and Susan Leigh Star put it in their pathbreaking *Sorting Things Out*, every classification system "valorizes some point of view and silences another."[13] That is, classification systems aren't neutral descriptions. They limit what it's possible to imagine and to do. Classification systems are made, made by people and institutions with their own interests and investments about who counts as what. They are often invisible infrastructures that surround us, that structure nearly every aspect of our lives: what diagnosis we get at the psychiatrist and how we're subsequently treated

---

xi   These are patterns in the literature, not explicitly stated criteria for classification as trans. In a representative example, in the opening of *True Sex*, Skidmore recounts the stories of two trans men who both changed their names, married women, frequented saloons, and earned respect for their hard work as men (1–3).

when it appears in our medical record; whether or not we're a member of a legally protected class; what toxicants are accepted as safe in the water we drink. In practice, classification systems rarely fit everyone, everything, or every eventuality. When people come into contact with classification systems, especially people who don't easily fit into the available categories, disjunctures between system and person often result in harm as lives must twist to fit categories that can never encompass the full range and complexity of existence. Classifying is a high-stakes enterprise.

The classification practices of trans history do just what Bowker and Star warn against. The traits that scholars use to find trans people in archival documents only work because they're supposed to be distinctive—even as many people in the past who cross-dress or work unexpectedly gendered jobs are not considered trans.[xii] The reliance on purportedly distinctive traits produces a view of the past in which some small number of people transgress gender to such a degree that they leave the category they were assigned at birth, but most people don't. Or to put it another way, as Susan Stryker does in *Transgender History*, being "'trans' is like being gay—some people are just 'that way,' though most people aren't."[14] This specific point of view—that trans people are "some minor fraction of the population"— assumes a happenstance, near-universal alignment between most sexed bodies and their gender identities.[15] If most people in the past are categorized as not-trans and don't require painstaking effort to locate, and trans people are a numerically small minority that need to be carefully searched for, then a cis/trans distinction seems natural and eternal. It hides how cisness has had to be constructed as a privileged way of being.[16]

No one is outright saying that everyone else has always been cis. After all, cisness only became a named concept in the mid-1990s, and it only began to see broader circulation in the early 2000s, so good historians would not call people in the past "cis." But that's the point: You don't have to bother saying anyone *wasn't* trans. "Cis," as Finn Enke has succinctly

xii   Sears's *Arresting Dress* illuminates this tension—Sears proposes a "trans-ing" analysis that "can reinvigorate and open up cross-dressing histories, without embracing every cross-dressing trace as indicative of a lesbian, gay, or transgender past" (6). On one hand, Sears's approach makes a very similar move to this book in resisting definitive categorization; on the other, it still implies that while everyone might cross-dress, some of those people are trans and some are not. Some women "used men's clothing . . . to challenge the limited social roles assigned to them," Sears writes, but "for the most part these women did not seek to become men" (64).

put it, "never needs to prove itself."[17] Herein lies the central contention of this book's methodological underpinnings. Most people *don't* just happen to fit the category they were assigned.[xiii] A whole apparatus churns along in the background, willfully ignored, to make it *seem* like they do. The power of cisness comes precisely from this hiding of its own invention, and it is there that trans history can intervene. Let me be clear—historians can argue for the importance of writing histories of trans people *and* refuse a cis/trans binary. As Eve Kosofsky Sedgwick put it, we need a "multipronged movement . . . whose minority-model and universalist model strategies proceed in parallel without any high premium placed on ideological rationalization between them."[18] Sedgwick discussed this coexistence of minoritizing and universalizing models of sexuality in terms of a fixed, distinct minority of homosexual individuals, on one hand, and a much more expansive model, on the other, in which bits and pieces of queerness attach themselves to a wider, more nebulous range of people of various identities. The same dynamic applies to simultaneously talking about trans people and the systemic incoherence of sex and gender that pertains to everyone. I take a universalizing approach to show how gendered power operates by sorting most people into normative categories even though they don't remotely fit that norm, and then pretending that sorting hasn't happened.

This book's approach to trans history builds on scholarship in trans history and historical work in trans studies, which have both shown the multitudinous possibilities of gender that have existed throughout history, and expanded the bounds of transness itself to encompass analyses of racial formation, the nonhuman, and opaque figures who resist easy sorting into a modern trans category.[xiv] The past is a very trans place, this

xiii   One is not born a woman, indeed. Simone de Beauvoir, Judith Butler, and a number of other feminist scholars have, over decades, established the constructedness of all gender, but the message seems to have gotten lost. The splitting of "gender" from "sex" soundly situated the body in easily identifiable, biologically based categories, in contrast to the mushy social and cultural stuff that feminist theory had the authority to make claims about.
xiv   I distinguish between trans history and historical work in trans studies to draw attention to somewhat divergent methods and especially divergent levels of institutional support. While the past few years have seen the publication of several new monographs on trans history, more trans history and trans studies courses, and a small handful of job ads for history and gender studies positions that mention trans history and/or studies as a preferred specialty for applicants,

work has told us.[19] I offer a corollary: History is far less cis than one might anticipate. I propose a model of trans history that focuses on the systemic absurdity of static and binary gender and sex classification. Such an approach addresses the history of structures that have produced transness as rare exceptions to a rule. Instead of responding to accusations of newness with assertions that we've always been here, this book presents the provocation that if you want to talk about newfangled subject positions, we need to talk about cis people.

Trans history already has the necessary tools to do so, and scholars have recently moved away from an outright distinction between trans and cis.[20] Scholars taking this approach have begun to outline how putting trans analysis together with questions of race, species, and age all throw the notion of well-constructed cis/trans *and* male/female binaries into disarray.[21] The recent collection *Feminism Against Cisness*, to which I contributed an expanded version of this section, is particularly exciting in this regard; the volume posits that for feminism to "address patriarchy without reifying the categories—woman and man" that it relies on, it must disavow cisness. The essays within treat cisness as a contingent political formation in service of white, colonial, bourgeois brutality. Even before this recent resurgence, there is a longer tradition of using trans analysis to demonstrate the constructedness of *all* sex and gender. In the late 1980s

---

it should be noted that there is far greater enthusiasm for trans studies *outside* of history departments than within them. See, e.g., several of the most recent works on the history of transness published by scholars working outside of history departments: Snorton, *Black on Both Sides* (now English, previously Africana studies); Gill-Peterson, *Histories of the Transgender Child* (English when the book was published); Heaney, *The New Woman* (English); and the essays in Chess, Gordon, and Fisher, "Early Modern Trans Studies" (English and various types of literature departments). This is not to say there is *no* trans history coming out of history departments in recent years. There certainly is: Skidmore, Manion, and Chiang are all publishing on trans history from history departments. Gill-Peterson and I traded places, maintaining the numbers despite my ascension to the tenure track: she is now based in a history department, while I have landed in a stand-alone history and sociology of science department. Regardless, it's striking to see how many of the most influential voices in trans history right now are working external to disciplinary history, and it is perhaps worth considering how this path of field formation has influenced what kinds of trans histories are being written. For a brief discussion of trans approaches being pushed out of disciplinary history, see Stryker, "Transgender History, Homonormativity, and Disciplinarity," especially pages 153–55.

and early 1990s, scholars argued that looking at the history of transness exposes how binary gender classification systems are made and how they unravel under scrutiny.

In 1987, Sandy Stone articulated this framing in her rebuttal of Janice Raymond's *The Transsexual Empire*, which had argued that "transsexualism" reinforced patriarchal gender stereotypes. Stone argued that it was doctors specifically who insisted that trans people adhere to gender stereotypes. This corrective, though, was not just an origin story for well-rehearsed narratives that Stone and others had been effectively forced to recount to access transition-related hormones and surgeries. "The origin of gender dysphoria clinics," Stone said of the institutionalization of requirements that trans people perform gender "correctly," "is a microcosmic look at the construction of criteria for gender."[22] Historicizing how trans women were assessed for femininity made clear the borders of gender norms writ large.

Susan Stryker built on this in "My Words to Victor Frankenstein Above the Village of Chamounix." Stryker, too, took aim against medico-scientific efforts to subsume trans experience into normativity: American doctors of the mid- to late twentieth century deigned to provide care only if trans people hid their transitions. Like Frankenstein's monster, Stryker spoke back to her medical "creators," explaining that she had exceeded attempts to produce normative womanhood through scientific expertise.[xv] "I offer you this warning: the Nature you bedevil me with is a lie," Stryker wrote, referring to the supposed "unnaturalness" of trans bodies. "Do not trust it to protect you from what I represent, for it is a fabrication that cloaks the groundlessness of the privilege you seek to maintain for yourself at my expense. You are as constructed as me. . . . Heed my words, and you may well discover the seams and sutures in yourself."[23] This is the promise of trans history: to trouble the line between "natural" cisness and "unnatural" transness and thereby make it abundantly clear that "non-trans" people's sex and gender are just as constructed as trans people's.

The sites of knowledge production I examine in this book are places where researchers and clinicians routinely encountered bodies that did not match paradigmatic forms of male or female and then had to decide what to do with them. Sometimes, those bodies were labeled as deviant,

---

xv   Subsequently, in an adjacent move, Meyerowitz and Gill-Peterson have examined how the development of "transsexuality" influenced the construction of "gender" at midcentury.

degenerate, and clearly other. More often, their existence was explained away as researchers shepherded them back into binary sex categories. The categories *male* and *female* became filled with bodies that did not match their definitions, and the definitions of male and female adapted in each sorting to receive their new contents. In other words, sex and gender categories do not simply fit or not fit; they are *made* to do so or not. Trans people, then, are not alone in exceeding sex and gender categories. People who don't imagine themselves as gender nonconforming and who are not viewed as such in their social worlds break the rules of sex and gender with shocking regularity, but many are either welcomed or forcefully yanked back into neat male and female categories.[xvi] If most people, according to a cisnormative historical imaginary, fit into the category they were assigned at birth, it's because the ways that the contents of those categories don't actually match what they're supposed to have been rendered invisible.

This is why I said earlier that even if you don't think trans history is relevant, it probably is. Given all the aforementioned work in trans history and trans studies, we are past the point where a rigorous study of gender can proceed as though the sex of only some bodies (i.e., trans bodies) has required negotiation, and that otherwise gender is the cultural baggage affixed to bodies that, for the most part, naturally fall into male or female categories.[xvii] Trans studies can no longer be imagined as distinct from gender studies; I'd even wager that studying gender *without* engaging trans studies can only produce partial understanding that ignores much of the most innovative work about gender currently happening. Considering the history of gender and sex while assuming that sex simply exists as an incidental fact of biology misses a crucial part of that history.

Take, for example, Sandra Eder's recent *How the Clinic Made Gender*. While Eder very successfully discusses the convoluted enactments of gender as simultaneously fixed and malleable among clinicians working on intersex and congenital adrenal hyperplasia, the book ultimately reinscribes the very sex/gender distinction that it seeks to historicize. Male, female, and intersex operate as prediscursive, stable objects, with the "normal" sexes and genders of clinicians assumed to require no manage-

---

xvi   Many, though, are not, and one of the underlying questions of this book is why some are punished for their deviance when for others, it can be overlooked.
xvii   The exception, of course, being scholarship on intersex, e.g., Reis, *Bodies in Doubt*, and Karkazis, *Fixing Sex*. Keep this in mind—it will be important later, especially in chapter 3.

ment. Because this makes the power dynamics of the actors' relationships surprisingly opaque, it's hard to tell *why* the clinic made gender. Engagement with recent work in trans studies and history on the construction of cisness suggests, though, that clinicians may have had their own investments and sense of normality affirmed as they constructed the idea of gender on intersex bodies.[24] The stakes of engaging with trans thought, then, are high for the history of science and medicine, as well as broader histories of the body, gender, and sexuality. "Sex" is necessarily part of analyses of the life and human sciences, sexual behavior, the construction of gender (on its own and in relation to labor, race, disability, and many other intersections), and countless other topics. Because it's one of the most basic ways that humans and nonhumans alike are categorized, its incoherence cannot be ignored.

Let us apply the same suspicion of subject categories that suffuses trans history equally to all forms of sex and gender. As Afsaneh Najmabadi has offered in provocation, we need to ask not only whether there were any lesbians in medieval Europe, but also whether there were any women. "That we ask the first question [about lesbians] with comfort," Najmabadi continues, "and presume the ease of the answer to the second (well, of course there were women, but defined differently) works on the presumption of naturalness of woman; that there have always been women."[25] The approach to trans history has been much the same: Of course there were trans people, but they were defined differently. Of course there were cis people, but they were defined differently. I want to reframe the conversation such that it's not only "there are more trans people in the past than we thought," but also "there are fewer cis people than we thought, and perhaps none at all." Which is to say, sure, "trans" is a historically contingent, invented category—but so what? "Cis" is a historically contingent, invented category. So is "male"; so is "female." It's invention all the way down.

Situating Knowledge

I've been told that some of the choices I've made in this book may not be in the best interest of me getting tenure. It's polemical, the footnotes are snarky, and it's equally invested in intervening in trans studies as it is in intervening in the history of science and science and technology studies (STS). Effectively, as many well-intentioned colleagues (who I'm sure are now grimacing in horror) have implied, the transness of it all is unsafe.

I therefore want to be explicit about where this book stands in relation to what Stryker has called "the micropolitical practices through which the radical implications of transgender knowledges can become marginalized."[26] Stryker notes that trans knowledges are often deemed "personal," less intellectually rigorous, and overly reliant on embodied, experiential knowledge, giving as an example the many occasions of her work being consigned to commentary sections of journals or regarded as popular rather than scholarly history. I've experienced something similar with accusations of bias against science and medicine (and against white men, in particular), advice to tread with extreme caution, and suggestions to prune references to my own political investments from my work.[27] There is, in there somewhere, a recognition of the transphobia of the academy and what it might mean for my career chances; yet, to achieve legitimacy by conforming to a thin slice of academic norms, I would necessarily have to leave out the particular trans insights and stakes that inform my work.

Instead, this book cashes the checks that feminist STS has been writing for decades. It is personal and political because all knowledge production is personal and political.[28] My experience as nonbinary and trans shapes the questions I ask and the conclusions I draw here because that is how knowledge works.[xviii] Histories of science that have been written by cis people are also shaped by individual experiences of sex and gender; they, too, are partial perspectives.[29] Knowledge production of all kinds depends on embodied practices, even when it is not addressed.[30] The question, as usual, comes down to who is considered a reliable knower, and by what standards.[31] Given the long history of trans people not being considered reliable knowers of ourselves, it comes as no surprise that our scholarship is suspect.[32] Part of my effort here is an insistence that knowledge about sex is not any less rigorous when it comes from a trans perspective (and might even be more so). Hence, a book that takes seriously its own trans perspective, as well as its entry into the world in the mid-2020s, when not just trans knowledge but trans life is under constant threat. In so doing, it unapologetically mobilizes feminist STS and what we already know about the politics of knowledge and classification.

xviii   An example of contradictory coexistence: I can both critique the category *trans* and also name myself as a trans person because what is useful to me in an analytic sense is not the same as what is useful to me in community membership and political legibility.

Bringing together the history of sexuality (especially queer history) and history of science/STS offers several additional methodological and intellectual interventions.[xix] These fields are poised at the edge of mutual benefit, but they mostly remain siloed from each other.[xx] Consequently, it's hard to find histories of sex that both attend to the specificity of category construction *and* do so in a way that isn't cis- and heteronormative.[33] Historians of science and STS scholars, on one hand, have extensively theorized classification and its enabling technologies.[34] Historians of sex and sexuality, on the other, have long been invested in the invention of nonnormative sexual categories, and they have paid close attention to how both sexologists and queer communities (and individual queer people) have come to know themselves as members of a distinct group.[35]

Here, I draw on both. I use methods derived from history of science and STS that privilege attention to on-the-ground practices of sorting and fact-making to build on decades of knowledge about categories of sex and sexuality. This enables me to track how sex categories coalesce and shift at a much more granular level than most histories of sexuality, which have largely—perhaps due to the borrowing of literary queer reading methods—approached sex science as a collection of texts, with less consideration of the networks and practices that produced those texts in the first place.[36] Turning to the practices of classification themselves, from paper-

xix Classification problems abound! I'm not interested in nitpicking the difference between history of science and STS. Increasingly, I'm feeling like the distinction is really just whether something is oriented toward social justice; if so, it's probably going to get labeled STS. I'm lumping them together here, partly because in drawing on both to do history of sex science, I don't think it particularly matters which is which (a hot take), and partly because despite all of these things ostensibly being my area of specialization, it remains unclear to me where anyone thinks my work falls. According to a very informal Twitter poll ($n = 17$), 23.5 percent of respondents said I do history of science, 29.4 percent said STS, 35.3 percent said history of sexuality, and 11.8 percent said queer and trans studies. Absolute chaos.

xx Scholars in the history of sexuality and trans studies are beginning to reach toward STS as an analytic: Gill-Peterson and Mak cite Annemarie Mol's foundational work on ontological multiplicity, and there is starting to be some overlap at the site of animality in trans studies (indicated by, e.g., the inclusion of chapters by STS scholars Mel Chen and Myra Hird in *Transgender Studies Reader 2*). On the whole, though, the fields remain separate, as indicated at the very least by the multitude of conferences I find myself having to go to in order to maintain a presence in both history of sexuality and history of science/STS.

work to funding structures to the relationship between abstraction and hands-on sorting, brings into sharp relief how sex categories have more to do with the power to classify than the bodies being sorted, and demonstrates the frequent divide between theory and practice in sex science. At the same time, I bring queer methods and concerns to the history of science. Building on decades of work in queer history and literary studies that have, with a careful eye, picked out the "queer presences and implications in texts that do not otherwise name them," as Siobhan Somerville has put it, I look for unintentional ends to which writers had no expectation of their work being used.[37] I do that in a different manner than the historians seeking traces of queer desire whose work informs mine—I look for failures of sex categories, rather than the presence of queer people.[38] Still, my propensity toward holding the archive upside down by its ankles to see what embarrassing scraps come out of its pockets is grounded in the reading practices that queer history taught me, applied here to the history of scientific knowledge production.

I also bring to the study of sex science a queer analytic. Histories of sex emerging out of history of science have tended to focus on the construction of binary sex.[39] Primarily, they have examined the patriarchal impulses that have led scientists to render women and men as fundamentally different from each other, usually for the sake of justifying women's (especially white women's) exclusion from political life and their social subjugation more broadly.[40] In essence, the field has thus far explained categories of maleness and femaleness always already conceived within a binary framework, with a goal of identifying how one category has been naturalized as better than the other rather than imagining liberation from the categories themselves. Even histories explicitly about the construction of a sex binary (or "two-sex model") itself concede that a binary was eventually successfully created.[41] I ask, instead, whether a binary ever coalesced, and I answer, not really.

While scientists in the period under study here often deployed distinct maleness and femaleness as research variables, and made social claims *as though* science wholeheartedly treated sex as a settled binary, a unified understanding of sex is hard to find in the historical record. What is often now referred to as "biological sex" is more of a cultural product than a scientific one.[xxi] "Biological sex" is far more conceptually unified

xxi A Google Ngram suggests that use of the term *biological sex* only really started to proliferate around 1970, increased steadily over the course of the

than what scientists themselves enact.[42] Yet, out of what I read as concerns of being taken seriously while making a feminist critique, historians of sex science who know biology is a construct nonetheless tend to defer to science as knowing the reality of sex. "I want to stress from the outset," Londa Schiebinger clarified in a foundational 1986 article on the history of visual renderings of "female" skeletal anatomy, a representative example of a broader tendency in histories of sex science, "that it is not my purpose to explain away physical differences between men and women but to analyze social and political circumstances surrounding the eighteenth-century search for sex differences."[43] With work like Schiebinger's having already laid the groundwork for disrupting biologized notions of essential, hierarchized differences between men and women, I do mean to suggest, to quote one of this book's epigraphs, "there is no such biological entity as sex."[44] Rather, as Geertje Mak has succinctly put it, "'Sex' is then not the physical thing, but the category to which a person belongs."[45] How something called "sex" came to function as though both humans and nonhumans are male and female because of some inherent physical state (and, therefore, men and women) is the central target of this book. The differences are the construct, and I'm here to explain them away.

This theoretical integrity of a category falling away upon examining actual scientific practice is by no means unique to sex science. Many other forms of knowledge production shift definitions, assume that objects rarely match prototypical ideals, and accept that the observed world is messier than their models.[46] Facts and objects of all kinds can be and often are incoherent and still do work—science, after all, is infrequently built on simple consensus.[47] Sex science differs, however, in the uneven distribution of harm that its incoherence causes. Sex science, imagined as an authority with access to a singular truth, informs who has access to medical care, job security, public space, a general sense of safety, and countless other quotidian needs. In the present, the violent, incoherent deployment of "biological sex" weighs most heavily on trans people, especially trans youth. So while sex science is not necessarily distinct in its knowledge production processes, their consequences distribute life chances particularly unequally.

1990s, and then rose precipitously from about 2012 through the present with only a brief leveling off around 2007 (potentially an artifact of the contents of Google Books).

While sex science has had particularly sweeping and dramatic effects, studies of historical (and contemporary) ontologies demonstrate the broader prevalence of multiple enactments of a given object. M. Murphy describes this analytic style as "accounts of how objects . . . came into being as recognizable objects via historically specific circumstances."[48] It builds on a central contention of STS: Facts are not facts because of things that really exist out there in the world, but rather become facts when people agree that they are true.[49] Likewise, discrete objects are not so because they exist, free-floating, waiting to be identified and described. Diseases, chemicals, species, and a whole array of other objects develop their existence *as things* as the result of contingent and specific practices. Simultaneously, as these things are enacted in multiple ways, they become an expanse of *different* things by the same name.[50] Combining historical ontology with the aforementioned attention to classification, STS is already quite methodologically queer in its dedication to questioning categorical stability far beyond sex and gender.

I take this mode of historical ontology as a point of departure to refuse the idea of sex as something that existed coherently before scientists began trying to figure out what it "really" was, and to consider it as an object that was and still is enacted differently by different people in different spaces for different reasons. I do not try to understand and re-create scientists' and doctors' assessments of where sex was located, or who was truly male or female, or who or what was hermaphroditic, or who was trans or cis—in essence, I do not imagine that there is a taxonomy that can be precisely understood. The logics I retrace are not the logics of successful delineation, but those of an often inchoate struggle to make knowledge out of complex bodies, reliant on tacit agreement to not look too closely at the contradictions contained within it.

There are two concepts worth parsing out before mashing them back together: multiplicity and incoherence. By "multiplicity," I invoke primarily the work of Annemarie Mol, who uses the example of atherosclerosis to show how objects are enacted through practices—they are constituted and re-constituted, depending on how people use and experience them. Mol demonstrates that the same thing can be enacted in different, often mutually exclusive, ways at different sites, and all of those ways really are *that thing*, not mere alternate perspectives on it. For the pathologist in the lab, atherosclerosis *is* a thickened arterial wall that can be observed and dissected under a microscope. For the clinician, however, atherosclerosis *is* the leg pain that a patient complains of.[51] In the process, Mol succinctly puts it, "reality multiplies."[52] I operationalize a similar approach to multiplicity here. Sex *is* multiple things.[53] There is no single object called sex; instead, many sexes exist simultaneously, each of them equally real.

I use *incoherence* to get at a particular aspect of multiplicity where enactments conflict and cannot be resolved.[xxii] While Mol's version of multiplicity contends with tensions between different enactments, it also envisions these alternate realities as interdependent and nonexclusive.[54] There is a certain internal logic, a consistency. Even when there is contestation and for practical reasons an internist must send a patient to a surgeon who enacts disease quite differently, actors tend to stick to their usual enactment.[55] Less so with sex. In the present study, we find researchers changing their enactment of sex on the fly and keeping a toe in multiple realities at once. Sex derives its power from both its multiplicity and its ability to contain direct contradiction that might otherwise produce conflict.

In this respect, sex aligns, in part, with Leigh Star and James Griesemer's explanation of boundary objects—objects that "inhabit several intersecting social worlds . . . *and* satisfy the information requirements of each of them." Their utility comes from their preservation of contradiction so that they can adapt to "local needs" even as they "maintain a common identity across sites," or in other words, so that scientists with divergent meanings can understand each other and work to shared ends.[56] Boundary objects enable what Star and Griesemer call the "fundamental tension of science": how "findings which incorporate radically different meanings become coherent."[57] Eventually, they're usually replaced by (attempted) standardization.[58] Sex, however, did not become coherent so much as settle into incoherence. Sex remained multiple in its local uses in service of the myth of its universality. Its findings, rather than becoming coherent over time, were ignored so as to avoid an integrated version of sex that would profoundly constrain scientists' idiosyncratic enactments. There is also an affective element in my imagining of "incoherence"—this book reads against the grain by taking seriously the moments when deployments of sex start to seem not just divergent but absurd, held together, notwithstanding its illogic, with so much metaphorical spit and paper clips.

What I term the incoherence of sex relies on a state of not-knowing that enables contradictory knowledge to exist.[59] The multiple, incoherent en-

---

xxii   To be clear, I don't mean incoherent as in incomprehensible. It's all quite comprehensible; it's just kind of ridiculous.

actments of sex in the period I focus on here coincided with a strong drive to structure social hierarchies, assumed biologically inherent, whether according to differences between men and women or to racial categories defined by differing levels of sexual dimorphism. For sex to be the basis of such social claims, it had to exist in multiple adaptable forms; yet, for one enactment of sex to avoid getting in the way of other contradictory enactments, vast swaths of sexual knowledge had to selectively be unknown.[xxiii] Exceptions to scientific rules—anomalies—became irrelevant rather than spurring a rethinking of knowledge systems as a whole.[xxiv] Jules Gill-Peterson has argued that by the 1950s "sex was in crisis" as a result of research on sexual plasticity in children, which had undermined the idea of sex as binary.[60] I argue that there had already been a multi-decade barrage of anomalies, some treated as noise and some made significant, almost all of which made professionalized sex research an important tool for solving these problems. Thus, an opening gambit of this book: Scientists have rendered the many exceptions to static, binary sex insignificant enough to keep them from overrunning the entire system.

I started this project with the romanticized idea that queer liberation would come from a rejection of classificatory structures. I still, in many ways, believe this: While ever-more-precise identity categories can have social and political utility, they also tend to demand frozen legibility, not to mention their vibe reminiscent of the table of contents of *Psychopathia Sexualis*.[xxv] They largely presume a preexisting set of categories, often biologized.[61] But incoherence is an incoherent thing. It can be part of a politics of refusal, of saying no to invasive questions and the need to prove

---

xxiii   Eve Kosofsky Sedgwick's remarks on "ignorance effects" in *Epistemology of the Closet* (4–8) apply: Not knowing, and incitements not to know, dictates the range of discursive possibility just as much as the creation of knowledge.
xxiv   See Kuhn, *Structure of Scientific Revolutions*, and Shapin and Schaffer, *Leviathan*, on how anomalies can be rendered noise or calibration errors in order to preserve an existing mode of thought. Kuhn, given his disinterest in reckoning with power and social forces, was an unlikely inspiration for this project. Anomalies, in his rendering, eventually pile up and can no longer be ignored and cause a crisis, which causes a paradigm shift. Meanwhile, I was finding sex anomalies everywhere and the stubborn persistence of binary sex despite them. Something, I realized, must have been preventing those anomalies from becoming meaningful, and here we are. Thanks, Tom!
xxv   Foucault, after all, warned at the end of *History of Sexuality* of the declaration of sexual identity enabling capture within the biopolitical regime of sexual knowledge production.

one's existence.[62] Embracing incoherence can be a way of mitigating the damage, or "torque," as Bowker and Star have framed it, caused by trying to cram lives into imposed classification systems.[63] An imagining of queerness as exceeding definition and impossible to pin down has been a rich analytic frame, both within the academy and outside of it as queer people—myself included—question narrowly bounded categories.[64]

At the same time, incoherence can be mobilized to cause harm and weaponized to enact violence with truly impressive agility, as many moments of this book show, as well as to avoid responsibility for solving problems.[65] Incoherence may produce feelings of risk that cannot be divorced from biopolitical decisions about who gets care and resources and who doesn't, who gets too much attention and who gets abandoned, and who is worth protecting from whom.[66] In the last few years, trans bodily plasticity, effectively a physical form of incoherence, has come under scrutiny by scholars concerned with how abilities to reshape bodies have, in part, their origins in the violence of slavery and eugenics.[67] Anti-transition rhetoric has relied substantially on the production of uncertainty to justify paternalistic "save the children" narratives.[68]

This book refuses the naturalness of sex categories by showing how scientists used incoherence to smooth over evidentiary conflict, make sex categories look natural, and establish science as the proper way to know things about them. To be clear, pointing out the use of incoherence by scientists is not a call for "better" science. While some STS scholars writing about sex have deployed a tactic of holding scientists accountable to their own standards of rigor, I am less interested in engaging with science on its own terms.[69] Nor do I want to appeal to some unified vision of what science—which has never had a unified vision of sex—says about natural diversity. I want, instead, to consider whether science can know sex without doing harm, if there can be a queer science beyond science *about* queerness, and what other methods we might imagine for knowing sex outside of science. The answers to these questions are outside the scope of this book, but I hope this can be a starting point for asking them.[xxvi]

xxvi   I've been teaching a Queer Science seminar the last few years, and depending on how it goes, sometimes students conclude that queerness and science are antithetical to each other, and sometimes they decide that hope for a

Finally, This Book

What follows traces studies of sex across zoology, eugenics, gynecology, statistical sexology, and early trans medicine. These knowledge spaces were tied together both conceptually and by a network of colleagues and institutions that, between the mid-nineteenth and mid-twentieth centuries, created a wide-ranging but well-connected domain of knowledge.[70] Crucial findings and theories about sex developed through these inter- and cross-disciplinary relationships and infrastructures. The same individuals show up again and again—the men (they were mostly, though not entirely, men) who studied sex in this moment saw themselves as a pioneering group on the cutting edge of science and respectability. They attended conferences together, they taught and learned from each other, they served on the same committees, and they stopped for dinner at each other's homes when passing through town. Between them, and sometimes even individually, they produced incoherent enactments of sex, often unintentional and unstated but necessary to their work. The tentacular reach of sex science through multiple domains is part of its power: Its incoherence emerged not only from different meanings of male and female and what sex was, but also from the deployment of different methods of knowing (and not knowing), epistemologically and in daily practice.

In the period under study, science became a privileged site for understanding sex, but its outsize cultural influence does not correspond particularly well to its role as only one of many manufacturers of sex. The law, state bureaucracy, and various social and cultural apparatuses likewise increased their interest in sex during this period, and they, too, created incoherent enactments. Sometimes they did so in conversation with science: For example, the boundaries between immigration policy and scientific investigations of sexual deviance were paper thin as efforts to prevent "public charges" from entering the United States integrated assessments of sex development.[71] Other times, science had little to do with it, as in the recognition of the third-sex "fairy" in working-class communities in New York City.[72] In these cases, an expansive range of sexed possibility enabled justification of exclusion in the former, providing a way to make sense of an obviously extant social role in the latter. Sometimes, clearly demarcated male

---

queer science remains. I remain agnostic: I'd love to believe it's possible to do science queerly but struggle to feel particularly hopeful about it.

and female categories served a greater use than more complex enactments. We'll see in chapters 2 and 4, for example, how enacting sex as binary and static frequently offered the path of least resistance for researchers processing large quantities of information.[xxvii] The matters of reproduction that most concerned agriculturists and heredity researchers also relied on a binarily sexed breeding pair of certain maleness and femaleness. Everyday usage, though, could undercut such stark categories. Life insurers, for example, classed applicants into two, stable sexes, intending to reflect divergent mortality rates between women and men. Yet using the same data, actuaries came to opposite conclusions about what sex meant for their bottom line, with some instituting policies that refused to insure or charged more for women, who apparently generated more risk than men, while others determined that women were in fact less likely to die in a given period and thus were cheaper to insure.[73] So, too, in sex science: Sometimes a binary was more useful, and sometimes it decidedly was not.

In this context, the researchers covered in the following chapters enacted sex. This book spans, roughly, the temporal bounds of the mid-nineteenth and mid-twentieth centuries, a century or so in which sex science began to coalesce into a legitimate topic of study and underwent considerable changes along the way.[74] Some scene-setting is therefore in order. On one end of that range, the study of sex in nonhumans presaged the development of sexology as historians typically consider it. In chapter 1, the United States was at the periphery of sex research. Historians have written extensively about the taxonomies of sexual personhood that had emerged across Western Europe around the 1870s and 1880s and subsequently proliferated throughout the first decades of the twentieth century. Especially in Germany and England, sexologists like Richard von Krafft-Ebing, Magnus Hirschfeld, and Havelock Ellis began using case studies of increasingly pathologized forms of sexual being to understand their etiologies and manifestations. Some were motivated by a desire to reduce sodomy's legal penalties, while others sought better scientific bases for social hygiene, colonial power, and various biopolitical regulatory schemes.[75] In the United States, however, research on sex had another trajectory: Expertise in sex and reproduction first accrued to the domains of zoology and agricultural science, supplemented by

---

xxvii I don't mean this in a technologically determinist sense; rather, researchers used a simple, commonsense approach rather than spend time and energy figuring out how to do something else.

European knowledge. Most American researchers, coming from both formal and amateur backgrounds, didn't conceptualize their work as sex science per se, and their research rarely centered on humans (though that knowledge often supported eugenic and racializing goals).[76] Their primary concerns were improving agricultural yields and quality, or contributing to a growing body of scholarship in the life sciences—also with a European center of gravity—that sought to understand the anatomy and physiology of sex.[77]

Chapter 1 turns to zoology and animal husbandry texts rather than the traditional source base of human sexology to explore the use of *sex* in this period. Unlike the rest of the book, it focuses exclusively on published sources to provide a grounding sense of where conversations and debates stood before the institutional consolidations that mark later chapters. These studies of animals produced two models whose contradictions would trouble sex research for over a century: In one, scientists could shore up their authority by identifying the "true" male or female sex of animal specimens; another framed hermaphroditism in "lower" organisms as more common than separate sexes. The former articulated stark differences between white women and men, while the latter supported theories of racial hierarchy based on degrees of sexual dimorphism. These parallel understandings of sex—sex was limitlessly knowable in a binary that allowed nothing outside of it, and also exceptions to that binary were constant and threatening—remained in tension throughout the period. This chapter close-reads nineteenth-century research on three problem animals that each show a facet of the struggle to make meaning out of sexual variation. Scientists established their own expertise by constructing hyenas' sexual morphology as a mystery; made sex itself malleable and sex organs effectively interchangeable in disagreements about freemartins; and added to confusion about what counted as maleness and femaleness in trying to reclassify worker ants and bees as female while popular science sources framed these insects' three-sex system as more advanced than sexual dimorphism. The failure of this research to successfully produce a stable binary stoked antimiscegenationist fears about a collapse of whiteness into animality. This chapter takes a broad approach compared to the deep anchoring in a particular individual or institution that follows in subsequent ones in order to highlight the multiplicity of sex manifesting across genres, fields, and methodologies, and its inseparability from evolutionarily informed racial politics. Sex science thus emerged not out of whole cloth and suddenly in the late nineteenth century but in fragments from studies of animality and race that undergirded colonial and white supremacist thought and expansion.

These connections wound tighter from the first decade of the twentieth century through the 1930s, as American sexology institutionalized through and alongside eugenics, now supported by philanthropic money and, by the mid-1920s, federal funding augmented by private fortunes. Agriculture continued to generate knowledge about better breeding, and its knowledge was supplemented by the increasingly professionalized fields of psychiatry, psychology, and social work, as well as new findings in genetics, Mendelian evolution, and biometry. Grounded in the first decades of the twentieth century, chapter 2 looks at conflicting approaches to sex research in two prominent laboratories run by Charles Davenport in Cold Spring Harbor, New York: the Station for Experimental Evolution (SEE) and the Eugenics Record Office (ERO). The former saw sex as malleable and viewed this malleability as something to be manipulated for eugenic gain, while the latter employed a binary framework that supported its studies of heredity. While researchers at the Station like Oscar Riddle and Albert Blakeslee worked on projects about sex reversal and sex differentiation, ERO fieldworkers mapped the heritability of desirable and unwanted traits through reproductive male-female pairs. The Cold Spring Harbor case illustrates how sex researchers' understanding of what sex was could shift to suit their goals, and how they made sex binary and not binary, a classification system and a bodily process, and variously defined across species. As American eugenics rose to global prominence, it operated as a key site for the development of sex science.[78]

Around the same time, birth control, gynecology, and sex hygiene were critical sites for research that would improve reproductive and marital success for racial betterment.[79] Chapter 3 investigates the contradictions held within early to mid-twentieth-century gynecological research and private medical practice, foregrounding the work of the clinician, sexologist, and eugenicist Robert Latou Dickinson. Dickinson bridged an era of sex research that used case studies to understand pathology and another that used large data sets to search for normality; personally, he was a close correspondent of Charles Davenport, a main figure in chapter 2, and a mentor to Alfred Kinsey, whose research forms the basis of chapter 4. Dickinson was a virtuoso of incoherence. He espoused a belief that sex manifested in degrees rather than in binary kind; that theory all but disappeared in his assessment of white women patients. In case notes, correspondence, and publications, Dickinson framed pathology and pain as components of normal white womanhood that need not trouble femaleness. Missing ovaries, menstrual insanity, and genitals supposedly transformed to a more masculine shape by masturbation did not, in practice, indicate that a patient might have strayed from her sex category, even as he asserted that "full sex endowment" was rare. The designations "female" and "woman" could be maintained as the very organs that were supposed to constitute them went awry. Alongside his expansion of the meaning of femaleness to encompass a tremendous range of bodily configurations, Dickinson's commitment to racial improvement led him on a quest to quantify and represent in visual form the ideal, eugenic female body. This chapter positions Dickinson's effort to identify that body—eventually rendered as the statue "Norma"—as a way to protect white sexual dimorphism even as he encountered an onslaught of evidence against it. The chapter also argues for a historiographic approach that does not always attempt to tease apart transness, homosexuality, intersex, and "normal" pathology.

By the 1940s, American sex research had ostensibly distanced itself from eugenics.[xxviii] On the cusp of this new world, entomologist-turned-sexologist Alfred Kinsey and his collaborators declared that they had brought the study of sex into the modern age. Chapter 4 examines incoherence in the largest and most heavily popularized study of twentieth-century American sex science. The case study was over, and so was an outdated understanding of sex as a spectrum. With university backing and considerable funding from the National Research Council Committee for Research in Problems of Sex, what came to be known as the "Kinsey Reports" used innovations in statistical practice to demonstrate that American sexual behavior was far more varied than previously believed. This narrative of novelty plastered over the many ways that the Kinsey studies drew on the past, especially their insistence on binary sex while presenting evidence to the contrary, linking of race with sexual deviance, and refusal of an Identitarian form of sexuality while foregrounding behavior over inherent type. In this quantitative behemoth, anomalies like those discussed in previous chapters became more formally noise. The Kinsey researchers privileged frequency and incidence as the most important facts about sex and treated anything they perceived as numerically uncommon as unimportant—especially the possibility of sex outside of a static binary.

xxviii   Spoiler: It hadn't.

At midcentury, the United States consisted almost entirely of clearly male or female people, with only two options to choose from, still defined and enacted in a mess of incoherence. Most deviances, whether lingering assumptions about racialized sexual difference, bodies that did not quite conform to ideal types, or nonnormative behavior, had been recaptured back into femaleness and maleness. Anomalies had been rendered exceptional and therefore not disruptive. Chapter 5 brings the narrative to a close by examining how the development of the category "transsexual" posed little threat to binary sex and cemented the distinction of transsexuals as unusually discordant in body and identity compared to the masses of people who simply were not. This final chapter focuses on Harry Benjamin and Elmer Belt, early practitioners of trans medicine, as they attempted to sort out eligibility for transition-related surgery in the 1950s and early 1960s. Benjamin defined transsexuality as an uncomplicated desire for hormonal and surgical transition, but, obsessed with risks to themselves if a patient regretted having surgery, he and Belt created a gulf between taxonomic clarity and quotidian action. They traded questions of who counts as female or male and woman or man for concerns about who might sue them, reject their authority, and interfere in their efforts to self-fashion as medical pioneers. Management of those fears established habits of assessment and views of transsexuals as dishonest and psychologically deficient, which eventually structured requirements for surgical access more formally. By the end of the book's arc the incoherent enactments of a male/female binary are joined by a nascent cis/trans binary, in concept if not in name.

Sex science both shaped and responded to developments in disciplinary formation and changes in the political economy of sex research. While this book is not about either of those things per se, such shifts over time provide a foil for the continuity that is my focus here. Though the contours of sex research and its relation to knowledge structures and governance transformed over the period under study, the incoherence of sex persisted. It persists today, visible in the frantic redefinitions of sex by those who wish to bar trans people from bathrooms, sports, and other areas of public life. The book concludes with a brief discussion of the contemporary ramifications of this history and an assertion that attempting to counter anti-trans rhetoric and legislation with better science is bound to fail. I propose my own incoherent approach: a simultaneous insistence that there are more ways to know sex than science and that we need to take seriously the possibility that sex is not a useful category at all.

A brief note on terminology: Throughout this book, I use *binary* with some regularity to mean "exhaustive and mutually exclusive." My use of it is consciously ahistorical—while I considered avoiding the term, *binary* is ultimately the most concise and legible way to put it. However, a crucial thing had to happen for this to be the case: The concept of "binary" as "exhaustive and mutually exclusive" had to develop. Centuries of sources use *binary* to mean "a combination of two things of the same or a similar type, a pair, two."[80] Notably, that now-obsolete definition does *not* include the connotations of exhaustiveness and exclusivity, and that definition has been used for much more of the English language than the current one. Old English used the prefix twi-, derived from the Latin prefix bi- (also the root of binary, "binarium"), to mean, essentially, something with two parts that belong to the same category, rather than mutually exclusive opposites—for example, *twibille*, a two-edged ax, or *twi-féte*, two-footed.[81] The *Oxford English Dictionary* (OED) dates the first known usage of "bynaries" specifically to 1464, when John Capgrave used it in the phrase "þink þat ȝe be mad of to natures, body and soule."[xxix] "Body" and "soule" are two aspects of the self here but not necessarily the only ones; indeed, the same manuscript refers to another binary of "love of God, and love of your neighbor," decidedly not an exhaustive list of lovable entities.[82]

This broader meaning of a matched pair continued on through the nineteenth century—an 1837 use refers to "the binaries of boats and Anubises," where two objects are again paired, but there are clearly more objects in the world than watercraft and Egyptian gods. In 1876, a book on color theory used "binary" to refer to colors made of two primary colors, yet green, for example, is not made of only two colors, nor is it a single distinct shade. It's not until the mid-twentieth century that "binary" comes to carry its present meaning, "consisting of two opposing or contrasting aspects."[83] Though somewhat speculative, my hunch is that the contemporary meaning of binary emerged alongside the development of electronic computing in the 1940s.[xxx] According to Google Ngram, use of *binary* increased substantially in the mid-1940s, further suggesting this

xxix   Or, more modernly, "think that ye be made of two natures." Thank you to Caz Batten for assistance with Old English concepts of sex/gender and vocabulary!
xxx   Thank you to Mar Hicks and David Dunning, who confirmed for me (not a historian of computing) that this makes historical sense.

relationship. The either/or mechanics of various forms of computing—a circuit is closed or not, a punched hole is present or not,[xxxi] code contains a 1 or a 0—provided a potent metaphor for thinking about twoness. No longer just related pairs, binary came to refer to two mutually exclusive options outside of which no additional possibilities exist. So the "binary" of "binary sex" is a product of the latter half of the twentieth century and not an actors' category, but I've nonetheless chosen to use it for simplicity.[xxxii]

"Sex," however, most certainly is an actors' category. When I say "sex," I refer to the vague, ever-mutating hydra with endless heads that is the object of study and research variable of my actors. Rather than attempt to impose precision on a category that was constructed without regard for consistency—and depended on a lack of consistency—I have allowed my terms to be somewhat slippery. I ask that you not take "sex" to mean "the biological" in contrast to a social or cultural gender. I mean something more akin to my actors' gesture toward the natural as a source of explanatory power for a particular axis of social classification, which might at any given time bundle morphology, social role, psychological feeling or identity, reproductive capacity, and sexual behavior, et cetera. I mostly don't use *gender* here, except in the final chapter, since it wasn't invented until the mid-twentieth century and my actors assumed that the above-mentioned generally went together. I use *woman/female* and *man/male* interchangeably, since, again, my actors mostly didn't separate them. This book is not an effort to decode precisely what historical actors meant when they talked about sex. Sex is a snarl of contradictions, and my goal here is not to untangle, or uncoil, or cut through Gordian-style. The knot is the thing. If this seems like a lazy analytic or use of terms, I ask that you take it up with the late nineteenth and early twentieth centuries, and join me in this exploration of what a refusal of coherence opens up.

xxxi   Notwithstanding hanging chads.
xxxii   With the risk of losing some nuances about the relationship of female and male. See Park, "Myth of the One-Sex Body."

1

———

Constructing Sexual Multiplicity
in Animal Research

If you're looking for yet another specimen of peculiar sexual configuration, peer into a zoology journal, popular science magazine, or natural history text from the nineteenth or early twentieth century. Sometimes these exceptional beings appeared as individual subjects of case studies. Other times, an entire species, utterly alien in its form, might be on display. Throughout this period, a veritable menagerie of hermaphroditic, sexless, triple-sexed, sex-changing, and otherwise category-defying creatures galloped, swam, and flitted their way across the pages of multiple genres of writing. Ostensibly, they were part of a project to make sense of the world, but more often they left confusion in their many- and un-footed wake. If scientists looked to the animal world for evidence that a clear distinction between male and female was the norm, they didn't find it.[i,1] Instead, they plucked from every land- and seascape an unruly diversity, and the more they

i   Following Harriet Ritvo's example, I use *human* and *animal* rather than *human* and *nonhuman* because it reflects the fragile dichotomy constructed by my historical actors, in which the two categories were easily blurred and often overlapped. Indeed, what counted as "human" was not a foregone conclusion, with full membership in that category typically reliant on adjacency to whiteness. More on that soon.

learned, the less they seemed to know.[2] Yet, even as scientists encountered sexual variations that didn't map neatly onto male and female categories, they staked both their claims to expertise and their ideal social orders on their ability to know sex.[ii]

Animals are everywhere in sex science. Animal research has served as one of the primary sources of information about human sex and sexuality: the theories of evolution that underpinned fears that homosexuality was a manifestation of degeneration, related evolutionary explanations of sex differences, agriculturally informed conceptions of heredity through sexual reproduction and the eugenic theories that came with them, and the idealization of monogamy all gained empirical support from study of animals.[3] Animal physiology and experimentation provided the initial basis for hormonal theories of sex differentiation, and animal bodies were the first testing sites of the medical technologies that would eventually, in the twentieth century, become available to trans people as possibilities for bodily reconfigurations.[4] Many of the researchers whose careers would eventually take them in more human directions began their work as animal scientists—perhaps most famously Alfred Kinsey, whose training and career in entomology preceded his shift from wasp phylogeny to the studies of human sexual behavior that are the focus of chapter 4.[5]

The animal body, purportedly untouched by the vagaries of society and culture, has served as proof for the naturalness of social arrangements like patriarchy, white supremacy, and colonialism over a long arc of history, with particular intensification during the period under study here. Over the course of the nineteenth century and into the twentieth, challenges from feminist and abolitionist movements, changing immigration demographics, and violently assimilationist encounters with Indigenous peoples threatened to expand the bounds of citizenship in the United States and Europe.[6] The study of animals offered insight into the epitome of nature and generated supposedly objective evidence about what life should be, unfettered by politics or sentiment. These studies conveniently tended to reproduce a vision of the natural world that aligned with patriarchy, white supremacy, and colonialism. This, the animal as source of normative claims, is old news in the history of science and the history of sex alike.[7]

ii A reminder: This book is about how "sex" contains multitudes. In this chapter, the historical texts I discuss often collapse anatomical difference, reproductive capacity, and sexual behavior.

For the most part, both historians of sexuality and scholars in science and technology studies (STS) have conceptualized the history of sex science as a relatively steady march toward increased disciplining of nonnormative forms, and an ever-smaller unit of unchangeable determination from genitals to gonads to hormones to, finally, chromosomes that can't be changed, no matter what one's gender identity might be.[8] Even though science underwent periods of interest in indeterminacy, models like universal bisexuality have tended to bubble up and then dissipate under the weight of conclusions that, this time, scientists had decided what sex was.[9] Much of this narrative aligns with Thomas Laqueur's much-cited, though perhaps not entirely accurate, argument about the two-sex model: that, by the end of the eighteenth century, "sexual difference in kind, not degree, seemed solidly grounded in nature" and that since then the "dominant view . . . has been that there are two stable, incommensurable, opposite sexes."[10] In this framing, sex, over time, resolved from mess into clarity.[iii]

Reexamining research on animal sex throws that history of sex science into disarray. As I argue in this chapter and throughout the rest of the book, while scientific and popular narratives often sought to use examples from the "natural" world to justify both sexual and racial hierarchies, that effort did not establish a cohesive order so much as a multiplicity of sexual possibilities. Instead of crystallizing sex into male and female categories, or indeed an array of traits and processes into an entity called sex, this research fractured any semblance of neat delineation. The incoherence of sex became increasingly apparent when scientists had to contend with not only differences of sex but differences of species. Features that might seem self-evidently male, for example, might not exist in all species, or they might appear in presumed females instead. Some species might not have easily recognizable sexes at all. Others might have more than two. Sexual traits in different species conflicted, and animals with strange morphologies or that reproduced parthenogenetically threw a wrench in scientists' efforts to theorize sex as a whole.[11] The result was that even within a single

iii   Laqueur himself at times mentions that aspects of the one-sex model persisted, especially in embryology, but never quite takes up the simultaneity problem. As he's cited, the one-sex to two-sex epistemic shift has up until recently often been taken by historians of sexuality (less so by historians of science— see Park and Nye's review of Laqueur, and Park's 2023 expanded retort) at face value. In a representative example, Jennifer Terry reports that "in this two-sex system, the sexes were not different in *degree*, as they had been in the ancient Aristotelian and Galenic models, but in *kind*." *American Obsession*, 33.

journal article, an author might define sex one way and then change the definition only pages later, undercut their argument with their own evidence, or even explicitly express the difficulty of making a sex determination. Comparative study of sex across species promised insight into the evolution of sexual reproduction and the development of different species traits, but it also brought to the fore an incredible diversity of sexual forms that could not be accounted for by a mere binary of male and female.

Things went further awry when scientists used research on animal sex to make racial claims about humans. Race science and sex science were mutually reliant on each other, both conceptually and structurally, at this time (more on this in chapter 2), and the animal was one of the most salient sites of overlap. Throughout the nineteenth century and well into the twentieth, biologized racial classifications relied in large part on sexed traits for their definition. Full sex differentiation, meanwhile, could only be achieved by the most evolved bodies—white human ones—while organisms that had not reached this pinnacle of developmental achievement tended toward hermaphroditism.[iv] This meant that humans racialized as anything other than white were framed as less than fully human and therefore closer to animal ancestors, as well as less than fully sexually dimorphic and therefore closer to a hermaphroditic evolutionary past. As scientists studied sex in animals in service of racial contrast, human sex became a racialized expanse beyond a clear binary, even as scientists tried to keep white dimorphism safe from the intrusion of the hermaphroditism commonly found elsewhere in the animal kingdom.

By studying animals, then, scientists produced opposing versions of sex that nonetheless coexisted. In one, all creatures belonged to male and female categories that could be correctly identified with a trained scientific gaze. In another, a hierarchical evolutionary order of life hinged on differing degrees of sexual distinction according to race and species. Such a bifurcation was a crack in the foundation of sex research that

---

iv   Today, *intersex* is the preferred term for what people in the past called "hermaphroditism." I use *hermaphroditism* here in part because it is what the texts I'm discussing use, and, more importantly, because the two are not fully commensurate. In the period under study, *hermaphroditism* referred not only to a specific type of bodily configuration associated with a particular identity category, but to a broader range of manifestations of sex that could include varying combinations and degrees of physical and behavioral failures of perfect sex dimorphism.

persisted through subsequent decades as later researchers built on it. To sketch out this multiplicity and the practical conditions of its production, this chapter proceeds in three parts. First, I consider how scientists created knowledge about sex through and with animal bodies by making evolutionary arguments that linked whiteness to sexual dimorphism. In the second section, I examine three types of animal—hyenas, freemartins (co-twins of bulls with ambiguous sexual characteristics), and hymenopterous insects like ants and bees—that particularly troubled the efforts of scientists to classify them as male or female.[12] Finally, I use the work of sexologist and ornithologist Robert Wilson Shufeldt to show how a feared return to an animal state fueled antimiscegenationist arguments, even as comparisons between humans and animals used for racial hierarchizing constantly blurred the line between human and nonhuman. Together, these three parts offer a new perspective on sex science that questions the success of efforts to construct a binary of male and female, and of human and animal, even as sex remained a privileged site of sorting bodies.

This chapter diverges from those that follow in its broad scope and focus on published materials. It is about sex science and animals, but it is also a general introduction to the looping investments in sex, race, and the evolution of humanity that would go on to shape both knowledge produced about sex and the practices of its production in the first half of the twentieth century, and therefore the rest of this book. The eugenicists we'll meet in chapter 2, for example, owed their attention to and knowledge of heredity and development to agricultural breeding practices. The racial politics of sex research, though generally less explicit over time, remained indebted to the evolutionary model of the development of sexual dimorphism established by the scientists who appear in this chapter. The same kinds of categorical reconfigurations that animal research deployed to determine if a particular specimen was male or female would appear again and again, coming to a head in chapter 3's study of the boundaries of womanhood in gynecology.

While subsequent chapters will zoom in on individual researchers or institutions, each builds on a series of entanglements already swirling through nineteenth-century sex research. This chapter makes visible some starting points of those entanglements by situating incoherent sex in a larger context of scientific sex across fields, ranging from agriculture to embryology, and it shows how a consistent inconsistency of meanings of sex and its categories was embedded in sex research even before more formal consolidations of sex science.[v] It also, by making published texts its source base, illustrates that incoherence underpinned sex science even in its most polished form. It can be found in highly respected, public-facing texts just as easily as in private correspondence and casual research notes.

Not all of the researchers here saw themselves as doing the work of defining sex; nor, as discussed in the introduction, would many of those who followed. As Adele Clarke has argued, work on what would eventually be known as the "problems of sex" and the "reproductive sciences" solidified and differentiated from sexology over the first decades of the twentieth century, and the United States became a scientific global power in that area only in the 1940s.[13] But one needn't identify as a sex scientist to produce scientific enactments of sex, and a field needn't be articulated as *about* sex in order to make sex a fundamental part of scientific inquiry and thereby play a role in its construction.

The techniques of knowledge production discussed here and the infrastructures that enabled them are not unique to sex science. Although this chapter is, on its face, about animals in sex science, it points also to much larger shifts in the political economy of American science. Over the course of the late nineteenth and early twentieth centuries, sex science shifted from being primarily organized around agricultural and small-scale natural history and zoological research, to being funded first through large philanthropic organizations, and eventually also supported by major universities and federal dollars. This chapter situates early sex science in the preamble to its later philanthropy- and university-supported state, demonstrating a consistent scientific interest in sex before Michel Foucault's 1870 mark and before the consolidation of "sexology" or "reproductive science." It illuminates an incipient social world through which the incoherence of sex emerged, in part, by way of a collective insistence that science needed to know sex to make claims about other aspects of life, and scientists themselves had the tools to know it.

v   There are, of course, many other genealogies of sex's incoherence. We can trace out the intersection of the fields and social networks discussed in this chapter as also emerging through the relations and epistemologies of medical experimentation on Black women (e.g., Owens, *Medical Bondage*; Snorton, *Black on Both Sides*; and Rich, "Curse of Civilised Woman"); travel narratives and other colonial forms of knowledge (e.g., J. Morgan, *Laboring Women*; Mitra, *Indian Sex Life*; and Levine, "Naked Truths"); and contraceptive research (e.g., Clarke, *Disciplining Reproduction*).

The Logics of Sorting

Sex and gender, as countless historians have demonstrated, have long functioned as organizing principles for social, legal, political, and economic relations. If, to name just a few examples, the type of labor one performs, the ability to own property, or the right to vote is determined in part by one's sex (always alongside race), then those delineations require the sorting of people into categories of male and female.[vi] The high stakes of those determinations, alongside the increasing cultural authority of science in the eighteenth and nineteenth centuries, made scientists and doctors the ultimate arbitrators of what was male and what was female.[14] That power, rather than any kind of inherent coherence of scientific approaches to making sex determinations, was what undergirded scientists' claims about their specialized knowledge and the classifications that they came up with. Internal logic, other than the logic of expertise, was not particularly necessary for scientists to make truth claims about sex.

The ways that scientists sorted out animal sex designations had little to do with the inherent classifiability of bodies. Instead, scientists made choices at every step of the way that privileged certain ways of knowing and outcomes of knowledge production over others. They actively constructed certain kinds of bodies as ambiguous and requiring a particular kind of expertise—expertise that only they could supply—to understand.[vii] They allowed for varying exceptions from the norm so that they could place certain kinds of bodies in normative categories of male and female, and they differed on what counted as one or the other. They held wildly different opinions on what any of their animal insights had to do with human experience. They constructed not a binary but a multitude, and their choices reflected much more about their stated and unstated investments in white supremacy and their own expertise than about the animals they studied. The result was that although scientists did frequently sort bodies into male and female categories, exceptions, uncertainty, and contradiction were much more common than binary consensus.

---

vi   Sometimes without regard for the body—prior to the nineteenth century, at least, picking one and sticking to it was more important than any kind of bodily truth. Foucault, *Herculine Barbin*; Brown, "'Changed into the Fashion of Man.'"
vii   Again, as noted in the introduction, this is not some kind of nefarious plot. This is just how science works, through the accrual of expertise and becoming trusted as a source of knowledge.

This section dwells in the ambiguities of three species that each show a different facet of the struggle to make meaning out of sexual variation among nonhuman animals, and therefore demonstrates three different ways of producing knowledge. First is research on the hyena, which enabled scientists to establish their own expertise precisely by constructing hyenas' sexual morphology as a mystery. Next, I turn to long-standing debates about freemartins, or what we might today call intersex cattle, which made sex itself malleable and sex organs effectively interchangeable. And last, I look at discussions about bees and ants that not only added to the confusion about what counted as maleness and femaleness but also framed their three-sex system as potentially more advanced than mere sexual dimorphism.

Over the course of the nineteenth century, scientists looked to hyenas to establish their own superiority over prior knowledge producers. Natural history texts, encyclopedias, travel writing, and other publications discussing hyenas reported that "the ancients," or occasionally Indigenous peoples in regions home to hyenas, believed every hyena to be hermaphroditic or to change sex on a yearly or seasonal basis.[viii] Modern science, however, knew better.[15] Dissection under a trained eye revealed that this belief in nondimorphic or malleable sex in a mammal was absurd—an outmoded idea that could be dispensed with now that better methods for getting to the truth of sex had been developed.

But in order for science to be the hero of the story, the hyena, especially the female hyena, had to first be constructed as confusing, its sex as ambiguous. An anatomist and medical school dean named Morrison Watson sought to cut through the mystery of the hyena and reveal the truth of its sexual morphology, and in the late 1870s and 1880s, he published a series of articles on the anatomy of, in the binomial nomenclature of Linnaeus, *Hyaena crocuta* that would become foundational texts in the study of hyenas.[16] Watson's articles, which repeatedly emphasized the oddness of the genitalia of the female hyena, suggested that only modern developments and his expert eye could unravel that mystery.

Watson was explicit about his confusion, turning the hyena into a problem to be solved. He immediately framed the hyena as strange, referring to

viii   The dismissal of Indigenous knowledge is about both the construction of a science of modern civilization and the dismissal of purportedly less-civilized forms of sex and gender, enabling Europeans to locate themselves within a modern scientific and properly gendered and sexed imperial order. Robin, "Platypus Frontier," and Pratt, *Imperial Eyes*.

the "peculiar characteristics" of female hyena genitalia and reporting that the colleague who had sent him a female hyena specimen did so because he was so baffled by it that he wanted a second opinion.[17] What was so mysterious? Watson initially believed his first hyena specimen was male due to "(1) the absence of any well-marked vulva such as one generally sees in the female (2) the presence of an elongated pendulous penis-like body . . . ; and (3) the presence of two projections which at once called to mind the appearance of the scrotum."[18] Essentially, female hyenas did not look particularly female.

Watson repeatedly remarked on how hard it was to derive any sure answers from the hyena's body. Referring to the female hyena as "deceptive," Watson complained that its "peculiarities . . . render the identification of the sex a matter of difficulty."[19] After declaring that it was nearly "impossible to distinguish between the sexes," Watson admitted that even with his considerable anatomical knowledge and careful inspection of both male and female hyenas, it was *still* difficult to find differences between male and female. "After a very minute examination of [the external genital organs] in both sexes," he wrote, "I have only been able to recognize . . . by no means very evident points of distinction."[20]

Nonetheless, Watson insisted, he, a well-trained anatomist, *could* tell the difference. The hyena clitoris, easily mistaken, apparently, for a penis, was "perforated by a single canal of so small a size that one is at first sight induced to believe that he is dealing with the extremity of the male urethra, an error only corrected by an examination of the internal organs."[21] In an article about the dissection of a male hyena, Watson asserted that a "very accurate examination" could reveal the differences between male and female hyenas: the slightly smaller diameter of the clitoris, the dilatability of the urethra, and amount of hair present around the genitals.[22] He concluded that one would not be able to "decide the sex of the animal in the absence of such an examination as is well nigh impossible so long as the animal is alive."[23] It was not just that the internal organs mattered more, in keeping with a definition of sex that was in this period moving away from external anatomy and toward internal anatomy, but internal organs required expertise to be able to access and decode. In these arguments, the need for dissection likely produced an even more specific kind of expertise than one defined by "objective" professional knowledge or more skills and resources than the nonexpert public. Over the course of the century, systemic zoology had undergone a slow transition (which by the end of the century had not yet fully settled) from taxonomies based on visible external characteristics to internal ones only accessible by cutting, principally informed by Georges Cuvier's comparative anatomy.[24] Declarations that sex could only be determined by dissection, then, were as much about signaling an affiliation with newer modes of classifying and their techniques of knowledge production as they were about placing the power to define sex in the hands of experts. Dissection was the ultimate solution to the mystery of ambiguous sex that Watson had so deftly created.

Certainly, Watson was engaged in an effort to shoehorn the hyena into a binary framework, and he went to great lengths to identify "female" structures where they were not particularly evident. At the same time, however, he actively constructed the hyena as *not* conforming to binary expectations, so that he could use his expertise to solve the problem. His outright descriptions of the mastery required to make sense of the hyena's confusing anatomy make his expertise even more apparent in contrast to the ambiguity he describes as existing in nature. Even as Watson tried to describe the hyena within a binarized category of "female," he merely emphasized how poorly that classification fit and how slight the distinctions between maleness and femaleness could be.[ix] Watson's writings on the hyena indicate that sexual ambiguity does not exist on its own but rather has to be manufactured in relation to the privileging of distinction. Any

ix   The framing of the hyena as both sexually aberrant and not disruptive of sexual order stands not on its own, but always in relation to the racialization of sexual deviance. While authors rarely drew explicit comparisons between hyenas and Black humans, the particular types of deviance that some authors claimed as the origins of the hyena/hermaphroditism link also tended to crop up in anti-Black writing. This took many forms—allusions to disgusting odors, loudness, the possibility of taming hyenas but their dislike of captivity, cannibalism, laziness, and the potential to vocally "pass" as human, among others—but is particularly interesting with regard to femaleness and purported genital abnormality. As Siobhan Somerville has argued, turn-of-the-twentieth-century anatomists argued that Black women and white lesbians possessed abnormally large clitorises, which signified a failure to achieve sexual dimorphism. That the female hyena, an African animal, became particularly notable for its penis-like clitoris does not seem coincidental. Writings from this period, especially vernacular texts (including story books for children), also frequently include stories of what were effectively hyena lynchings, in which a hyena had preyed on a woman or child and was then killed by an angry mob in retribution. For examples, see T. Jackson, *Stories About Animals*, 69–70; Lang, *Red Book of Animal Stories*, 254–55; Nott, *Wild Animals Photographed*; Bingley, *Animal Biography*. For a similar discussion of the nonbinary hyena as allegory for Jewish uncleanliness in medieval bestiaries, see DeVun, *Shape of Sex*, esp. chap. 3.

attempt to binarize sex will thus necessarily produce the opposite effect of demonstrating just how inadequate that system is for describing the world.

Truth claims about sex like Watson's thus tended to veer into the nonsensical, bringing more incoherence into attempts to neatly classify bodies the harder that scientists tried to identify sexual difference. Watson's determination to make the hyena—which he insisted was extremely confusing—conform to male and female categories pales in comparison to the way that studies of freemartins, or twins of bulls who exhibit a variety of ambiguous sexual characteristics, repeatedly shifted what particular body parts might be classified as male or female. Though the freemartin had long been known to cattle breeders and butchers, it made its debut in scientific literature with the 1779 publication of Scottish surgeon John Hunter's "Account of the Free Martin" in *Philosophical Transactions of the Royal Society of London*. Further articles describing additional dissections by other authors soon followed, making the freemartin one of the most frequently discussed examples of hermaphroditism in the animal world and a site of great interest for learning about sex determination and development.[25] Almost all of them made some kind of claim about the freemartin's "real" sex, declaring it either a malformed male or a modified female. To do so, they had to make choices about what counted as male or female. One scientist's choices rarely mapped onto another's, providing another example of how to argue about sex, even in this either-or binary frame, and emphasizing that sex is not self-evident.

From the beginning, Hunter noted that freemartins appeared more like females, reporting that one of his specimens displayed "external parts [that] had more of the cow than the bull" and that it did not truly have a mix of "both sexes." For this reason, although he found organs resembling testicles upon dissection, he concluded that they were "only imitations of such."[26] James Simpson's oft-cited chapter on hermaphroditism in the 1839 volume of *Cyclopaedia of Anatomy and Physiology* discussed the freemartin under the heading "transverse hermaphroditism with the external sexual organs of the female type," and concurred that the freemartin seemed more female than male.[27] Other writers were more willing to accept a neither/nor approach: An 1851 animal husbandry text referred to the freemartin as "not, in any case, either a bull or a cow" and noted that all freemartins had "the generative organs of both sexes combined" to differing extents.[28]

The sex classification of the freemartin started to shift toward maleness in the second half of the nineteenth century. By 1876, Francis Galton described the "modern view" of the freemartin as an understanding that while a freemartin was "neither a complete female nor male," the male characteristics "predominate" over the female ones.[29] The designation of the freemartin as male, however, produced a jumble of characteristics within that category. An 1859 article on the rarer ovine freemartin by W. S. Savory captures the idiosyncrasies of that sorting process well. Savory argued that his "hermaphrodite sheep," which the author proposed was the same type of being as the freemartin, was male despite its many female characteristics. Yet, implicitly, Savory opened up the possibility that sexed features were not inherently sexed at all. The specimen testes were "precisely similar to those of an ordinary male," such that they were similar but, in needing comparison, not the same, or in other words, not male.[30] The sheep also had a uterus and vagina that were not female: The "uterus is altogether smaller than that of the female" and "the vagina is exactly half the length of that of the female."[31] Savory specified that the case was particularly remarkable to him because it demonstrated "the complete union of such perfect male and female organs."[32] He concluded, however, "of course this animal was really a male."[33] The sheep in Savory's case is a male (of course!) even with its uterus and vagina. In Savory's writing there is a delightfully pre-trans sensibility that suggests that sexual biology is not destiny, and of course there are male vaginas. This is not to say that maleness and vaginas are mutually exclusive, but rather to illustrate Savory's efforts to hold on to a self-evident maleness, while using the presence of "perfect" female organs to make claims about the specimen's peculiarity, *while also* insisting that those organs are not necessarily female. These contradictions were perfectly acceptable science, absurdities created by the very system intended to maintain order.

By the turn of the twentieth century, the most up-to-date conclusion shifted again. In 1909, gynecologist David Berry Hart referred to "the nature of the free-martin" as having been, to that point, "one of the unsolved problems of anomalous sex" and claimed that the mystery had finally been solved.[34] The freemartin was a sterile male: Hart had examined Hunter's 140-year-old specimens and found no ova present in the indeterminate sexual glands, but he noted their "characteristic testicular tissue" and identified spermatozoa in one such gland.[35] He likened these findings to "the condition usually found in undescended testes in man" and noted that both the freemartin and its normal sibling came from one male zygote that had split.[36] The two were thus "identical male twins except in their genital tract and secondary sexual characters," implying, in the process, that true maleness could still be found, regardless of sex characteristics.[37] Indeed, for Hart, freemartins were not hermaphrodites at all, merely males with a uterus and a vagina. Hart was not the only scientist to make this claim—geneticists Leon J. Cole and William Bateson and gynecologist Otto Spiegelberg had also begun to articulate the possibility that the freemartin was a male that resembled a female.[38]

The freemartin question took on an even greater significance in the early twentieth century, in the heyday of the theory of universal bisexuality (i.e., the idea that all organisms contain both male and female attributes) and amid a surge in an inversion model of homosexuality.[39] It was of particular import in the development of the endocrine theory of sex as evidence that hormonal exposure in utero could affect embryonic development of sex characteristics, including in humans. Frank Lillie, whose work on the freemartin served as a foundation for these new endocrine theories of sex differentiation, argued that the freemartin "could not possibly be interpreted as a male"—the opposite of Hart, Cole, Bateson, and Spiegelberg.[40] Nearly all of the cases of bovine twins that he had examined were dizygotic and therefore did not emerge from a single male zygote; rather, the twins emerged from one male and one female zygote that had been connected through a large arterial anastomosis and, crucially, shared blood and hormones.[41] Lillie reasoned that even though "the general somatic habitus inclines distinctly toward the male side," including a phallus and gonads that most scientists had identified as testes, the freemartin was in fact a female that had been modified by these "male hormones."[42] The freemartin, then, might be a male with female genitals or a female with male gonads, and that apparent contradiction need not interrupt claims about "true" sex. Moreover, the results of Lillie's study indicated the malleability of sex; he compared his findings to those of Eugen Steinach's experiments on gonad transplantation in rats and the rodents' subsequent changes in sex characteristics.[43] Sex characteristics, of course, were not the only things that could shift in studies of the freemartin. Maleness and femaleness were neither self-evident nor closely guarded categories. They fluctuated over time and across the work of different scientists in efforts to make sense of indeterminable bodies.

Arguments such as these about the classification of animal sex were not abstractions discussed only among academic and medical elites. American agriculture played a significant role in the development of theories of sex via practical issues relating to the raising of crops and livestock.[44]

The requirements of feeding a rapidly proliferating urban population incited a new, scientific approach to food production, with the funding to match.[45] Nutritional volume would, agriculturists hoped, increase through the manipulation of heritable traits that privileged fertility and size, the improvement of reproductive efficiency, and innovations in selecting for the sex of offspring. This research and its infrastructures, especially the American Breeders' Association (founded in 1903 by both academic researchers and animal husbandrists, and a key player in the following chapter) laid the groundwork for much of American sex science. Work on animals demonstrated techniques and outcomes of good breeding, eugenic organizations grew directly out of agricultural research on heredity, and the human reproductive sciences built themselves on decades of animal investigations.[46] Agricultural science fundamentally shaped the path that American sexual science would eventually take; as such, scientific debates over the classification of animals as male or female were imbricated in the political economy of daily life, not mere intellectual curiosities.

Agriculturally inclined texts beyond the pages of zoological and medical journals served as another space for the attempted working out of sex classification. Guides for breeders of cattle, horses, and other livestock frequently featured descriptions of hermaphroditic animals so that breeders could identify them. Texts published for an audience of cattle breeders, especially, took interest in freemartins and the broader phenomenon of hermaphroditism. These books often dedicated pages and sometimes entire chapters to the identification of freemartins and explanations of their genesis.[47] Butchers even apparently regarded the freemartin as a source of superior meat.[48] For horses, making identifications in instances of ambiguous sex became an issue in the context of certificates of description, which enabled owners to establish a horse's identity in the event of sale or theft. Because sex was one of the key attributes recorded on the certificate, breeders had to be able to decide based on sight if the horse exhibited ambiguous genitalia.[49] Though these texts intended to provide sexual clarity to their readers, and frequently offered instruction in making binary distinctions, they also had the effect of introducing sexual uncertainty into their readers' lives as a fact of nature and taught them to expect it.

This flow between academic science and agriculture—the arterial anastomosis connecting two zygotic fields, if you will—is particularly apparent in an ongoing disagreement regarding the sexual system of hymenopteric insects, especially ants and bees.[x] The long-standing debate was rooted in disagreement about the sex of worker insects, which, according to a variety of writers, were variously neuter, female, or technically female but also effectively a third sex in their incomplete development.[50] The worker as neuter largely won out until the 1880s, until that designation was tentatively and unevenly replaced by an insistence that the worker was an underdeveloped female.[51] This was more, however, than a simple reclassification in which a binary sex system won out over time. Fights over what made femaleness, and an exaltation of a three-sex system in utopian imaginings of the future, interrupt a simple narrative of the collapse of multiplicity into a binary. As with hyenas and freemartins, in order to force the worker insect into a female category, the meanings of male and female had to flex in order to fit their new contents—and this time, those might not be the only available categories.[xi]

The sexual system of bees, in particular, brought discussions of sexual variation to the farm or backyard. Magazines like the *American Bee Journal*, geared toward members of the apiculture industry and amateur beekeepers, invited readers to send in specimens of ambiguously sexed bees that they found in their hives—and were told to be on the lookout for them—and find an expert analysis of its true sex in the following month's issue. Lectures from scientific organizations about the two-sex or three-sex issue were reprinted and sometimes even translated for an English-reading audience.

x   Insects of the order Hymenoptera include bees, ants, sawflies, and wasps. The order is one of the most frequently discussed examples of what are known to contemporary scientists as "eusocial" insects, meaning that they have a social system involving "cooperative care of juveniles" and "reproductive division of labor," among other characteristics (Plowes, "Introduction to Eusociality," 7). Bees and ants are by far the members of the order most likely to be found in the kinds of archival sources I discuss here, perhaps due to their usefulness to humans for the former and low likelihood of stinging the people studying them for the latter. A major exception to this is Alfred Kinsey (the subject of chapter 4) and his work with gall wasps.

xi   While I argue that this issue was particularly meaningful in the case of sex, shifting taxonomies that widened the possible definitions of their categories were not limited to sex alone and highlight a broader move in comparative anatomy from external to internal characteristics as defining ontological properties, even if an animal's new taxonomic home might be counterintuitive. Graham Burnett demonstrates this process at work in the reclassification of whales from fish to mammals in the early nineteenth century, and he cites the example of the mammalian redesignation of bats (formerly birds) as well in *Trying Leviathan*.

One could read a critique of a new theory of parthenogenetic bee reproduction[xii] and find suggestions for further reading. As beekeepers sought
to optimize the production of their hives to increase profits from honey
and wax, understanding how bee sex and reproduction functioned was,
in their eyes, crucial for raising and managing a healthy swarm. Like
texts for livestock breeders, beekeeping literature put complex theories
of sex into circulation for a much broader audience than purely academic
texts.

Questions of ambiguous sex, then, were not intellectual exercises limited
to a small circle of elites based in urban universities.[52] Problems of sex
classification were an everyday consideration of farmers and agricultural
hobbyists, whose texts served as a location for the transfer of information
from academy to farm and back again, aided by the new land-grant agricultural colleges that dotted rural America, which gave further authority
to farm-based conceptualizations of good breeding and, of course, animal
sex.[53] Professors found themselves dissecting specimens sent to them by
curious farmers, and farmers read excerpts of Spencer and Huxley. Some,
like Albert John Cook, who handled much of the incoming specimen mail
for the *American Bee Journal*, had a foot in each world. Bees, then, provide
another example of how agriculture, in conversation with academic animal sciences, produced knowledge about sex before sexology and the state
developed much interest in regulating human sexuality.[54]

Perhaps unsurprisingly, entomologists and apiculturists introduced additional incoherence into the late nineteenth-century enactment of sex as
they tried to figure out how many sexes these insects had, let alone which
individual specimens might be classified as which sex. Even less surpris-
ing is the way that some writers framed the aforementioned transition
from the neuter worker to the female one through a progress narrative
that emphasized the advances of modern beekeeping. Just as scholars of
the hyena mocked the "ancients" for their misguided understanding of
hyenas as hermaphroditic, late nineteenth-century writers frequently
derided work on bees from previous decades, allowing contemporary
beekeepers to envision themselves as more enlightened and intelligent
than their counterparts of the past. At the eleventh annual North American Bee-Keepers' convention, held in 1880, A. J. King highlighted such
progress in a lecture on the recent systematization of the beekeeping in-

xii Beeproduction.

dustry. Before the introduction of modern beekeeping theory and methods, he noted, "Superstitions the most foolish were held, and practices the most unreasonable prevailed. . . . The workers were regarded by some as males—others as females—others without sex, and still others as about equally divided in this regard."[55] Now, however, apiarists had exited what King called the "Dark Ages of bee-culture in America" and could produce more honey with their greater knowledge.[56] Two years later, in 1882, the *American Bee Journal* reprinted under the title "The Ignorance of the Past Ages" an article originally published in 1860 in *British Agriculture* and introduced it as a "relic of ignorance and superstition" presented to "amuse and interest as well as disgust our readers." One of these archaic beliefs was the concept of "neuters, without sex," which had since been replaced with the idea of worker as underdeveloped female.[57] Some blurriness persisted—an 1887 article described worker ants as both "imperfect females" and "neuter ants" and also called a particular subset of larger workers in some species "a second, or rather a third, form of female," indicating further confusion about exactly how many sexes, or perhaps more accurately subsexes, existed among ants—but writers increasingly described the neuter worker as a thing of the past.[58] To some extent, then, the understanding of the worker as an underdeveloped female took hold, but the acceptance of that designation was uneven and partial at best.

Regardless of its success, the reclassification of the worker as female was not as straightforward as King and his colleagues imagined. It required a new understanding of femaleness as not necessarily attached to reproductive capacity and instead defined by behaviors of labor and kinship. One article in the *American Bee Journal* suggested that a lack of reproductive capacity didn't make the worker any less of a perfect female than the queen, because the worker had mental qualities that helped with the "preservation of the family."[59] Another, part of a multi-issue, multi-journal debate about what pronouns to use for workers argued that the worker should be regarded as female because of their role in food preparation and rearing of young.[xiii, 60] Yet another argued against the neuterness of the worker because it was their task "to keep house, nurse the babies, and hustle around to get something to put on the table."[61] For these thinkers, sex

---

xiii Eugene Secor, general manager of the National Bee-keepers Association, even contributed a poem on the topic, which declared that "from this time on, IT ought to be HE" because the worker bee "was ALWAYS A MAN" (capitalization original in "The Worker-Bee—He, She, or It?," 216).

could be conceptualized irrespective of reproduction—gendered behavior could fill in when the body proved ambiguous. The body itself mattered far less than what it did.

In their efforts to understand the sex of worker insects, these texts also opened up possibilities for advantageous nonbinary forms of life. While expert scientists and professional beekeepers largely rearranged the meaning of femaleness to make room for worker bees to become female, vernacular publications like *Popular Science Monthly* and *Scientific American* allowed for the possibility that worker bees, as well as worker ants, were neuter—and that these insects' three-sex system made them superior to species who had only two sexes. Over the course of the nineteenth century, hermaphroditism came to represent an undifferentiated state of primitivity or degeneration, with dimorphism signaling a more evolved state (more on this in a few pages, so hold that thought). In that model, some writers reasoned, the existence of three sexes meant that a species had differentiated, and therefore evolved, even further.

In light of ants' complex social structure of "males, females, and neuters," entomology writer E. R. Leland wondered in the pages of *Popular Science Monthly* "if man, or any other order of the vertebrata, is destined to remain forever the higher animal!"[62] The splitting of ants into three, rather than two, types was the result of a social economy with a complex division of labor, "which in mankind," Leland noted, "always marks a high state of civilization." That division of labor was so clearly demarcated in ants that it had led to physical modifications that freed the neuter worker ants from "the gentle functions of maternity" so they could perform other kinds of labor for their community (an intriguing departure, to be sure, from the beekeepers' insistence that nurturing behaviors revealed workers' true femaleness).[63]

Another *Popular Science Monthly* article, titled "Our Six-Footed Rivals," discussed at greater length the evolutionary advantages of a three-type sexual system. The specialization for particular tasks indicated that certain insects had evolved to a higher state than humans. "The polymorphic species is higher than the dimorphic," claimed the author, "just as the dimorphic is higher than the monomorphic."[64] This higher stage brought better distribution of resources and greater efficiency with it, effectively positioning ants and bees as post-Malthusian success stories. Bees and ants were, in fact, so much better at resource distribution and, on account of their sexlessness, unattached to the reproductive family unit that they would be able to succeed at communism where humans had thus far been unable to.[65] These arguments for the advanced sexual trimorphism of bees and ants do not seem to have been taken up by professional scientists and beekeepers, and even the unnamed *Popular Science* author was not enthusiastic at the prospect of humans becoming three-sexed. While a neuter class was of great advantage to insects, a group of neuter humans might not be "entirely desirable," because, as the author cautioned, celibacy and castration showed that de-sexing enfeebled the human body and mind. Nonetheless, they indicate a popular willingness to think more broadly about the possibilities of sex—specifically, nonpathological and advantageous forms of nonbinary sex, even if not in humans—than a mere two-sex system would allow.

Writings on hyenas, freemartins, and bees and ants suggest that the study of animal sex did not merely discipline diverse manifestations of sex into a neat binary. Rather, it produced endless questions and problems that could not be entirely resolved. Scientists invented ambiguity to prove themselves experts, and they constantly modified what counted as male, female, or something else. Considerations of the potential benefit of a three-sex system even suggested possibilities of the advanced nature of nonbinary sex. Without doubt, the scientific study of sex enabled the pathologization of certain bodies and the normalization of others, as will become abundantly clear throughout this book. But that process had little to do with internal consistency of theories of sex or the inherent classifiability of bodies, and everything to do with the power to classify. Looking more closely at the place where discourses of animality, race, and sexual differentiation meet demonstrates forcefully how power dictated the logics of sex classification.

The Racialization of Hermaphroditism

With the advent of new evolutionary theories in the nineteenth century, studying animal sex became a key facet of studying species development.[xiv] In studies of human biology, this quickly translated into interrogations

---

xiv   See Milam, *Looking for a Few Good Males*. Both Lamarckian and Darwinian theories of evolution depended on sex as an engine of change and therefore invited increased scientific attention to not only sex itself but the evolutionary role of maleness and femaleness. As Kyla Schuller has argued, the Lamarckian framework of change in response to a particular environmental stimulus generated a fear of white overimpressability that would create vulnerability to degeneration. Theories of sexual differentiation soothed that anxiety by locating sentimentality in women and rationality in men. Darwinian approaches toward

of racial difference defined by sexual difference, all within a spectrum of natural life ordered from the most primitive to the most evolved.[xv] This built on existing approaches in race science. Studies of animals and concepts of animality itself have long structured racial hierarchies in the Euro-American context, or, as Asian American studies scholar Claire Jean Kim has put it, race is "a classification system that orders human bodies according to how animal they are."[66] Eighteenth- and nineteenth-century science, especially, located whiteness farthest from animality, while Blackness and Indigeneity rendered a person more bestial and closer to the wilderness itself. Those forms of animality also structured sex and sexuality: White Euro-American researchers thus conceptualized, for example, Black men as hypersexual apelike rapists, Black women as physically distinct from white women and closer to simian ancestors, and Indigenous people of all sexes as incapable of adhering to "modern" (that is, monogamous and patriarchal) sexual arrangements and gender roles.[67]

Much of that reasoning took the shape of a narrative of sexual development in which hermaphroditism was the natural condition of less-evolved animals, while dimorphism occurred in only the most advanced. In that model, hermaphroditism was remarkably common, a starting point of life that required tremendous leaps to get away from. Tucked within these

---

natural selection as a function of random variation, meanwhile, inspired the idea that males were more variable and the driving force of changes in a population, while females conserved the general shape of the species. The two models blurred and informed each other throughout the nineteenth and into the early twentieth centuries, as scientists infused Darwin's theories with Lamarck's focus on progress and orderly change due to their discomfort with the randomness of natural selection. Schuller, *Biopolitics of Feeling*, 16; Russett, *Sexual Science*, 92–93; Bowler, *Evolution*, 202.

xv   I refer to sex and race by those terms, but I do want to trouble their separation. While scholars like Kimberlé Crenshaw ("Mapping the Margins") and Roderick Ferguson (*Aberrations in Black*), among many others, have provided important tools for considering how race, gender, and sexuality are inseparable from each other, in practice historical analyses still tend to talk about sex and race as if they are parallel and co-constitutive but ultimately distinct. For the scientists I write about in this chapter, however, there was little distinction. Sex could not be separated from race because race dictated sexual attributes and indeed sex was the way, hereditarily and evolutionarily speaking, that race worked. "Sex science" was nearly always "race science," and vice versa, and often structurally so, as will become clear in chapter 2. It is hard to tease out race and sex in historical analysis because they were never separate things.

arguments, then, was the implication that normatively sexed bodies and behaviors were a rare find not only in nature more broadly, but among humans across racial categories. Arguments about racial difference thus interrupted the neat splitting of sex into male and female.

Some of the most influential scientists of the age openly declared that clear-cut dimorphic sex categories were rarer than hermaphroditism. "Male and female are the exception rather than the rule in nature," claimed a chapter heading in George Briggs Starkweather's widely cited 1883 book, *The Law of Sex*.[68] The geneticist Thomas Hunt Morgan, too, reminded his readers that while many of the animals with which humans interact on a regular basis have separate sexes, "we are apt to forget how often in other groups the sexes, or more accurately the sex-cells, are united in the same individual."[69] Morgan noted, too, that even in species with separate sexes, it could be difficult to identify their sex from external examination alone.[70] He also named parthenogenesis as a "not infrequent occurrence" and, perhaps in a nod to universal bisexuality, suggested that while parthenogenesis typically results in female offspring, parthenogenic eggs contain "latent, or in a potential condition, the male characters which at any time may become dominant."[71] In other words, for both Starkweather and Morgan, tidy, mutually exclusive categories of male and female were only one possible configuration of sexes among many.

As noted, for many scientists the apparent rarity of total sexual dimorphism indicated that only the most exceptionally evolved could achieve clear sexual differentiation. Already used to thinking hierarchy of life in a "chain of being" framing and increasingly adopting theories of evolution steeped in narratives of progress, nineteenth-century scientists were quick to point out that sexual dimorphism was a product of success, reserved for the highest orders of life.[72] Dimorphism as an extraordinary accomplishment could be a positive, in that it made binary sex seem special. But it could also cause trouble, because it meant that clear and stable binary sex was not the norm at all.

Often, an argument for the commonality of hermaphroditism could be safely sequestered in the nonhuman world. Starkweather, for example, though he pointed to the commonness of hermaphroditism, noted that "as the development of life is traced to higher forms, the functions are gradually specialized." In the same vein as triple-sexed bees, sexes emerged from this increased specialization.[73] The more evolved a species, then, the more differentiated its sexes would be, and humans were the most differentiated. No need to worry. Morgan, meanwhile, supported his claim that hermaphroditism was common by referring to sponges, flatworms, mollusks, and other relatively simple forms of life, and while he allowed the sexual similarity of some mammals, he offered as examples mice and pigeons.[74] The frequent appearance of hermaphroditism, it seemed, had little to do with humans.

Nonetheless, examples of hermaphroditism among higher species, including humans, filled the pages of medical and zoological books and journals. To cope with this threat to evolutionary order, authors frequently split the kind of hermaphroditism found among lower animals from the kind that occasionally manifested among the higher.[75] As John Hunter had put it in his influential 1779 work on the freemartin, there was a difference between natural hermaphrodites, who "belong to the inferior and more simple order of animals, of which there are a much greater number than of the more perfect," and "unnatural uncommon or monstrous" hermaphrodites.[76] This model persisted into the nineteenth and early twentieth centuries. Generally speaking, hermaphroditism in lower animals was their proper state, given both their place in the evolutionary hierarchy and the needs of their lives, while hermaphroditism in the higher was indicative of some kind of malformation in an individual that was supposed to have two sexes. The hermaphroditism of oysters, for example, was conceptualized as necessary for reproduction, because as stationary organisms they had no means of encountering differently sexed fellow oysters and thus needed an independent means of reproduction.[77]

This careful splitting of expected hermaphroditism in lower animals from unexpected malformations in higher ones corresponded to the racialization of sexual difference in humans. As Lisa Duggan has argued, scientists framed sexual inversion—a close cousin of hermaphroditism that ranged from an incongruously sexed psyche and body to various physical manifestations of homosexuality—as a collective tendency of people of color indicative of evolutionary failure, but a symptom of exceptional pathology in a white individual.[78] The "natural" hermaphroditism designation, then, largely aligned with the argument that nonwhite humans were less evolved than white humans on the basis of a lack of sexual dimorphism.[79] Hermaphroditism could again be sequestered, as with Morgan's mollusks.

In practice, this proposed binary that clustered human/white/sexual dimorphism on one side and nonhuman/nonwhite/hermaphroditism on the other proved distinctly unstable. It couldn't, for example, assuage fears that the conditions of modern living might cause white degeneration, and it couldn't support scientists' endeavors to identify the true sex of organisms.[80] With racial hierarchy at stake, binary and nonbinary models of sex coexisted. Human sex was binary, except for when it conveniently wasn't. Lower animals tended toward hermaphroditism, yet an expert eye could somehow overcome this with the tools of dissection and determine which of two, mutually exclusive sexes a given hyena or freemartin belonged to. In essence, efforts to circuitously define race and sex according to each other added even more contingency to the multiple ways of sorting sex discussed in this chapter.

Despite an insistence that hermaphroditism was to be located at the bottom of the evolutionary hierarchy, scientists working at the nexus of race and sex relentlessly blurred the lines between human and animal that served as the basis of racialized sex difference to begin with. Part of that, of course, was the designation of some humans as naturally more hermaphroditic, which turned humanity itself into a racialized spectrum of animality. Most humans, in fact, were not fully human because they were not fully sexed. Among historians, two examples are cited particularly frequently: first, British sexologist Havelock Ellis's *Man and Woman* (1894), which argues that pelvises of European women were wider, and thus more feminine, than those of Andamanese women because European women had achieved a higher degree of evolution and stood more erect, and second, the sexual objectification of a Khoikhoi woman named Sarah Baartman as she was exhibited as the so-called Hottentot Venus.[81] Andamanese women as a whole and Baartman specifically were, tautologically, more primitive because they were sexually aberrant when compared to the white pinnacle of evolution, and they were sexually aberrant because they had not evolved as far from ancestral life. These instances, and others like them, reflect a tendency within and adjacent to science that linked racially "lower" orders of humanity with physical failures of proper sexual development, in an intellectual milieu in which hermaphroditism was the mark of less evolved animals. Sex made a mess of sorting human from animal.

This boundary troubling occurred much more immediately than the distant conceptual as scientists positioned cases on the page and reasoned comparatively. In an oft-cited 1827 article, "Description of an Hermaphrodite Orang Outang Lately Living in Philadelphia," Richard Harlan asserted that hermaphroditism was a "vegetable attribute" and said that "the disposition . . . to hermaphroditism, is more rare, as we advance in the scale of perfection, or rather to a more complex organization."[82] Harlan then immediately proceeded to describe a human case quite similar to his titular simian example.[xvi] The "human individual in Lisbon," whom I presume to have been white given the absence of a racial marker, "unit[ed] both sexes in apparently great perfection," just like the primate in question.[83]

Even the image (figure 1.1) included with Harlan's text emphasized the human qualities of the "Hermaphrodite Orang Outang," which was drawn standing upright, holding a walking stick. Simpson's chapter on hermaphroditism mentioned above, which was cited as an authoritative text for decades, similarly transitions back and forth between animal and human cases with little fanfare, as if the comparison should be self-evident.[xvii] In Simpson's chapter, a paragraph on dogs is followed by one about a twenty-three-year-old human, and a section on goats by the example of a French soldier.[84] These are only two of the most frequently cited examples, reflective of the seeming majority of writing about hermaphroditism, especially in human medical contexts, that included reference to cases in animals. Even if the authors' intent was not to frame human hermaphroditism as a particularly animalistic trait (and, thus, in the context of the period, a particularly racialized trait), they made that connection in the textual juxtaposition of human and animal.

Confusions embedded in the taxonomic organization of racialization via sexual difference could also be more explicit. Patrick Geddes and J. Arthur Thomson, whose 1890 book *The Evolution of Sex* would go on to influence sexologists like Havelock Ellis (himself Geddes and Thomson's editor), claimed in their very first sentences, sounding much like Starkweather and Morgan, "That all higher animals are represented by distinct male and female forms is one of the most patent facts of observation. . . . In lower animals, the contrast, and indeed the separateness of the sexes often disappears."[85] Soon after, however, they admitted, "Apparent exceptions occur, it is true, among the higher animals."[86] This included degrees

---

xvi   Harlan's use of "Orang Outang" reflects the term's nineteenth-century status as an inexact category derived from the Malay "orang hutan," or "person of the forests," rather than a specific reference to what is now known as the "orangutan" (Harlan names the primate's species as *Simia concolor*). Rubis, "Orang Utan."

xvii   A multivolume edition of Simpson's selected works, released in Scotland in 1871 and the United States in 1872, contained a reprint of the chapter. The reprint joined the original 1839 version as a frequent reference in subsequent scholarship on hermaphroditism in both humans and animals.

SIMIA CONCOLOR.   Harlan.

H. M. Bird. del.   H. Kearny. sc

FIG 1.1. Plate IX, "From a drawing of the animal, taken after death," from Richard Harlan, "Description of an Hermaphrodite Orang Outang Lately Living in Philadelphia" (1827). Illustration by R. M. Bird.

of sexual dimorphism that they claimed differed among humans according to race; like their contemporaries, Geddes and Thomson imagined full sexual dimorphism to be achieved by only white Europeans and their descendants.[xviii]

Geddes and Thomson thus found themselves in something of a bind. As they made claims about the failed sexual dimorphism of a large swath of humanity, it became necessary to brush aside nonwhite sexual traits as insignificant to the understanding of sex as a whole. Geddes and Thomson assured their readers that neither exceptional animals nor exceptional Africans required the revision of statements about humans being clearly sexually dimorphic. To that end, Geddes and Thomson directly compared exceptional animals and members of exceptional—in this particular case, African—human races. Anomalies, they effectively reminded their readers, could be safely ignored. Geddes and Thomson turned to bees as their example. Drone bees, they claimed, were anomalous in that they were "male members of a very complex society, in which what is practically a third sex is represented by the great body of 'workers.'" While males were normally the most active members of a society, drone bees were notably passive. The astute reader might thus use the drone bee to counter the two authors' arguments about natural sex differences—perhaps they weren't as inherent and widespread as the authors claimed. Geddes and Thomson preemptively combatted any such reading by rendering the exceptional drones unimportant. Drones, they argued, "are no more fair examples of the natural average of males, than the hard driven wives of the lazy K . . . are of the normal functions of women."[87] Geddes and Thomson could, then, argue for the less-evolved state of Africans on the basis of failed sexual dimorphism without threatening the sexual dimorphism of whiteness. The apparent whole of the animal kingdom had no bearing on the true sexual classification system and its binary pinnacle, even as adjacency to animals at the site of sex constituted supposed racial inferiority. It is notable, too, that Geddes and Thomson not only chose bees as their

xviii Another example of the intellectual and social connections among scientists interested in sex: In his article on the freemartin discussed above, Hart thanked Thomson for sharing "valuable references" ("Structure of the Reproductive Organs," 240). Morgan, meanwhile, critiqued Geddes and Thomson's theory as not "having in any degree even a proximate solution of the problem of the determination of sex" (*Experimental Zoology*, 388). Brutal.

example, but fell closest to the "third-sex" camp on the question of how to categorize worker bees. Implicitly, they suggest that humans were also a three-sex species, with Africans and other racialized peoples directly comparable to the neuter worker.[xix]

In sum, in order to make claims about race, scientists relied on a narrative of the rare evolutionary advancement of sexual dimorphism and the more common tendency toward hermaphroditism. While many historians have described nineteenth-century and early twentieth-century sex science as a defense of the pregiven social positions of clearly delineated men and women, that understanding of a binary framework begins to break down when interrogated through the lens of species and race.[88] In other words, because of the frequency with which hermaphroditism and other forms of difficulty taxonomizing sex and gender appeared in arguments about racial hierarchy, there was no rigid sex binary to defend, at least not in the scientific literature.[xx] Rather, a binary was used or disregarded, depending on which best suited the needs of white supremacy. Sex research aided in the process of separating the human from animals, with sexual abnormality shunting human subjects down an imagined evolutionary ladder. While those conclusions supported scientific arguments about racial hierarchy, they also made the distinction between human and animal utterly precarious. Throughout, a simultaneous acceptance of binary and nonbinary sex joined the flexing, contentious definitions of male and female in the production of an all-important and incoherent classification system. Within those nonbinary possibilities, it is tempting to see a retrospective otherwise, a path that may have led to a more inclusive enact-

---

xix   Their reference to a "complex society" echoes Leland's and the unnamed *Popular Science Monthly* author's contention, discussed above, that social complexity leads to the development of not just a third sex but a *laboring* third sex. It's unclear if Geddes and Thomson were positing that Africans were inherently suited to work while Europeans could focus on reproduction, but it doesn't seem that far of an intuitive leap.

xx   In terms of social and legal position, there are some ways that that binary was more fixed, if one is considering middle-class whites: Voting rights and other political opportunities, property ownership, divisions of labor, and so on were starkly divided between male and female. However, adding race and axes of social classification to the mix instantly complicates that. Is voting a specifically male prerogative in which one is either binarily a man who can vote or a woman who can't if Black men cannot vote? Not really.

ment of sex. Queer life, however, was neither the goal nor the outcome. On the contrary, the blurring of lines would have violent consequences.

The Failure of Taxonomy

The stakes of these discussions were high. White supremacy rested on claims about difference, but those were claims rooted in sexual classification systems that could be described as incoherent at best. On a personal level, scientists' own expertise required that they wade through ambiguity to find the definitive truth within it. The work of Robert Wilson (R. W.) Shufeldt, ornithologist and sexologist (among other kinds of -ologist), provides an example of the violent panic that could result when taxonomies begin to crumble, and when experts realized that the choices they had made had very little to do with inherent difference. Shufeldt was more willing to take male and female categories for granted than many of his colleagues. Even so, binary sex could not save white humanity. For Shufeldt, an attention to reproduction made clear the fragility of efforts to taxonomize whiteness as more human. As lines between race and species faltered, Shufeldt responded with genocidal force. This section, then, is testament to the cracks that could form in a scientist's capacity to tolerate incoherence.

Historians are most familiar with Shufeldt as the author of "Biography of a Passive Pederast," a 1917 case study published in the *American Journal of Urology and Sexology* that described a sexual invert named Loop-the-Loop who was apparently quite illustrative of a newly defined category, homosexuality. George Chauncey spotlighted Shufeldt for the historical gaze in early 1990s in his now-canonical *Gay New York*, when he used the Loop-the-Loop source to discuss relationships between feminized sexual inverts known as fairies and what Chauncey termed "normal men," who presented as masculine and took a penetrative role in sex.[89] Shufeldt's article about Loop-the-Loop has since appeared in several histories of sexuality as an archetypical instance of how sexologists framed homosexuality as a matter of sexual inversion in the years before a shift to object choice.[90] From this angle, Shufeldt becomes just one among many turn-of-the-century sexologists engaging in the invention and dissemination of new sexual categories.

Shufeldt, however, like many of his generation, both dabbled and attained expert status in multiple scientific and medical realms. Though he published widely on topics like masturbation, impotence, and sexual inversion, he was also a US Army surgeon, recreational anthropologist, scientific photographer and taxidermist, massively prolific and institutionally supported ornithologist, and writer of violent antimiscegenation screeds. Expertise in one area flowed neatly into expertise in the other, and techniques of seeking out taxonomic clarity served him in all of them. A closer examination of Shufeldt's sprawling career helps reframe the history of sexology as inextricably bound up in animal and race science: Shufeldt's work hinged on the threat to human—by which he always meant white—evolutionary development. Indeed, it shows how in the nineteenth and early twentieth centuries, sex, animal, and race sciences were not separate from each other at all. Shufeldt thus poses as a microcosm of the field of sexology as an interdisciplinary enterprise, in which its practitioners self-fashioned as superior knowers specifically because their work drew on so many kinds of knowledge, including that of the life sciences. Or in other words, his work shows that the power to classify in sexology emerged out of knowledge practices that themselves contested boundaries, not merely out of the coherent articulation and consolidation of a new discipline.

Shufeldt's biography maps out the entanglements of colonialism, militarism, white supremacy, the life sciences, and sex science specifically around the turn of the twentieth century. Born in New York in 1850, Shufeldt served as a ship's clerk and signal officer for the Union in the Civil War, and at the war's end he enrolled in medical school. By the mid-1870s, he had joined the US Army as a surgeon, and he provided medical services for the military during the Great Sioux War of 1876. Throughout, he collected interesting zoological and botanical specimens, which eventually gained some notice: In 1882, he was assigned a post as the curator of the then-new Army Medical Museum and soon after that obtained an honorary curatorship at the Smithsonian. It appears that he became increasingly interested in the similar collection and study of humans when stationed in New Mexico in the late 1880s and 1890s, where he spent his leisure time photographing Indigenous people against their wishes and stealing their bones.[91] This included the skull of a Navajo woman that he used to decorate a bookshelf in his office.[xxi] By the onset of World War I

xxi Shufeldt wrote in the article cited in the previous sentence, and I quote at length because, yikes, "A few years ago I remembered very well the danger that attended my efforts to secure a few Navajo skulls for Professor Sir William Turner, of the University of Edinburgh. It came to the ears of these Indians in

he had written more than 1,100 articles on ornithology. He continued to publish well into the 1920s within and beyond that scope, including articles on sexual deviance and racial degeneration, photography, Indigenous social and economic practices, and taxidermy. He died in a sanitarium in 1932 following a long illness.[92]

Beginning at least as early as the 1870s, Shufeldt positioned himself as an expert on classification, first as it pertained to birds, later mammals, and eventually humans. He saw himself as having a particularly specialized knowledge that enabled him to see what he called the "true relations" of species that existed in nature but were not necessarily obvious to the untrained eye. "Personally," he said in a 1920 *Medical Record* article, "I have studied all the habits, history, and structure of mammals continuously for the better part of half a century," which allowed him to understand the relationships that "exist among all forms that ever have or ever will appear upon this planet." That determination of "true relations" was the highest purpose of science.[93] Shufeldt was so invested in classification that he once wrote a letter called "The Distinction Between Anatomy and Comparative Anatomy" to the editor of *Science*, taxonomizing taxonomy.[94]

In his ornithological studies, Shufeldt repeatedly insisted that distinctions between kinds not only existed in nature but could be found with the right kind of scientific expertise. "That a real relationship exists among certain and various tribes of birds . . . is a fact that it is feared those who attempt their taxonomy do not always keep impressed with sufficient strength upon their minds," he lamented in an 1898 article on classification within ornithology. "Consequently we often hear of this classifier's arrangement, and that classifier's arrangement or scheme, just as though no real affinities existed."[95] This imagining that taxonomists were essentially making things up and imposing their will on a chaotic world

---

the vicinity, and I was repeatedly cautioned not to make the attempt to carry out my designs. . . . With a large bag rolled up under my arm, and my ambulance awaiting my return at the entrance of the gorge, I climbed up to the place in a blinding snowstorm. Notwithstanding all my precautions, however, my reputation had gone ahead of me, and I found armed Indians posted in several localities, evidently there to resist my depredations at any hazard. They showed their agitation upon my approach, and I returned unsuccessful. Skulls of these Indians were, nevertheless, secured by me at a later date, and are now in the anatomical museum at the Edinburgh University." Of a particular Navajo woman, he said, "A year afterwards I secured her skull, and at this writing it adorns the top of one of the bookcases in my study."

was all wrong, Shufeldt thought.[xxii] Species *could* be accurately identified and their evolutionary development traced.[96] Classifications *were* based on resemblances of structure, and while species constantly undergo development, their differences are nonetheless rooted in physical reality.[97] Understanding those sometimes subtle differences required significant training in morphology, he felt; taxonomizing was not merely the sorting of appearances by "any observer with common powers for comparison of objects," but a careful, studied observation of ontological difference.[98] This was not merely an intellectual concern: In 1906, he decried the American failure to maintain an interest in comparative anatomy, which had been overshadowed by financial obsession. The "mad race for money," he lamented, had made him "count our vaunted civilization a miserable failure." English scientists, especially those studying evolution, had "revolutionized the entire thought of the modern world." America was falling behind, and only increased support for comparative anatomy would save the country and its "national morals."[99]

Yet as Shufeldt's career wore on, his confidence in ontological difference began to falter, especially as he increasingly oriented his work toward sexology. By the latter decades of his life, Shufeldt came to believe that "the study and investigation of the matter of sex is many times by far the most important field of research in the entire range of science."[100] Shufeldt wanted to save white American civilization with sexological knowledge, for a lack of it had "been responsible for the greatest amount of human unhappiness."[101] Good information could prevent a whole range of sexual and racial ills. Parents, for example, would be better able to rear healthy white children with a strong knowledge of heredity, physiognomy, and morphology, so that the children might be raised in accordance with their "nature" and "organization." This was important, he said, so that the child could "become a desirable factor in the make up of the nation and the race as a whole."[102] Armed with that intent, shared with many contemporaries concerned about miscegenation in the wake of Reconstruction as well as "race suicide," Shufeldt wrote books and articles and lectures, letters to the editor and reviews, bibliographies and guides, all in the hope that greater sexual knowledge would lead to social improvement. As indicated

---

xxii   Ironically, Shufeldt's obituary noted that several of the species designations he had identified based on bird fossils later had to be revised. "Because of the pioneer character of his work," the obituary apologized, "he was liable to make a few mistakes." Lambrecht, "In Memoriam," 361.

by his specific investment in white progress, studying sex was a means to an end and the key to solving the problem that he called "the general nonprogressiveness of the race" that had resulted from centuries of failing to maintain the taxonomic purity of whiteness.[103] Shufeldt was less interested in sorting male from female than some of his fellow scientists, and more interested in sexual behavior and reproduction. Nonetheless, his writings indicate another aspect of the classification problems that suffused sex research in the late nineteenth and early twentieth centuries, as his study of sexology rendered other human taxonomies suspect and sent him scrambling to reinforce them.

Shufeldt feared the effects of the mixing of populations as white humans encountered racialized others in the course of new territorial conquests.[xxiii] He had seen it firsthand as an army surgeon on the front lines of Western expansion, and his father, a navy commodore also named Robert Wilson Shufeldt, had sailed around the world as an agent of American colonialism. When differently racialized populations meet, Shufeldt the younger cautioned on the basis of this experience: "It matters not a whit whether one of these races is highly developed in every way, and the other a savage."[104] Racial hybridity would occur in all such cases because the "demand to satisfy the sexual appetite in the male, even if he be an individual endowed with great refinement and learning, often completely blinds him, . . . and he will have carnal intercourse with almost anything in the shape of a woman."[105] American slavery, he added, had proven exactly that point—even though the "imported" race "came nearer the anthropoid apes than any other known group of people," the "negresses of that race have borne untold thousands of children to representatives of the white race in this country."[106] This, not the violence of forced labor and sexual exploitation, was apparently the true tragedy of slavery.

Shufeldt's fears about racial hybridity belie his apparently prodigious talent for identifying distinction. Traces of insecurity plagued even binary sex, reflecting the conflicting models of sex present in sexualized racial difference: The "negresses" were not, in fact, women, but merely "in the shape of a woman" and indeed required a word other than "woman" to mark them. Despite his vocal confidence in his own abilities to tease out the minute

---

xxiii   Shufeldt noted that the "crossing of species of men" has always taken place, but increasing populations, modern methods of travel and communication, and widespread migration had led to a decided uptick in frequency. *The Negro*, 85.

differences between species, Shufeldt's fears about racial hybridity rested on both the uncertainty of distinctions between groups and the desire to position Black women as other-than-women while panicking about their fecundity. Shortly after the turn of the twentieth century, Shufeldt's obsessions with taxonomy, white supremacy, and animality culminated in an intense focus on miscegenation in his writing, most notably in his 1907 book, *The Negro, a Menace to American Civilization*. He hoped that this book would convince the white American public and especially the federal government to transport all African Americans to Liberia—although he didn't particularly care where they ended up and said he would prefer to see anyone of African ancestry "at the bottom of the Pacific Ocean" rather than in the United States, where they would "jeopardize by race inter-mixture" the entirety of American civilization.[xxiv]

Shufeldt dedicated *The Negro* to Edward Drinker Cope, the paleontolo-gist and comparative anatomist, and one of Shufeldt's favorite taxonomists, for "the sincere effort he once made to maintain the purity of the race, of which he was so distinguished a representative."[107] Though Shufeldt began with this tribute to taxonomy, the order of the world quickly began to break down. Even with his powers of comparative anatomy, he re-ported that he found it difficult to draw clear lines of racial distinction—reminiscent, perhaps, of Watson's trouble with his hyena specimens. Some of the frightening racial hybrids he thought would destroy white society were, he wrote, "so white, it takes a very keen eye to detect the Ethiopian blood in them" and were particularly dangerous because "they may pos-sess all of the vicious and sensual traits of the negro, without the color of the latter's skin as a warning flag to the unwary."[108] This was not alto-gether different from his take on birds—finding true relations again de-pended on the discerning eye—but a note of panic began to creep in. The birds were, fundamentally, different from each other, and Shufeldt merely had to sort them out. Racial hybridity literally born of binary sex chipped away at those differences one generation at a time, such that they would eventually disappear and no eye, no matter how well trained, would be able to see the difference.

xxiv   Shufeldt, *The Negro*, 161–62. Notably, the elder Shufeldt made multi-ple trips to Liberia during his naval career and was a staunch supporter of the "repatriation" of free people of color from the United States to Liberia. See Shufeldt Sr., *Liberia*.

This called into question everything that Shufeldt held dear in the realm of classification. "Nature cares nothing about the species, lions and tigers have mated and offspring hybrids have resulted; and misfortunes of all misfortunes," he lamented, "the white and black races are interbreeding in the United States."[109] Indeed, this crisis of categories was so great that it transcended species, such that, quoting T. H. Huxley, "no structural line of demarcation . . . can be drawn between the animal world and ourselves."[110] As much as he wanted to believe that the fundamental separations that drove his taxonomical work existed somewhere in the material world, Shufeldt seemed to understand that he was the one imposing boundaries on nature.

Shufeldt concluded that managing humans like animals might be the solution. "The strong tendency [of nature] is toward the unification of all the races," he cautioned, but nature could be pushed along in the right direction.[111] Humans could be separated from each other, their traits administered like the artificial selection that produced domestic breeds of dogs and pigeons.[112] Thus, Shufeldt's plan for artificially selecting for whiteness in the United States: After toying with castrating all of the Black men in the United States and deciding that was an inadequate solution, he determined that the only way to stop the "horrors of their crossing continually with the Anglo-Saxon stock" was to deport anyone of African descent.[113] In his justification of this plan, he came back to birds. "On the same principle that were I building up a fine aviary," he explained, "I certainly would not tolerate the presence of a lot of black, undesirable, unrefined, flesh-eating vultures among my sweet-singing, educated and attractive bullfinches—that is all."[114]

Shufeldt feared the slide of whiteness back to a more animalistic state as a result of sexual practices. But in the process of making his racial arguments, Shufeldt himself constantly blurred the line between human and animal. When he decried hybridity based on its potential for atavism, for example, he used the coat patterns of wild horses as his example for how atavism functioned before segueing into humans.[115] His hopes for racial betterment came directly from his understanding of the breeding of domesticated dogs, pigeons, chickens, and horses.[116] He viewed the United States as a giant aviary awaiting expert manipulation. Enhancing the humanness of whiteness involved not a distancing from animality but an embrace of it: the recognition that the supposedly most human of all humans was just like any other animal, beholden to base sexual whims that would destroy taxonomies. Shufeldt's anxieties demonstrate that sexual incoherence was not only about sexual classification per se. If racial difference depended on sexual difference, and if sexual difference depended on evolutionary development, then these ways of sorting life, too, rested on an unsteady foundation. Incoherence enabled the persistence of binary sex in the face of a racial hierarchy dependent on a spectrum of sex.

Researchers who used animals as test cases for learning about sex produced two hierarchies of life rooted in sexual difference or lack thereof. In one, which took a clear sex binary as axiomatic, scientists could shore up their expert authority by identifying the "true" male or female sex of animal specimens, which also performed the cultural work of articulating stark differences between white women and men that could then be mobilized in arguments about the role of white women in politics, labor, and public life more generally. In the other, researchers described hermaphroditism as a common mark of "lower" organisms. This second model, in which sexual dimorphism was the infrequently attained standard by which racial progress and civilizational development were measured, denied the humanity of anyone racialized as not white and established sex as a nebulous continuum in which one could be more or less male or female, or potentially neither, according to one's race or species.

Neither of these hierarchies was particularly stable. Looking closely at animal science demonstrates that incoherence persisted throughout the study of sex and its use across fields, even when nature was supposed to provide evidence for the correctness of human social orders that privileged maleness and whiteness. This unintentionally left space for alternative ways of being and belies the narrative that all bodies fit into certain kinds of categories or that most bodies are normal bodies. At the same time, that space allowed for and justified racial violence, reminding us that sex "beyond the binary" contains possibilities for brutality as much as for freedom. Classifying sex was a fraught enterprise, requiring far more effort than it would need if all creatures belonged self-evidently to obvious categories. The history of sex research in this period, at least where animals are concerned, isn't one of hegemonic disciplining so much as a failure to achieve order. An understanding that sexual taxonomies could not contain the variation found in nature endured throughout the nineteenth and early twentieth centuries, to such an extent that they had to be actively managed and, for some, caused a sense of panic about the future of white humanity. Through this lens, binary sex—and as the next chapter will show, sex as a category entirely—comes to look less like a systemic truth taken for granted and more like a fleeting aspiration.

2

Conflicting Sexes at
Two Eugenics Laboratories

Cold Spring Harbor, tucked away in an inlet on the north shore of New York's Long Island, was a whaling hub until the industry declined in the 1860s and it refashioned itself as a resort town on the water. When Charles Davenport founded the Station for Experimental Evolution (SEE) there in 1904, he gave the town a new use for nonhuman bodies: eugenics.[i] Pigeons and chickens, dogs and mice, beetles, corn, and fungi would all have a role

---

i   Context for this chapter: I grew up a straight line south from Cold Spring Harbor as the crow flies and a somewhat circuitous path as the car drives, on the opposite shore of Long Island. First in elementary school and again in high school, I attended genetics summer camp at the Dolan DNA Learning Center, the educational arm of what is now Cold Spring Harbor Laboratory (CSHL). From these formative experiences, I learned a lot about pipetting technique and how to make *E. coli* glow in the dark, but no one ever mentioned that CSHL was, having been rebranded, the same institution that had led the country in eugenics research many decades previously. And so when I kept stumbling across references to Cold Spring Harbor in my early archival research for this project, I had to follow the breadcrumb trail back to the place I knew. In another twist of fate, one of my PhD advisors, Joanna Radin, had also attended the same genetics summer program. There is, I think, a lot to be said about how that program didn't turn us into geneticists, but we did become science and technology studies (STS) scholars.

in what Davenport called "the analytic and experimental study of the causes of specific differentiation—of race change."[1] The study of humans joined the effort with Davenport's founding of the nearby Eugenics Record Office (ERO) in 1910. These two laboratories made Cold Spring Harbor an epicenter of eugenics research and, as it would turn out, of sex research.[ii]

In the early decades of the twentieth century, eugenics researchers dreamed of achieving white perfection through a scientific approach to breeding. They aimed to make a better future for themselves and their descendants in the form of a physically, mentally, and morally robust population. Anticipated generations would be free from traits like blindness, feeblemindedness, alcoholism, and other negative qualities that, left unchecked, would be a draw on the coffers of those they had decided were genetically destined to lead society. To do so, these researchers set out to control heredity. Meticulous tracking and innovative experimentation would show them how traits were transmitted through generations. With that knowledge, they could make—and encourage or force others to make—the good reproductive choices that would bring eugenicist desires to fruition. Reproduction and therefore sex were, after all, the primary mechanism of eugenics, the way that human traits would be emphasized and eradicated through the crossing of genetic material, much like Gregor Mendel's peas.[iii] Scientifically augmented, able-bodied white supremacy, then, would

ii   I refer to the ERO as a laboratory in part because I see it as an extension of the SEE—though not experimental in nature, it was, as I will describe in more detail later, an effort to produce as much of an understanding of the same concepts of heredity as possible without turning directly to human experimentation. Moreover, it was intended to do the *work* of a laboratory, in the way that Bruno Latour has discussed it: taking a chaotic world on the "outside," distilling it down to something more controlled and knowable on the "inside," and then publicly demonstrating that what happened "inside" is useful "outside." Much more could be said about this; here, I will use "laboratory" as a shorthand insistence that the SEE and ERO were comparable institutions. See Latour, "Give Me a Laboratory."

iii   Gregor Mendel's mid-nineteenth-century experiments with crossing different types of pea plants had shown that discrete traits—in his case, things like smooth or wrinkly seeds, or yellow or green pods—were generationally transferable as coherent characteristics, rather than ever-increasingly blending together, and that there were dominant and recessive forms that appear in predictable ways. Mendel's work was largely ignored in the 1860s but taken up in 1900 by several scientists and would go on to be a key component in the development of modern genetics. Davenport's 1902 proposal for Carnegie Institution funding to

require a deep understanding of how these mechanisms worked. This research on sex would have tremendous practical implications: The fate of what they called "the race" (typically a shortened form of the "human race" or the "race of man," presumed to be, in its ideal form, Anglo-Saxon) depended on the study of sex.[iv] Sex came to suffuse every aspect of eugenic research at Cold Spring Harbor, from Davenport's high-level mission to control and improve breeding down to the collection and recording of data.

For the scientists working at the Cold Spring Harbor labs, though, sex was not a preexisting, unified object of study. In the course of analyzing the transmission of traits, the manipulation of bodies through experimentation and targeted breeding, and how sex and reproduction even worked, these scientists created their own versions of sex.[v] As with the animal research discussed in chapter 1, sex at Cold Spring Harbor became a constellation of traits and processes that bunched together even as they often conflicted. While something called "sex" sometimes functioned as a binary variable of male or female used for making sense of vast quantities of eugenic research data, sex could also be malleable bodily morphology, reproductive capacity, hormonal makeup, metabolic rate, and behavior. This coexistence enabled eugenic researchers to deploy sex in a wide range of arguments and situations, where a single meaning would have limited both theory and policy implications within a confining internal consistency. Eugenicists in the early twentieth century depended on an incoherent conception of sex in order to make their claims to scientifically backed racial superiority work.

I propose that this use of incoherent sex in the service of eugenic racial violence ("racial improvement," as the eugenicists would have it) was an important part of the development of twentieth- and twenty-first-century sex science. Historians have overwhelmingly demonstrated a long-standing connection between sex science and race science that

---

support the SEE (Davenport, "Biological Experiment Station") reflects precisely these principles in its statement of methods and aims.

iv   Francis Galton's initial writings on what would become eugenics primarily use "human race" or "race of man," and refer also to "our race." Subsequent eugenic writers often simply said "the race" as if it needed no specificity—and, for them, it probably didn't. Galton, "Hereditary Talent and Character."

v   A reminder: Scientific research is not merely about observing what already exists. Through an articulation of an object as a thing *to* study, and through practices and narratives that emphasize particular aspects of existence, scientists bring those objects into being. See the introduction.

reached far beyond Cold Spring Harbor.[2] Scholars explicating that connection have attended particularly closely to sexologists' portrayal of homosexuality and sexual inversion as symptoms of racial degeneracy.[3] However, that conversation has tended to be limited in its scope by its focus on sexology specifically and in its use of mostly published sources. In one of the bedrock texts of this discussion among historians of sexuality, Siobhan Somerville argued that imagined bodily differences of both African women and white lesbians were located in genital excess.[vi] The apparently failed sexual dimorphism of white inverts and Africans (and, by extension, African Americans) was used in the nineteenth century to make claims about the primitivism of both types of people.[vii] But Somerville made an interesting caveat, one that is fairly representative of the conversation about race and the construction of homosexuality that she helped spawn. The construction of categories of race and sex was not "a mere historical coincidence," Somerville argued in her introduction, but her "goal, however, is not to garner and display unequivocal evidence of the direct influence of racial categories on those who were developing scientific models of homosexuality." Instead, the article would focus on "racial ideologies, the cultural assumptions and systems of representation" that informed both race science and sex science."[4]

Since then, this is largely how historians have approached the co-constitution of racial and sexual categories: as a matter of ideology, assumption, and representation. Scholars have shown in great detail the attention that sexologists like Richard von Krafft-Ebing and Havelock Ellis paid to theories of degeneracy in their work as they attempted to determine whether homosexuality or sexual inversion was a mere individual pathology or a wholesale racial descent.[5] Recent literature on the relationship between race and bodily plasticity, too, has largely analyzed racialized sexual categories as emerging from a series of fairly canonical sexological texts.[viii] With this framing, sex scientists were primarily people

vi   Foundational, too, to my own scholarly career—this was the article that blew my senior-in-college mind so much it convinced me to go to grad school.
vii   While certainly not the first person to make claims about the co-constitution of racial and sexual categories, Somerville remains widely cited among historians of sexuality. Interestingly, the article doesn't cite people like Hortense Spillers or Kimberlé Crenshaw—but clearly Somerville is building on preexisting conversations among Black feminist thinkers. See also chapter I.
viii   See, e.g., Snorton, *Black on Both Sides*; Gill-Peterson, *Histories of the Transgender Child*; Schuller, *Biopolitics of Feeling*; and Amin, "Trans* Plasticity," which all

who wrote books and articles, and those books and articles manifested a shared language in which sex could be defined through racial terms, and vice versa. In other words, these histories (see chapter 1) focused primarily on the product of sex and race science, rather than on the science itself.

Here, I build on the links that these scholars made between sex and race science to examine the connection between eugenics and sexology— and their inseparability—on a granular level.[6] This chapter zooms in on eugenics specifically and tracks the research processes that scientists engaged in as they sought to both use theories of the malleable body to their advantage and cope with what the plasticity of sex would mean for social orders that seemed to demand self-evident categories of life. As eugenics and sexology developed into institutionally supported fields, the relationship between race science and sex science was not merely a matter of shared terms or overlapping interests and investments; rather, the two were mutually constitutive in the early decades of the twentieth century.[7] There is much to be said about the structural entanglement of funding resources in American eugenics and sex research. As just one example of this entanglement, a substantial amount of the funding for SEE and ERO research came from the National Research Council Committee for Research on Problems of Sex (CRPS), best known for its generous funding of Alfred Kinsey's work two decades later.[8] The CRPS was, for over two decades, headed by the primatologist Robert Mearns Yerkes. Yerkes, meanwhile, was Davenport's former student; indeed, Davenport had introduced him to eugenic research in the first place.[9] While efforts to have Yerkes set up a lab in Cold Spring Harbor never panned out, he nonetheless supported Davenport's enterprise throughout his career: He contributed his expertise to the analysis of heritable psychological traits, enrolled his own students to submit family data to the ERO, and served as chairman of both the ERO's Committee on Inheritance of Mental Traits and the National Research Council's eugenics committee.[10] Davenport, too,

—————————

attend to racialized narratives of which bodies are sexually malleable, whether Black women's bodies forcibly de-gendered as in Snorton or white bodies excessively responsive to their environment as in Schuller. Notably, while these scholars have made important strides in articulating the connections between sex science and race science, in a sense the field remains within Somerville's approach favoring representation, ideology, and assumption, where the crucial conceptual and theoretical insights of such scholarship are not necessarily matched by a fine-grained investigation of practice or the exact relationships that allow for and create those discursive effects.

took part in several CRPS studies.[11] While tracing that out is beyond the scope of this chapter, the gist is that throughout the period of time covered here, sexologists and eugenicists were in active, self-conscious conversation, regularly seeking each other out for input and aware of the alignments between their work. Each field offered the other legitimacy, money, and tools and data to work with as they grew not merely alongside each other, but inseparably entwined through both material infrastructure and social and intellectual networks.

This attention to the on-the-ground practices that take eugenic research seriously as sex research adds to the work of scholars writing at the nexus of the history of genetics, eugenics, and sex. Sarah Richardson, for example, has written about how the designation of the "sex chromosomes" was a turning point in the history of chromosomal theories of inheritance more broadly, and Jules Gill-Peterson has shown how research at the SEE contributed to an imagining and practical use of sex as malleable. I offer an additional argument: Some of the messiness of the category of sex can trace its roots back to the everyday practices of eugenic research about inheritance and genetics. Sex, that wily creature, was made in the laboratories of eugenics.[12]

This chapter traces, with Cold Spring Harbor as exemplar, a trajectory of the history of sex in which its utility as a mechanism for race "improvement" enabled its elevation as a topic of serious science. It examines the research undertaken at the SEE and the ERO at its most quotidian, from paperwork to training guides to administrative correspondence, and shows how these twin laboratories came to rely on virtually opposite definitions and enactments of sex. Ultimately, sex was built into every aspect of eugenic research at Cold Spring Harbor, but in consistently inconsistent ways. At the SEE, researchers emphasized scientific control over sexual malleability, and they investigated the wide range of forms that sex could take beyond the human. At the ERO, data collection and analysis depended on a fixed binary model of sex that was useful for tracking heredity of traits and making copious amounts of information both sortable and intelligible. These disparate meanings of sex did not coalesce into one; rather, in the eugenic research of the early twentieth century, there was no singular entity of sex. On the contrary, this research could take place because multiple contradictory sexes existed at the same time, with no need for ideological coherence between them. While it's not particularly unusual for researchers at the same institution to vary in their approaches—I can tell you with the utmost confidence that I have an extremely different view of sex than many of my colleagues—the case of the Cold Spring Harbor labs demonstrates with abundant clarity that different research called for different sexes.

A note on race: It might seem odd that, for a chapter about eugenics, there is almost no discussion of race itself here. This absence is just as odd in the sources themselves. While Davenport and his colleagues did examine the heredity of skin tone and hair type among people of different racial classifications in Jamaica, and investigated the outcomes of different "race crossings" like, as their notation put it, "Caucasian x Mongolian" and "Caucasian x Eskimo," the vast majority of their human subjects were white or members of national groups that later became white (Irish and Italians, for example).[13] It appears that for the most part, the eugenicists I discuss here had already made their decisions about who was racially worthy of better breeding practices, and who could not be saved and should merely be sterilized, and the distinction warranted little discussion or research.[14] The researchers I discuss in this chapter rarely named what race they wanted to improve, and as noted above, they wrote more generally of "the race" or "humanity." Eugenic research had a rather muddled approach to what today we might separate out as race, class, able-bodiedness, sexuality, and other now purportedly distinct axes of difference. For example, Nicole Hahn Rafter has pointed out that the racial anxieties held by researchers associated with Davenport often latched on to the white rural poor, particularly those living outside the disciplinary bounds of industrial labor.[15] Meanwhile, eugenic aspirations and practices wholeheartedly linked racial improvement to a weeding out of disability of all kinds, from epilepsy to flat feet to the conveniently open-ended "feeblemindedness."[16] Eugenic research—sex research included—thus targeted a wide range of people considered undesirable, while nonetheless being grounded in a racial logic that centered white supremacy.[17]

A Brief Institutional History

Though this chapter focuses on Cold Spring Harbor, Davenport's labs were not the sole headquarters of the American eugenics movement. Several other organizations, including the American Breeders' Association,[ix] the Race Betterment Foundation, and the American Eugenics Society,

ix   Sound familiar?

among others, dotted the eugenic landscape and operated in conversation with the Cold Spring Harbor labs. The fantasy and action of eugenic control over the US population held by these organizations and plenty of nonexperts and unaffiliated scientists was diffuse in its ideology, chosen practices, and geographic situation, all of which changed over time.[18] Within that context, the ERO and the SEE had a tremendous influence on the scientization of eugenics and genetics in the early twentieth-century United States, and the ERO especially was an important force in both proselytizing a eugenic worldview and getting nonprofessional individuals to think about themselves and their families in eugenic terms, as transmitters of hereditary traits across generations.[19] As such, the ERO's voluminous collection of case files, family pedigrees, and research files offers a lens into a widely influential strain of eugenics. The SEE, meanwhile, was a crucial location for genetics research in the United States in this period, eventually becoming the Department of Genetics at the Carnegie Institution of Washington (CIW).

Cold Spring Harbor researchers knowingly participated in a broader reconfiguration of sex happening in this moment. In the early decades of the twentieth century, scientists around the world increasingly looked to research in gonad transplantation and hormones as evidence of universal bisexuality, or the idea that human and nonhuman bodies alike were all located on a spectrum of sexual characteristics—and could potentially move around on that spectrum—rather than being locked into binary categories.[20] At the same time, state demands for adherence to binary sex largely increased, often with the backing of scientific and medical authority.[21] As in Cold Spring Harbor, different desired outcomes created different enactments of sex: Cutting-edge, maverick scientific experts made sex something counterintuitive that they could manipulate in the lab, while government agencies seeking control made it something easily administrable with bureaucratic tools. Sex did not linearly change over time as much as take a range of shapes concurrently that shifted, depending on location and purpose.

Given the centrality of Cold Spring Harbor and especially the ERO in the history of eugenics in the United States, much has been written about it, and curious readers can find extensive descriptions of its founding, day-to-day functioning, and role within the greater eugenics movement if they so desire.[22] The general overview is as follows. Before founding the labs, Davenport, a zoologist by training, earned a PhD in biology at Harvard before securing an assistant professorship at the University of Chicago with the help of his wife, Gertrude Crotty Davenport.[23] He initially aligned himself with the burgeoning method of biometry (essentially, applying statistical methods to biological research), and his early publications helped introduce US scientists to the quantitative and statistical approaches developed by Francis Galton and Karl Pearson. Davenport first encountered the space that would become the labs' campus in 1898, when he directed the summer school of the Biological Laboratory of the Brooklyn Institute of Arts and Sciences, located in Cold Spring Harbor for easy access to marine life and generally varied ecology.

The SEE and ERO owed the possibility of their founding to a sweeping shift in the political economy of science. Like many other areas of science and medicine at the turn of the twentieth century, eugenic science found its funding in the new largesse of the philanthropic organizations that allowed recently enriched magnates to engage in moral capitalism, fight social decay, and improve their public image.[24] In 1902, Davenport made a proposal to the CIW for a new laboratory at the familiar Cold Spring Harbor site, which would use experimental methods to learn about species development and the inheritance of traits, in light of the rediscovery of Gregor Mendel's work on heredity, and to re-create Darwinian selection in the laboratory by artificial means.[25] By 1903, he had successfully secured the grant—underwritten by the accumulated wealth of Andrew Carnegie and his steel empire—that allowed him to the establish the SEE the following year. Davenport, of course, would be its director.[26]

Over the next several years, Davenport's interest in the inheritance of physical, psychological, and behavioral traits in humans, and his desire to figure out how to breed for the best ones and eradicate those that supposedly led to crime, economic waste, and other forms of social degeneration, continued to develop. Gertrude Crotty Davenport, a prolific eugenic researcher in her own right, urged her husband to pursue the study of heredity for racial betterment—she had been invested in the cause long before Charles turned his attention to it.[27] During this time, Davenport joined the American Breeders' Association (ABA), which had been founded the same year he obtained the CIW grant. The ABA brought together a range of constituents interested in developing rationalized breeding methods in the context of the quickly accelerating study of evolution and genetics, strengthening the link between agricultural practice and the burgeoning biological sciences. The ABA's focus on improvement through science ensured its place as a cornerstone of American eugenics.[28] In 1906, the ABA established a Committee on Eugenics; by 1909, at

Davenport's prompting, it had promoted the committee to a full section of the organization.[29] Davenport became the new section's secretary, and eugenics became the primary focus of Davenport's work at the SEE.

It soon became apparent that the ABA eugenics section did not have the resources to pursue scientific research to the extent Davenport desired on its own, so, building on his success with the SEE and in connection with the ABA, he began the process of founding another institution—which, so that he could closely oversee it, would be located just down the road from the SEE. In 1910, Davenport secured funding from Mary Williamson Harriman, widow of the railroad magnate Edward Henry Harriman, to create the ERO.[30] This influx of dollars first allowed for the transformation of an old home into an office complex, complete with a fireproof vault for storing records, and later financed staff salaries and the massive infrastructure of filing cabinets, index cards, and form-printing required to amass and organize the data that Davenport felt necessary for understanding trait inheritance in the American population.[31] With this infrastructure in place, the ERO was poised to join the SEE in making Cold Spring Harbor a focal point of eugenic research and its incoherent construction of sex.[x]

Experimental Sex

At the SEE, researchers' investigations of how sex worked focused on its manipulation in the service of controlling reproduction. Their studies began in the bodies of nonhumans, from crops to chickens and, Davenport initially hoped, chimpanzees. Such bodies could be easily modified and closely observed in the controlled environment of the laboratory, with the proliferation of generations made possible on a timescale much shorter than that of human reproduction. Davenport was explicit about his desire that this research be transferable to human eugenics. "Neither physically nor psychologically is there a sharp break in the animal series where it culminates in man," Davenport reflected in a 1907 report to the CIW. "Consequently the discovery of the laws of organic evolution is . . . a study of human evolution. Since when we know the law we may control

---

x   As mentioned, later funding would come from the CRPS, itself funded by the Rockefeller Foundation, as well as from the CIW once the ERO got absorbed into the CIW Department of Genetics. Big snore, I know, but a useful fact to whip out when you get questions about what "real" geneticists were doing at this moment. Eugenics research, it turns out!

the process, the principles of evolution will show the way to an improvement of the human race."[32] The study of the laws of heredity and evolution that Davenport so valued quickly led SEE researchers to study sex.

Davenport's involvement with the ABA had emphasized for him the relationship between scientific practice, improvements in breeding, and sex. In 1910, the year of the ERO's inauguration, Davenport penned an editorial for *American Breeders Magazine* titled "The Relation of the Association to Pure Research." In it, he noted that even though "to the scholastic biologist of our universities the work of the 'Breeder' has for long been regarded with contempt," "the future of research in breeding is bright," thanks to new experimental methods that had begun to scientize and systematize breeding and thus lend it the legitimacy it deserved.[33] Sex, Davenport reported, was chief among the topics being pursued in this research.[34]

Over the first two decades of the SEE's life, investigations into sex and reproduction were the backbone of its operations. Davenport retrospectively characterized the institute in a 1925 letter to Robert Latou Dickinson as essentially "devoted to a study of the phenomena of reproduction and development in animals and plants." He described the scope of that research: "For some years," he wrote, "we have been working upon sex differentiation with special reference to its basis in the chromosomes and in catalyzers found in the body of the adult." He then listed several lines of inquiry being undertaken at the SEE, which included "the basis of the ability to call forth male production in a parthenogenetic species," "the phenomenon of exclusive male production in certain matings of some of the insects," "regeneration of ovary," "ovulation in twin production," "intrauterine deaths in mice," "control of primary and secondary sex characters in pigeons," "the chemical basis of sex distinction where there is no morphological distinction, as in the mucors," and "the study of sex intergrades and their hereditary basis." Davenport again emphasized that this animal research was of broader relevance. "We have no doubt at all," he concluded, "that general principles secured from a study of the lower organisms will be applicable to man." Davenport's list neatly dovetailed with contemporaneous progress in human sexology: The investigations ongoing at the SEE ran "very close to and, in a way, coordinate with the [recently established, Rockefeller-funded] National Research Council for Research on Sex Problems."[35]

From their earliest instantiations, long before Davenport enumerated the SEE's sex-related project to Dickinson, SEE researchers' efforts to pin down what determined sex in individual development and what sex itself was relied on an elastic enactment of sex. Sex, in this research so central to eugenic progress, was both modifiable in the lab by expert hands and incredibly varied in nature. SEE scientists' work conformed, particularly in the first regard, with a broader research agenda in sex science, especially among their colleagues in Europe. There, building in part on the interest in hermaphroditism discussed in chapter 1, studies of sexual inversion provided evidence that an "intermediate sex" might exist and that sexual pathologies could emerge from a somatic mixing of masculine and feminine characteristics.[36] In the early 1910s, European scientists, most notably Eugen Steinach, also began experimenting with castration and gonad transplantation and demonstrated that sex characteristics could be changed.[37] At the SEE, this apparent fluidity would be of great service; shifting the sexual characteristics of animals became key to innovations in eugenic breeding. Two of Davenport's scientists, Hubert Dana Goodale and Oscar Riddle, used similar methods as their European counterparts to determine whether gonads and their secretions dictated what sex characteristics a body developed, and how that sexual development might be controlled. Both used easily available poultry as their experimental subjects.

In 1912, Goodale, at the time a graduate student working on genetics at the SEE, removed the left ovary of a female chicken that subsequently assumed "most, but not all, of the characters of the male." The experiment, Davenport reported, "seems well calculated to demonstrate the essentially hermaphrodite nature of the female fowl."[38] Goodale himself quibbled with this assessment—he specified in his own publication the result that the female chicken was a *pseudo*hermaphrodite—and was unsure as to whether his experiment made much of an impact on how scientists should understand the process of sex determination.[39] He wrote, sarcastically, that his findings did not reveal anything new about sex determination, "unless we wish to extend the idea of internal secretions by assuming that all individuals are hermaphroditic and that at some period after fertilization a mechanism comes into operation that partially or wholly suppresses the opposite sex."[xi] Whether or not one was willing to go to such lengths, Goodale specified that the experiment *did* indicate that scientists "need no longer erect separate categories for characters

---

xi   "Internal secretions" here refers to a hormonal model of sex, which, as you may recall from the discussion of Frank Lillie's research on freemartins in chapter 1, had only very recently started to popularize among biologists. The hypothesis that Goodale mocked was precisely the point of universal bisexuality.

that appear in one sex only, but may classify a given character in the female with the corresponding character in the male, even though the two characters actually appear very unlike."[40] Particular feather patterns, to use one of Goodale's examples, were not actually *male* or *female* patterns but were instead outcomes of other bodily processes like the circulation of internal secretions, which again gave evidence to the possibility of editing an individual body to improve its traits. "The determining factor," as Goodale put it, "is not in itself sex."[41] Sex, then, could not change (which would require, in Goodale's view, a change in type of gamete production), but *bodies* could. Goodale's avian gonadectomy experiments did perhaps more than he realized: They demonstrated that experimental expertise could be wielded in order to modify sex characteristics.

Though Goodale soon departed the SEE for the Massachusetts Agricultural Experiment Station, by 1916, other projects enabled Davenport to make even bolder statements about how sex truly functioned.[42] "To the Station," Davenport told the CIW in that year's annual report, "seems to have fallen the opportunity of demonstrating that the current view that sex is determined solely by the sex chromosome is too narrow."[43] Davenport based his claim primarily on Riddle's work. Riddle had been a research associate at the SEE since 1912, and remained so until his retirement in 1945. He spent his career developing a metabolic theory of sex, which held that sex was plastic and subject to changes in caloric use. With this theory informing his research, he argued that he could determine the sex of future offspring through breeding practices, thus controlling male/female sex ratios.[44] Perhaps most thrilling to Davenport was Riddle's work on what he called "sex reversal" in pigeons: the transformation of male pigeons into female ones, and vice versa (see figure 2.1). Through injections of testicular and ovarian extracts, Riddle was able to change birds' sex behavior and so-called secondary sex characteristics.[45] Riddle's work, along with that of his colleagues like Goodale and Arthur Banta, who worked on sex intergrades in crustacea, led Davenport to conclude that "sex in general is a much less fixed and precise state than is commonly supposed."[46]

Transforming a pigeon's sex was not a laboratory exercise for its own sake. Such experiments would hopefully have tremendous practical bearing on understandings of heredity. The implementation of that work would be rooted in the discovery that sex could be modified. Riddle, like Davenport, saw these aspects of his research as beneficial to the mission of eugenics: Control sex, and you could control more aspects of breeding. As Riddle concluded in an undated lecture, "The complete transformation of sex in adult

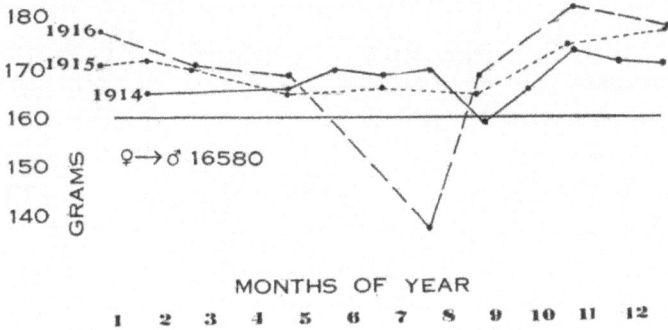

FIG 2.1. Graph from Oscar Riddle, "Sex-Reversal in the Adult Pigeon" (1924). The original caption reads "Curves showing body weight of ♀ → ♂ [Specimen No.] 16,580–1914 to January, 1917." The three lines demonstrate an upward trend in body weight over time after tuberculosis destroyed the pigeon's ovary (birds possess only one), as well as seasonal fluctuations. Before No. 16,580's death, the pigeon had begun to crow and force "her *male* mate to function as a female in copulation." Upon autopsy, Riddle located testicular tissue in the remains of the ovary. He took this as evidence, along with the results from his experiments with testicular and ovarian extracts, that sex was modifiable.

higher vertebrates has now been demonstrated" through his work with pigeons. "The demonstration of the complete transformability—not a mere modifiability—of one truly hereditary and chromosome-determined character [i.e., sex] now makes it quite illogical to assume that any hereditary character similarly founded in the germ may not be transformed in and during the life of the individual." He predicted that "this aspect of our knowledge of the heredity and development, when adequately developed and applied, should prove of much interest and of very real practical value to mankind."[47]

For Riddle, compared to Davenport, it seems that the practical outcome of his research may have had less to do with specific changes that might be made to individual bodies, and more to do with a fundamental shift in imagined possibility.[xii] Riddle insisted that scientists' understanding

xii   As it would turn out, Riddle's experiments *would* have practical use by contributing to the groundwork for hormonal therapies for trans people. This was, ah, probably not what he had in mind. See Gill-Peterson, *Histories of the Transgender Child*, 54–56.

of sex had been limited by an assumption that sex was fixed and that scientists actually had much more power over sex than they believed. His work had revealed a bodily malleability, which "now enables us in the case of some animals to force this character to develop into its alternative or opposite form, a hitherto unsuspected measure of experimental control over numerous, possibly all, similar hereditary characters becomes theoretically possible."[48] Sex had seemed unshakeable; Riddle's work demonstrated the power of scientific expertise.[49] Riddle put it another way in a 1917 article: As a result of his proving that sex can be experimentally controlled, "territory hitherto labelled 'impossible' is open to investigation."[50]

Riddle reiterated the practical importance of research on the malleability of sex for eugenics when he complained about his stagnant wages in 1921.[xiii] That Davenport wasn't paying him enough, he said, "cannot rest upon any lack of importance or relevancy of my work to the chief purposes of a Station for Experimental Evolution; for, sex, the modifiability of sex in particular, is of central interest here; it is moreover now being widely recognized as a practiceable [sic] mode of attack on the fundamental aspects of heredity and evolution." The emphasis was Riddle's own, a reminder to Davenport that his work could not be dismissed as an abstract curiosity. "It has also long been so recognized by the Station," Riddle continued, "and much of the past and present work of the Station is a study of one or another aspect of sex."[51] This vitally important sex research and vision of sex as something one could edit with the right expertise would unlock the power of science to sculpt humanity.[xiv]

Some SEE research, then, demonstrated that sex was not a static thing. Other research expanded the definition of sex itself. Like the animal researchers in chapter 1 who reinvented maleness and femaleness, botanists Albert Blakeslee and Sophie Satina found sex where it was not immediately

---

xiii   It appears that Davenport was generally rather stingy when it came to his employees' paychecks, and there are countless instances in the archives of researchers asking for pay increases. Beyond Riddle's appeal for increased salary, Sophie Satina, Mabel Earle, and several of the ERO fieldworkers whom we'll meet later in this chapter also wrote to Davenport about insufficient pay.

xiv   The New York Times reported on Riddle's work with this framing several times throughout the 1920s, as did additional newspapers across the country, including the Chicago Defender, the Cincinnati Post, the New York Tribune, and the Pittsburgh Press. Headlines included "Female Pigeon Is Changed to Male by Scientific Act," "Holds Superman Can Be Developed," "Science Learns How Animals Change Sex," and "Finds Modern Life Lessens Sex Fixity."

apparent.[xv] The pair worked on what Blakeslee called "biological investigations of fundamental problems in sex," with fungi as the target of their experiments.[52] But none of the familiar markers of sex existed in fungi—there were no genitals, gonads, or secondary sex characteristics to even attempt to sort out.[53] Blakeslee and Satina had to take a different approach in deciding what would count as sex.

Blakeslee had been working on the problem of sex in mucors, a genus of molds, since the early 1910s. In these initial experiments, Blakeslee sought to figure out which of the "sexual races" of the mucors, which at that point he referred to simply as (+) and (−), were in fact male and female. This was a difficult task, because they were "nearly indistinguishable each from the other."[54] Ultimately, he determined that the "vegetatively more luxuriant race may be considered female," while the "less luxuriant" fungi were male. Indeed, "whenever a difference exists in vegetative luxuriance," the "more luxuriant" one is female.[55] Blakeslee came to this conclusion by demonstrating a "sexual reaction" between fungi with gametes of two different types that he had defined either as male because they were smaller or female because they were larger, which aligned with their differing luxuriances.[56] Sex was not malleable here in the same way that it was for Goodale and Riddle. Blakeslee was more interested in sorting fungi into stable sex categories than demonstrating that sex or its characteristics could be modified in the lab. Yet Blakeslee's work demanded its own version of sex itself, thanks to the species on which he performed his research. In the world of fungi, sex was simultaneously gamete size, vegetative luxuriance, and, he would eventually come to claim, chemical difference (this chemical hypothesis holding that sex difference might actually be a matter of a difference in male and female proteins).[57] The question here was not whether sex could be changed, but what sex was.

Questions of how to determine and understand sex in fungi continued to interest Blakeslee through the 1920s. In those years, he was joined at the SEE, by then a subset of the CIW Department of Genetics, by research associate Sophia Satina, a botanist who had fled her post at the Women's University in Moscow during the Bolshevik Revolution. Satina and Blakeslee continued working on the problem of how to identify sex in fungi. Throughout, the two maintained an expansive understanding of

xv   Sophia Satina's name is written in different sources as Sophie Satina, Sophia Satin, and other variations, perhaps a result of different transliterations of the Moscow-born botanist's name.

sex, one that Blakeslee saw as a virtue: When funding problems struck, he wrote to Davenport that their work was of great import and deserved to continue because "few students of sex, I believe, have taken so broad a view of the subject and worked with such a diverse series of sexual forms."[58] With their capacious understanding of what sex could be, Blakeslee and Satina looked to the smallest differences in cells for insight into male and female distinctions.

They articulated this project as a mission to "test the biochemical differences between the cell contents of males and females."[59] This was a problem with wide-ranging relevance, because even though "sex is well nigh universally present in both plants and animals," as Blakeslee put it, "little is known . . . of the fundamental differences between the sexes."[60] To tease out these details, they conducted various tests at the microscopic level. Instead of looking for vegetative luxuriance as Blakeslee previously had, Blakeslee and Satina ground up their mucors, applied alcohol or distilled water, and centrifuged them, winding up with a liquid extract to test.[61] Unlike in sex research conducted on birds or mammals, the "body" of the subject was no longer needed. In Blakeslee and Satina's research, the determination of sex had nothing to do with morphology.

Mucor extract in hand, Blakeslee and Satina attempted to identify sex according to the smallest markers. They looked for the presence of various enzymes in the extracts, and how much of the extract was required to oxidize a sample of potassium permanganate. They tested the pH of the extracts. They applied the Manoilov reaction—which could apparently "distinguish female blood from male"—to their fungal specimens.[62] They assessed individual gametes for the presence of a yellow substance they presumed to be fat in one experiment, and gauged the gametes' reactions with tellurium salts in another. The outcomes matched what Blakeslee had determined over a decade previously: In their biochemical reactions, "the (+) sex of Mucors corresponds to the female and . . . the (–) sex corresponds to the male."[63]

The point I want to emphasize here is that because of the specific kind of life that they were studying, when Blakeslee and Satina decided what their markers for sex would be, they were *vastly* different markers than what colleagues like Goodale and Riddle would have used, or what the researchers from chapter 1 used. Sex, for Blakeslee and Satina, could be abstracted from a classification system in which there were certain kinds of bodies and reduced to a series of male and female chemical reactions. It was not only the two of them who viewed this as a legitimate way of defining sex.

While the CRPS apparently struggled to see the relevance of their experiments to a research agenda that privileged investigations of human sexuality and threatened to discontinue funding their mucor studies, Davenport was ready to enthusiastically defend them. In light of the CRPS hesitance to fund Satina's presence at Cold Spring Harbor, he said it would be "a calamity to science" for her to "abandon her work at this time."[64]

Putting Blakeslee and Satina's research alongside Riddle's indicates that sex was malleable at the SEE not just in the sense that female pigeons could be transformed into male ones. Sex itself could be different things for different research purposes. What we see here is not a winnowing of sex into a more specific entity. Rather, Blakeslee and Satina's approach to sex as disembodied yet identifiable enabled their work in the same way that Riddle's coeval metabolic sex supported his research. Meanwhile, friends of Davenport's like Frank Lillie were deploying a hormonal model, much like Goodale had, and countless researchers continued to, in practice, assess sex on the basis of external morphology. Multiple sexes coexisted, from the early days of the SEE through its later reincarnation as the CIW Department of Genetics. What sex *was* was ontologically variable. It could, concurrently, be the presence or absence of certain body parts, chemical reactions, or different kinds of hormones, and an attribute of a body that could be changed or a fixed state to be identified. At the Cold Spring Harbor campus's other research institute, the Eugenics Record Office, sex was something else entirely.

Recording Sex

Less than a mile down the road at the ERO, sex research had long taken yet another form.[xvi] While experimentally breeding humans to see what happened was apparently beyond the pale of even eugenicist research ethics (or logistics), tracing the results of human reproduction that had already occurred promised to illuminate what traits could be inherited according to which patterns. The ERO relied on this technique from its founding in 1910 onward. Compared to the SEE's expansive enactment of sex, the ERO's investment in analyzing reproductive outcomes generated

---

xvi   The ERO was located at what is now 1628 Laurel Hollow Road—Google Maps reports that, with the modern road layout, it's a fifteen-minute walk from the former SEE headquarters, now the Carnegie Library Building, at 6 Lower Road.

another approach: Sex took a more stable form to answer a different set of research questions. At the ERO, sex was simply male or female, a classification system to be deployed in making sense of tremendous amounts of information.

The ERO's primary mission was the collection, organization, and dissemination of data regarding heredity, with the express purpose of using that information to eugenic ends. As a 1918 report put it, the ERO's number one goal throughout its life span had been thus far "to serve eugenical interests in the capacity of repository and clearing house," followed by, number two, "to build up an analytical index of the inborn traits of American families."[65] The ERO chased the first of these goals by accumulating its own library; hiring researchers to comb through medical journals and other potentially useful literature and abstracting their contents; and publishing pamphlets, books, and, beginning in 1916, its own journal, the *Eugenical News*. The latter it pursued primarily through fieldwork. In its summer program, which ran from 1910 until 1924, the ERO trained a corps of 258 fieldworkers, most of them women who had a background, sometimes at the doctoral level, in biology or medicine.[66] These fieldworkers then traveled to institutions like the Kings Park State Hospital for the Insane, the Massachusetts State Industrial School for Girls, and Sing Sing Prison, and to communities known to have particularly large and distinctive families or isolated populations. There, using a combination of interviews and institutional records, they documented extensive individual and family histories with an eye for anything that might be a heritable trait. The ERO also teamed up with other institutions, independent volunteers, and the eugenic family contests that popped up around the United States in this period to accumulate information beyond the reach of its own fieldworkers.

The ERO's campaign to gather as much information as possible about the heritability of traits was a massive undertaking. By 1935, a few years before the ERO shut its doors for good, it had accumulated "nearly 1,000,000 index cards" that consisted of "perhaps 35,000" individual records.[67] The almost one million index cards were only a small subset of the mass of paper that constituted the ERO's records. That mass further included endless stacks of multipage Record of Family Traits forms that organized information about characteristics that seemed to crop up within individual families, ranging from feeblemindedness and left-handedness to musical talent and thalassophilia, or love of the sea. The pedigree charts that fieldworkers made to visually represent the data they collected as they interviewed subjects, often multiple sheets of paper fastened together to form a multifoot record, further contributed to the volume. Perhaps the best indication of the scale of the operation are its relocation receipts. When the ERO, after its closure, was preparing to ship its records to the Dight Institute for the Promotion of Human Genetics at the University of Minnesota for safekeeping in 1948, their moving company wrote them an estimate for 9,000 pounds of files alone, and a second quote included books, which added another 2,000 pounds.[68] A letter to the CIW reveals that the movers had woefully underestimated: The ERO's clerk wrote that he needed approval for an additional $682.61 in moving costs because "we now find that the estimate submitted by the movers was 7230 lbs. under the actual weight."[69]

The ERO needed a way to make sense of all the information contained within these tons of paper and to make their research usable for the crafting of eugenic policy and educational resources. That required the development of an extensive organization system and attendant paperwork. Scholars in science and technology studies and media studies have closely analyzed how the form—literally and figuratively—that information takes has tremendous power in what knowledge gets made. Paperwork is not neutral. Any form to be filled in places normative constraints on what can go on its lines and bounds what can be considered possible, even as fillable blanks help provide bureaucracies with an air of objectivity.[70] Even considerations like whether materials are stored in bound books or as movable pieces of paper in filing cabinets change the way that information is conceptualized and used.[71] The data collection and organization systems that the ERO used both relied on and helped produce a stable binary model of sex.

That process is clear at the ERO in its method of keeping track of findings on the heritability of discrete traits. To cope with the overwhelming amount of data, Davenport developed an intricate system for distilling fieldworkers' case histories into "trait files." All relevant information about a single trait would be placed in a folder (or series of folders, if necessary), then organized according to a scheme that Davenport had created and formalized in the rather obviously named *Trait Book*. The classification system therein relied on a decimal system for indexing traits, beginning with ten classes corresponding roughly to systems of the body (skeletal system, sense organs, circulatory system, et cetera), with each further broken down into smaller and smaller component parts (reproductive system, movement, criminality, and so on). This filing system provided the scaffolding for the ERO's research results. But the trait files, indexed according to the *Trait Book*, were a final product. In order to get to the point of indexing information about traits, the ERO staff first had to turn fieldworkers' interviews into more manageable data sets. It is here where the ERO's data collection and organization method came to utterly rely on sex as a self-evident, straightforward research variable.

To begin with, their primary means of data collection was the family history, which meant that every single piece of information that the ERO collected was filtered through normatively sexed family and reproductive roles.[xvii] For those who wanted to donate their family histories to science, whether students in genetics classes, members of eugenics organizations, or friends and family of fieldworkers, the ERO had a set of standardized, preprinted forms that interested parties could receive in the mail, fill in at their leisure, and send back for processing. All of these forms were organized by family role. On the Abridged Record of Family Traits form, for example, the first section of the questionnaire concerned the "Father's Father," and it provided several lines for recording things like his birthplace, occupation, medical history and cause of death, "number of sons who grew up," and "special tastes, gifts or peculiarity of mind or body." Spaces for the same information about "Father's Mother" followed further down the page.[72] One could not fill in the form without explicitly classifying each family member by sex through their family relationship to the person inputting the information on the form. While each form allowed the person filling it out to choose which traits to focus on, depending on what they thought was most common among their family members, how to organize that information was *not* up to them. The goal was specifically to track traits through reproductive family lineages, with binarily sexed breeding pairs as the focal point, such that it was not possible to voluntarily send in information beyond that using the provided form.

Sex was equally important for representing data in the form of pedigree charts. These charts allowed for the visualization of an entire family through multiple generations, with specific kinds of notation to indicate whether or not a family member possessed a certain trait. Genealogical pedigree charts—basically family trees—had been around for centuries, but the use of pedigree charts for mapping the inheritance of traits rather

xvii   Though, as Rafter has noted, precisely what constituted a "family" was a matter of some debate even as eugenics researchers presented families as self-evident in their pedigrees. *White Trash*, 18–19.

than the relationships of individuals, as well as a set of standards for doing so, emerged out of the development of eugenics and genetics around the turn of the twentieth century.[73] For fieldworkers and individual contributors alike, pedigree charts were the primary method of rendering long narrative family histories and trait forms coherent, and their form was highly standardized. The ERO's approach to pedigrees was first settled on at a meeting of the eugenics section of the ABA in 1910, in a session devoted to "arriving at standards for fieldworkers," including a "system of nomenclature."[74] Notes in the ERO's files on several possible methods for tracking heredity in use at various institutions indicate the importance of having a systematic, yet flexible, method for doing so. The New Jersey Training School at Vineland, for example, initially used a preprinted form for taking family histories that measured a mere five inches by eight inches. Soon, however, researchers found that in practice the "printed form was almost the same as useless because of its inelasticity."[75] Blank paper that would allow workers to create their own charts as needed allowed for a wider variety of representation—perhaps a tacit acknowledgment of the diversity of family formations that workers might encounter. A pedigree chart that a fieldworker could draw on blank paper enabled the representation of far more complex family histories than a printed form could. But the pedigree charts needed to be comparable from one fieldworker to another, and how the individual and their sex would be represented on the chart was a matter of some contention.

Since at least the mid-nineteenth century, there had been two main methods for doing so. One used a Mars or Venus symbol next to a circle or name represented each individual on the chart; the other merged sex designation into its notation of an individual and used a square to represent someone designated male and a circle to represent someone designated female (see figure 2.2). While the square and circle sex markers had been in use in pedigree charts at least since the physician Pliny Earle began to use them in 1845 to chart colorblindness at the Bloomington Asylum for the Insane in New York, and his system remained popular in France, other researchers, especially those informed by German methods, preferred the separate symbols. The ERO used both notation methods in its early years but eventually standardized to the circle and square system.[76] Not everyone appreciated this choice: Even as late as 1929, biologist Raymond Pearl argued strongly in favor of using Mars and Venus symbols instead of circles and squares, which he referred to as "horrid and intrinsically meaningless" and likely only in use "because some eugenic uplifter was shocked

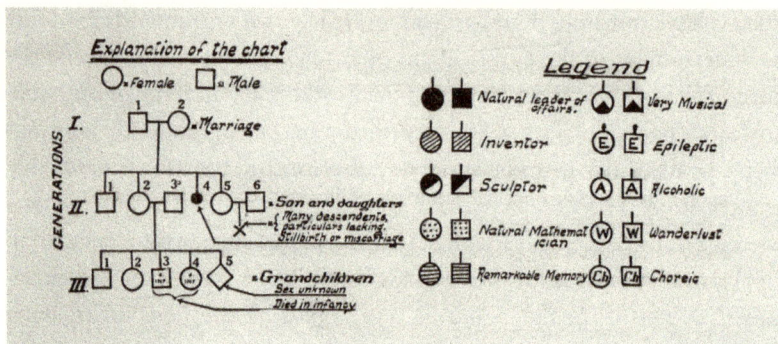

FIG 2.2. Eugenics Record Office instructions for drawing a pedigree chart. A circle for a female individual and a square for a male individual are foremost in its explanation of symbols. The guide included the pictured sample of how to represent relations between and within generations, which also shows a diamond for "sex unknown." The diamond was specifically used to indicate a lack of information rather than ambiguous sex. Courtesy of the Cold Spring Harbor Laboratory Archives.

by the phallic significance of the old traditional sex signs."[77] According to Davenport, the ERO had settled on squares and circles simply because they were easier to read and resulted in less clutter.[78] The ERO also had an interest in determining which traits were sex-linked; those patterns would be more discernible on a chart that clearly noted sex.[79] The squares and circles, however, had implications beyond mere visual legibility.

The entire structure of the pedigree chart relied on sex. Consider, for example, the "Brief Instructions for Making a Scientific Family Study," located inside the cover of the ERO's Family Distribution of Personal Traits form as well as in a free-standing bulletin, "How to Make a Eugenical Family Study" (figure 2.2).[80] These instructions aided the informant in drawing a pedigree chart based on the sum of the information they had collected about the individuals in their family. In the Distribution of Traits form itself, the section for recording information about an individual family member only provided a blank line to record sex as one of many biographical facts. But the "Brief Instructions" show how binary sex was necessary for organizing one's family data, not simply additive. The instructions provide a model chart to use as a template. The first item listed in the key to the sample chart's symbols is that circle = female and square = male.

At the same time that pedigree charts allowed for more flexibility than a prewritten form, then, they began to emphasize sex as the core attribute of an individual person, rather than a mere descriptor. In the days of the five-by-eight-inch form, each person on a given pedigree chart was represented by a circle, with sex marked next to the circle alongside any other traits. With the turn to the handwritten method, the circle and accompanying sex annotation was replaced by a square (male) or circle (female) to represent each person.[81] As a result of paperwork choices, the individual, in effect, became their sex, and sex a stand-in for the individual in eugenic analysis. While a diamond could be used for someone whose sex was not known, it was functionally a dead end on the chart, for use when a fieldworker or informant finally reached a family member whom they did not know enough about to continue drawing out the line of inheritance in that direction.[82] The square and circle notation visually highlighted the assumed importance of a binary sex framework for assessing the results of various reproductive couplings, and indicates that same binary's structural role in data processing.

Furthermore, each chart had to be hand-drawn, and while some of the fieldworkers possessed square and circle stamps to do that aspect of charting for them (and the ERO offered free stamps to the public), others carefully drew squares and circles.[83] Even those who had the stamps at their disposal would have had to continually switch back and forth between them, using the tools for each and every individual they recorded on a pedigree chart, which could number more than a hundred for a single family. This routine would have constantly reinforced the importance of sex in eugenic research. While fieldworkers received training in ERO summer courses and received general directions about how to organize their findings, each fieldworker developed their own distinct style for recording the outcome of their interviews, prioritizing different kinds of information according to their own judgment and interests. But the representation of an individual by their sex, and an organization of pedigrees based primarily on the male/female breeding pair, remained consistent among them all.

Recall, though, that in addition to his role at the ERO, Davenport—the one ultimately making the decisions about these notation systems—served as director of the SEE. He received detailed yearly reports from, and was regularly in close contact with, researchers like Goodale, Riddle, Blakeslee, and Satina, whose work actively argued *against* the existence of stable binary sex, or at least questioned how easy it was to identify. Davenport clearly supported that work, or he wouldn't have pushed for its being funded, or hired and kept on staff the researchers who were doing it. As an early twentieth-century race scientist trained in the late nineteenth century, he also would have been familiar with the arguments about the supposed degrees of sexual dimorphism informing American racial hierarchy (articulated in chapter 1). Yet none of these ideas made it into the ERO's data collection or organization methods, which relied entirely on the assumption that humans were split into obvious categories of male and female.

In the broader scope of research, Davenport was willing to forgo the complexity of nature in favor of being able to make his trait classification scheme work. At a meeting of the fieldworkers held in 1915, one of the fieldworkers asked Davenport when they should mark a person on a pedigree chart as "Fm." for "feebleminded." Davenport replied, according to the notes, that there was "no set line. Nature recognizes no such thing as feebleminded. Man tries to make laws of distinction but can draw no hard and fast line, if such a classification is artificial." Fieldworkers, he advised, should "not diagnose," a direction that a meeting attendee underlined three times in their notes. Yet classification was the entire goal. Case histories would be analyzed for their patterns at the ERO. The goal of fieldworkers was to gather as much information and write up a case history thorough enough so that further classification could be made later. If nature recognized no such thing as feeblemindedness, did nature recognize sex? No one asked, it seems, and Davenport did not directly advise the fieldworkers in that respect. Fieldworkers were, however, instructed to ensure clear communication around binary sex categories. "When Christian names for both boy and girl are alike, as 'Jesse,' 'Jessie,' 'Francis' and 'Frances,'" read a fieldworker's notes on how to record information, "always use the sign [Mars or Venus] after the name to avoid confusion."[84] Perhaps neither nature nor naming conventions dealt in sharply demarcated categories, but proper recordkeeping could help avoid confusion.

The mismatch between the complex, malleable model of sex used at the SEE and the binary, stable one used at the ERO could, if one were so inclined, be chalked up to a distinction between nonhuman and human research. After all, we saw in chapter 1 the way that many animal researchers sought to locate hermaphroditism in the animal world in contrast to the sexual dimorphism of humanity. Perhaps it was okay for pigeon sex to be malleable if the assumption was that it had nothing to do with people. Yet in this very moment, scientists *were* using animal research to try to modify sex characteristics in humans.[85] Even at the SEE, the point was that animal research would be translatable to human eugenic policy. This deep dive into the ERO's paperwork therefore demonstrates a larger problem: Collecting and sorting large quantities of data with sex as a binary variable impacts sex *as a thing*. The ERO's family history methodology and pedigree system simply would not work if sex were more complicated than a binary male/female system.[xviii] Although the classification and interpretation methods of the ERO relied on binary sex to make sense, the ERO's binary sex model was not purely an outcome of the demands of their paper tools. They, of course, could have designed their tools differently.[86] That organization constantly reenacted binary, stable sex, such that the more drawers of filing cabinets the ERO filled, the harder it became to conceive of sex being anything other than it was in their organization system.

Paperwork, however, could only do so much to stabilize sex. As the ERO sought to promote eugenic sexuality and campaigned for sterilization of the imagined unfit, concerns emerged about the malleability of sex. Though their data collection and organization seemed not to question their simplified version of sex, field reports and conversations among researchers at the ERO indicate a simmering anxiety about its fixity. Like zoologists and agriculturalists, fieldworkers often noted in their investigations that exceptional forms of sex and sexuality tended to proliferate widely.[xix] Even worse, research on the impacts of sterilization raised concerns that the removal or modification of gonads would affect the sexual characteristics of humans, reminiscent of how gonadectomy affected birds. Multiple imaginings of sex at the same time, then, did not merely occur in conflict between the SEE and the ERO, but within the ERO itself. The eugenic dream to control sex opened up the uncomfortable possibility that scientific intervention would cause more problems than it would solve.

Eugenic Nightmares

We already know that Davenport had a vested interest in sex as a biological process or state at the SEE; he was also interested in the sexual behavior of humans. Described by historian Daniel Kevles as someone who "bridled at the merest hint of sexual indulgence," Davenport was, like many of his contemporaries, concerned about the social impacts of what he perceived

xviii   Nor would their strictly Mendelian model of heredity, organized around the distinct contributions of one male and one female parent, but that's another story. Thank you to Yingchen Kwok for inspiring this realization.
xix   Despite efforts to limit sex to a binary, life, uh, found a way.

as sexual immorality and degeneracy.[87] Fieldworkers' attention to prostitution, extramarital sex, and sodomy in their field reports and pedigree charts in part reflects these concerns. By tracking the incidence of what they called "sex offenders" within a given family, the ERO could see whether such social ills were as heritable as feeblemindedness or epilepsy and, as with any other supposedly negative trait, prepare to breed them out.[88] But sex research at the ERO went far beyond investigations of the heritability of sexual behavior. As Davenport wrote to Robert Latou Dickinson, "We are also interested at the Eugenics Record Office in the human applications of general principles, particularly those of sterilization, of marriage, fecundity, with special reference to quality rather than quantity."[89] In these investigations, the ERO got more than they bargained for: bodies and behaviors that clashed with the tidy reproductive binary that their research demanded and threatened their vision of a world in which scientists could mold sex to fit eugenic needs.

Fieldworkers regularly solicited an expansive range of information about sex from their interviewees. They cared about behavior, about the body itself, about the psychology and morality of sex, and about reproductive capacity, all of which were crucial to the project of eugenics. Fieldworkers reported on behavior like masturbation, prostitution, incest, sex out of wedlock (which they termed "immorality"), and occasionally homosexuality—though given the eugenic focus on reproduction, fieldworkers were primarily concerned with the possibility of hereditary heterosexual behavior. Assessments of these dysgenic sex behaviors were of particular note, since they might lead to the reproduction of such unseemly traits. Fieldworkers marked individuals who demonstrated those behaviors with an "Sx." in pedigree charts; some even had a special stamp for this purpose. In the process of ferreting out these so-called sex offenses, fieldworkers routinely came across and documented bodily imperfect sexual specimens: pubescent boys with "infantile genitals"; women brought closer to death by menopause; a rarely talked about relation who was a "hopeless idiot" and "never wore men's clothes but an old woman's wrapper"; girls who became insane during menstruation; a brother who was "somewhat 'sissified'"; a boy with "female type of hair."[90]

While comments about "infantile genitals" and the like raised an intersex specter, and while the obsession with menstruation suggests a concern with the perilous state of being female, ERO fieldworkers were not particularly concerned with the violations of a male/female sex binary that they found in their travels. They were, on the whole, less attuned to the intricacies of sex categories than their colleagues at the SEE. As a result, these instances were treated largely as anomalous and not worth devoting much effort to describe beyond a quick reference on a pedigree chart or in an interview summary. The fieldworkers were on the lookout for dysgenic forms of sexual behavior and reproductive trouble far more than instances of cross-dressing or ambiguously sexed bodies.[xx] Unlike the animal researchers discussed in chapter 1, and unlike their colleagues at the SEE, ERO fieldworkers spent little, if any, time hashing out the meanings of maleness and femaleness, or struggling to sort the two categories out. Even so, fieldworkers' narrative records of the many kinds of sexual lives that they encountered as they moved from one institution to another stand at odds with the distilled versions that ultimately made it into the ERO's records as filled-in forms and pedigree charts. Fieldworker records reflect brushes with bodies and behaviors that were far more expansive in their variation than the research methods of their home office allowed for; narrative records reflect, too, how that expanse was easily confined into something smaller. Despite unearthing copious evidence of sexual diversity, fieldwork encounters did not lead to a wholesale questioning of their research system, or of sex. This disjuncture exemplifies a persistence of the routine articulated in the previous chapter as evidence of common manifestations of variability were again rendered meaningless exceptions.

Sterilization initiatives, however, produced a different kind of threat: Attempts to create a better population by modifying their sexual organs might instead create pathologically sexed and de-sexed bodies. In addition to its role as collector of family history data and clearinghouse of eugenic knowledge, the ERO was also heavily involved in the quest to pass eugenic sterilization laws throughout the United States. But sterilization posed a problem. While Davenport and his allies imagined sterilization of the "unfit" to be an important component of limiting the persistence of bad

---

xx   This corresponds with Margot Canaday's assessment of both immigration officials and military doctors often stumbling across, but not intentionally looking for, sexual inverts and homosexuals in the early twentieth century. In Canaday's account, the US federal government had not yet conceptualized "the homosexual" as a type of person to seek out for exclusion in the 1910s, and the same dynamic may have been at work at the ERO. This is especially likely given the close relationship between the ERO and US immigration policy—ERO Assistant Director Harry Laughlin provided extensive expert testimony to the federal Committee on Immigration and Naturalization. See Canaday, *Straight State*, and Laughlin, *Eugenical Aspects of Deportation*.

traits, sterilization was, through its connection to reproductive anatomy and physiology, a sexed procedure, with sexed consequences. This was unsettling at a moment when the American public was particularly concerned about the fragility of not only masculinity and femininity but male and female bodies threatened by "overcivilization."[91] Between that broader context and the sex reversal research going on at the SEE, the modification of reproductive organs came with significant perceived risks. Harry Laughlin, the ERO's superintendent, reported that castration was the only sure way to sterilize men, but that "when young boys are thus operated upon it appears also to inhibit the development of their secondary sexual characteristics as well as to destroy the procreatory functions"—a major drawback.[92] The castration of adult men, he said with apparent relief, "seems to be unaccompanied by any great physiological change other than sterilization." He then added "nor does ovariotomy often have any apparent untoward effects upon adult women."[93] This presented high stakes, though, if sterilization statutes were to be the eugenic way forward.

Fearing the outcome of such operations, the ERO researched the effects of surgical sterilization and found that perhaps they would get more than they had bargained for. In 1914, Davenport hired a research assistant named Mabel Earle to trawl through medical and scientific journals looking for any cases of hereditary traits and diseases that contained family histories and compile them in abstract form, to save himself the time and effort of maintaining his own bibliography. Earle's first assignment from Davenport was to research "the effect of the germ gland [i.e., gonads] of both man and woman upon the mental and particularly the emotional characteristics of the individual, especially the influence of removal of the germ glands upon the behavior and any difference in its effects dependent on the removal at an early age or at a later age." Davenport further advised, "Possibly you can get some information upon this subject by looking thru the Index Medicus under the title castration or eunuchoidism."[94] He reiterated in another letter that he was specifically interested in the "physiological" effects of sterilization.[95]

Davenport never specifically articulated a concern to Earle about the masculinization of women or feminization of men as a result of sterilization, but the literature that Earle sent him was decidedly worried. Throughout 1915 and into the rest of the decade, Earle took extensive notes, often several handwritten pages, on medical texts that documented the outcome of sterilization by "castration," which could refer to the removal of ovaries or testes—though mostly Earle provided notes on articles about women.[xxi] While these articles would occasionally emphasize that there were no real negative side effects of such sterilizations, Earle primarily transcribed reports of cases in which something went quite wrong after surgery. From Earle's attention to these articles, we can get a sense for what the ERO, with Earle as both representative and filter of what new knowledge would reach Davenport's desk, found important, as well as broader anxieties about ovariotomy among medical men.

A significant question in this literature was whether women whose ovaries were removed would become masculinized. Many of them said, emphatically, no. Earle quoted Robert Battey's landmark 1873 article on "normal ovariotomy": "of the loss of outward feminine graces, of mammary atrophy, change of voice, growth of beard, etc, it may be said that such occurrences after double ovariotomy are exceptional and not the rule."[96] Earle subsequently wrote twice, perhaps accidentally, perhaps for emphasis, "Thus double ovariotomy, as a rule, is not followed by any loss of the special characteristics of woman—the only decided physiological change being a final cessation of menstruation, as well as of ovulation."[97] Earle's report on a 1914 article by F. E. Walker reflected a similar take on ovariotomy and included Walker's line, "I am convinced that a masculine type in any form does not develop from the removal of any of the female sexual organs, nor does any abnormal condition supervene other than would obtain in a perfectly natural menopause."[98] Indeed, Earl reported that one article noted that the doctor's patients became "splendid examples of womanhood, enjoying the most perfect health, and retaining all their former attributes of mind as well as of body, and with undiminished sensory capacity in their matrimonial relations."[99] On the surface, at least, Earle's research would suggest that the effects of sterilization by removal of the ovaries were nothing to worry about.

Yet the articles Earle reported on protested perhaps a bit too much. Battey, to be sure, had a personal investment in proving the success of

xxi   The research would continue throughout the decade. In 1918 alone, researchers at the ERO located and wrote case histories of 777 people who had been sterilized as a result of eugenic sterilization laws. According to an ERO report penned by Laughlin, these case studies constituted "a practically complete record of such operations legally performed." Laughlin, "Official Records Under the Carnegie Institute of Washington," in "Official Records in the History of the Eugenics Record Office," (1939), 5, RGI: Eugenics Record Office, H. H. Laughlin 1907-1940, series 3: H. H. Laughlin—Publications, 1912-1940, Box 6, Cold Spring Harbor Laboratory Archives (hereafter EROC, CSHL).

ovariotomies, as he was the physician who had first suggested that they might be used to cure a variety of women's ailments.[100] On the whole, the masculinization of women was clearly a concern based on the intensity of insistence that it had not happened—language like "perfectly" and "splendid" suggests an apprehensive overcompensation. Moreover, the kinds of poststerilization symptoms that Earle attended to when she found instances of negative outcomes emphasize that the removal of ovaries would cause specifically sexed and gendered problems. Earle noted cases in which the removal of the ovaries results in "atrophy of the breasts and arrest of general physical development."[101] A report on the sterilization of girls at the Kansas State Home for the Feeble-Minded likewise stated "breast atrophy noted all cases" and added that among the boys who had been sterilized at the same institution, "one assumed feminine type voice, breasts increased in size. Loss hair on face. Change body contour. All desire lost."[102] Women's voices might become "harsher and more masculine," and they could develop a beard thick enough to require shaving.[103] These reports suggest that the confidence of some of the doctors that Earle was reading was perhaps unwarranted; either way, Davenport would have read them alongside tales of vanishing breasts and sprouting beards.

Beyond noting physical indications that sterilization might result in the modification of sexed characteristics, the vast majority of Earle's abstracts reported on the gendered neuroses that might occur. "Oophorectomies were followed by decided nervous disturbance of a various character," reported one article, which included neurasthenia, anxiety, insomnia, "obsessions of various nature," fears of death and "impending calamities," and the intriguingly masculine symptoms of "restlessness, discontentment with everybody and everything, tendency to quarrel, to contradict, desire to control, to dominate."[104] While some women lost their interest in sex after ovariotomy, in one case, a woman "not only retained her sexual powers, but in an exaggerated degree, and actually became aggressive in her demeanor."[105] Another simply underwent "a complete change in her entire being."[106] Not all doctors believed in such severe outcomes: Some argued that these neuroses were either already present before surgery or might be the result of the operating specifically on women who "live lives of debauchery and intemperance" (and thus would not occur in "ordinary" patients). Others, however, reported that they had seen at least one patient who had been "in perfect health mentally" prior to surgery but afterward "became hysterical, then maniacal" and "developed suicidal tendencies."[107]

Earle's notes paid a particular attention to women's ability to mother post-ovariotomy. Sterilization would, of course, prevent additional conceptions, but some of these women already had children. Earle provided more than two full handwritten pages about a woman who, following the removal of her ovaries, had either attempted to kill herself and her two children or had accidentally nearly killed herself and her children (an uncertain event that either way did cause one child's death). The woman, Lina K., had moved a bed into her family's kitchen and then turned on the gas for the stove. Lina and her children were found unconscious with an empty flask of port beside them on the bed. A note in Earle's handwriting attached on a separate piece of paper to the report reads, "Castration and attempted [word "attempted" inserted with arrow] suicide. Reported fully in part because it showed mental condition after castration."[108] Another report noted that after the removal of ovaries, several women's "homes were neglected, their children were abandoned to themselves, former clean habits changed radically, untidiness, uncleanness, nonchalance became conspicuous."[109] In women who did not yet have children, the "hope [of maternity] is lost and she is very likely to become morbidly introspective and chronically neurotic."[110] Depression could also result "as most women feel that they are sacrificing the greatest blessing of wifehood and motherhood" (though it was, apparently, quite noticeable in women who do not desire a family that a "complete and radical operation never depresses").[xxii] Sterilization by ovariotomy, then, might damage that fundamentally feminine quality, maternal instinct.

xxii   Earle notes on Walker, filed March 22, 1915. This is, to be sure, not mere turn-of-the-century discourse. Twenty-first-century studies also suggest that (presumed) cisgender women who undergo oophorectomy are at greater risk for cognitive impairments, can experience negative feelings about loss of fertility, and tend to use antidepressants at greater rates than a control group. In a moment when hormone replacement therapies did not yet exist and surgical techniques were in their infancy, suddenly being thrust into postsurgical menopause was likely an even more traumatic experience than it is today. Of course, this contemporary research is also shaped by gendered expectations about womanhood. Surprise, surprise, there's little research on the effects of this kind of surgery in trans people who actually want it—or even cis women who do. While I'm writing a particularly discursive footnote, let's take it a step further. As Johanna Schoen has argued, sterilization could allow some women to exercise control over their own reproductive lives at the same time that it was weaponized against others to deny their reproductive autonomy, usually along classed and racialized lines. Similarly, the technologies of sterilization and hormonal

It should be no surprise that researchers would be concerned about the effects of removing ovaries, the organs that were supposed to make women *women*.[iii] Of course women who had been deprived of their gonads might become neurotic, attempt to kill their children, and develop beards and deep voices. These imaginings were part of a larger fascination with the role of the gonads and anxiety about the malleability of the body. They were also part of the process of that malleability being brushed away: Women might become neurotic or appear more masculine, but they were not *not women*. Women without ovaries joined the ranks of freemartins and bees as they forced the meaning of female to stretch around them, as we'll see again with the gynecology patients foregrounded in chapter 3. In Earle's reports, an understanding that sex was malleable *and* a commitment to stable binary sex coexisted.

In addition to Earle's secondary research, ERO fieldworkers conducted their own study of the problem. H. H. LeSeur was assigned to investigate the effects of sterilization by interviewing people who had either chosen sterilization themselves or been forcibly sterilized while incarcerated. All the surviving examples of these sources document sterilization outcomes in men; they ask about demographic information, the sterilized person's attitude to sterilization in general, how the sterilization was carried out and if any anesthetic was used, pain levels during and after surgery, and the immediate and long-term effects of sterilization.[xxiii] The bottom of the form that LeSeur used to collect information

-----

shaping of sexed characteristics were used coercively for eugenic ends *and* closely guarded such that they became nearly impossible to access when people actually desired them. Indeed, by considering trans people within the history of reproductive choice and coercion, it becomes clear that the bodily autonomies allowed and denied include not only access to birth control or the capacity to give birth if one so chooses, but also the very kind of sexed body one has to exist within. Sexual malleability could be forced on those who did not want it, even as it made space for trans possibilities. Sexual stability, on the other hand, was imposed on those whose control over their own bodies required more fluidity, at the same time that it remained out of grasp for those whose maleness and femaleness were denied on the basis of their class or race status. In the present, fights for reproductive autonomy and transition autonomy remain inextricably linked. See Schoen, *Choice and Coercion*; Rocca et al., "Increased Risk of Cognitive Impairment"; Farquhar et al., "Prospective Study"; Abildgaard et al., "Use of Antidepressants in Women"; Reilly et al., "Barriers to Evidence-Based Transgender Care."
xxiii   H. H. LeSeur's records were noted as missing when transferred from the ERO to their second home at the Charles Fremont Dight Institute for the Pro-

reads, "Note: The purpose of this record is to secure some reliable first hand data that will contribute toward records numerous and accurate enough to justify accurate generalization concerning the physiological, mental, and social effects of each of the several types of sterilization on both males and females of different types, in different conditions, and at different ages.... Facts are wanted."[112] In other words, the ERO did not know what the effects of sterilization were, and they were worried about their lack of data in that respect. Certainly, proponents of sterilization had some apprehension that removing a major consequence of sex (i.e., pregnancy) might lead to an increase in "immoral" sexual behavior, and the fieldworkers were likely in part attempting to figure out if that was a real effect.[113] But sterilization was also sexed in a way that might result in blurred categorical lines. LeSeur's notes on an article by C. V. Carrington about the sterilization of criminals include the quote "Young degenerate, who even at reformatory where he was sent when quite young, was a notoriously lusty boastful sodomist and masturbater [sic]. When he came back to the prison a second termer and even more confirmed in his bestial habits I unhesitatingly sterilized him. He has since repeatedly thanked me for this operation and said it made a man of him."[114] If sterilization could make a man, perhaps it could unmake one, too.

Whether fieldworkers were encountering a diversity of sexual forms and experiences in their travels, or considering the possible risks of sterilization, these challenges to binary order did not bring the whole system down. Anxieties about the outcome of eugenic sterilization stood at odds with the ERO's concurrent data collection method and framework for understanding the hereditary transmission of traits. On one hand, members of the ERO feared the possibility that gonadal interference would change the sexual characteristics of those who were operated on, in line with a developing understanding of sex as a spectrum that individuals could be

---

motion of Human Genetics at the University of Minnesota, as were Jaime de Angulo's. Both had been working on the sterilization project. While I was at the American Philosophical Society, Head of Manuscripts Processing Valerie Lutz heroically located one folder each from LeSeur's and de Angulo's research files that had been misfiled in other parts of the collection. While it's certainly possible that the remaining missing LeSeur and de Angulo files were simply lost, I would also not be surprised if, as the most sexually explicit and pro-sterilization records, they were destroyed at some point before they arrived at the APS. Many thanks to Val for her extensive knowledge of the collection and commitment to the search.

pushed along by medical intervention.[115] On the other, the ERO depended on a male/female binary for determining the familial relationships they used to track the movement of different traits through generations, and it needed a clear-cut sex-classification system to organize and analyze their massive quantities of data. These contradictory models coexisted throughout the ERO's working life, reflecting a broader theme in the history of sex research: Sex could and did mean multiple things at the same time.

Sex, Which Is Not One

To borrow a phrase from Annemarie Mol, when seen through the lens of eugenics, sex multiplies.[xxiv] At Cold Spring Harbor, sex was both a trait that an individual could have and a classification system for organizing research. It was a fixed designation as male or female, and also a malleable aspect of the body subject to environmental, surgical, and metabolic influences. Sex was not a single thing but an object enacted differently depending on the particular research question at hand. These disparate meanings of sex did not have to coalesce into one. On the contrary, the research that took place at Cold Spring Harbor relied on incoherent sex. A complex version of sex was perfectly well and good in the labs of the SEE. In that context, as Riddle had hoped, scientists' ability to modify sex characteristics in their experiments emphasized the power of science over bodies. Different subfields of biology would eventually deploy their own specific expertise and jockey for importance by defining sex according to their own terms, whether gamete size, metabolic function, reproductive capacity, or endocrine profile.[116] In the abstract, more scientific control over sex—regardless of exactly what sex meant—was good for eugenics, because it allowed researchers to imagine a world in which there could be more control over breeding.

When it came to the collecting and sorting of thousands upon thousands of family histories, though, the SEE's enactment of sex simply would not do. How would a researcher draw a pedigree chart with a more nuanced understanding of sex? How would eugenicists pushing for mass sterilization proceed if they took seriously research on how gonad modifications changed sexual characteristics? Their research aims and questions demanded a different sex than those of their colleagues, just as the enactment of sex today

---

xxiv Mol's version in *Body Multiple* is that if you "understand [objects] instead as things manipulated in practice," then "reality multiplies" (5).

varies between geneticists and neurologists, and differs again from that of trans theorists. Moreover, the resulting binary version of sex that allowed the ERO to do its work had self-reinforcing effects. The kinds of practices engaged in by eugenicists wound up reinforcing a binary model in conversations, as people filled out forms, as eugenicists painstakingly drew circles and squares on their pedigree charts. That repetition, and particularly the way that ERO fieldworkers brought these "expert" perspectives into nonexpert communities, probably helped spread and perpetuate the idea that science backed a singular, binary understanding of sex.

The ERO's turn to the binary as a way of making sense of massive quantities of data, alas for them, could not sustain their research. A 1935 review by a special committee of the CIW found the ERO's entire organization system lacking (ironic) and reported that the records "do not appear . . . to permit satisfactory use of the data."[117] The review went on to suggest the use of a Hollerith punch-card machine to process them and see if anything was salvageable.[xxv] It nonetheless raised concerns that most of the records were "unsatisfactory for the scientific study of human genetics," due to both flaws in research methods and their assessment that eugenics was not itself a pure science so much as a practical application of science for social reform. After struggling along for a few more years, the ERO finally shut down at the end of 1939.[118] Today, the SEE, later reconfigured as the CIW Department of Genetics, continues on as Cold Spring Harbor Laboratory, a foremost site of biomedical research. Researchers there continue to enact a medley of sexes.[119]

In the discussion of animal sex research in chapter 1, researchers routinely swapped out one definer of sex for another—what body parts constituted maleness and femaleness could shift, and the definitions of maleness and femaleness, too, could flex to encompass wildly different kinds of bodies. Eugenics as practiced at Cold Spring Harbor did not have a coherent understanding of sex either, and this was not a problem for the people doing research there. Eugenics research thus shows in action how sex is multiply, incoherently enacted at different sites by different actors. At Cold Spring Harbor, sex was not an assumed binary to be protected but a question mark, a mystery yet to be fully understood, and

---

xxv   The ERO's exact approach became obsolete as ways of dealing with large amounts of data evolved in the first half of the twentieth century, but their insistence on quantity of data as key to knowledge production would have a much longer life. We'll return to this in chapter 4.

nonetheless crucial to classificatory practice. For Davenport's team, there was no unified thing called sex. Rather, sex functioned as an umbrella category for a constellation of things that were and remain assumed to clump together in sometimes contradictory ways: the body, the mind, reproduction, behavior, hormones, gametes, morality, family role, a square or circle on a form. What mattered was what would best serve the eugenic mission. There is not, for scientists in the past or historians in the present, a preexisting thing called sex whose boundaries can be precisely mapped. There is only something made by a set of research practices, responding, in the case of eugenics, to a desire for white, middle-class, able-bodied domination.

3

Maintaining Womanhood in Gynecological Practice

Sometime in the early twentieth century, Robert Latou Dickinson, venerated don of American gynecology with terrible handwriting, scrawled a page of sentence fragments as he organized his thoughts on deviations from sexual normality. In blue ink on rough paper, he scratched, "Inversion—homosexuality—rare in my experience," followed by "infantilism of various degrees common."[i] Further down the page, he concluded, "Small vulva—scanty periods → male type girl." Encapsulated in these disjointed

---

i   By "infantilism," Dickinson referred to a state of arrested development in which male or female sex had not fully developed. Robert Latou Dickinson, handwritten note, "Inversion—homosexuality," Box 6, Folder 54, Robert Latou Dickinson Papers, Center for the History of Medicine, Francis A. Countway Library of Medicine, Harvard University (hereafter RLD). Just like the Eugenics Record Office, Dickinson generated massive quantities of paper in his quest to understand the human body. His papers contain folders upon folders of informal handwritten notes that are untitled and undated—it seems that he liked to think on paper, which makes it possible to track his thought process but difficult to cite it. These untitled and undated notes make up a considerable amount of the primary source material in this chapter. So, a brief note on notes: I've indicated the first phrase on each sheet of notes in lieu of a title. Rather than repeating "n.d." incessantly, I've excluded date information for these notes. Based on surrounding dated papers and publication dates of material Dickinson discusses

phrases is Dickinson's theory of sex, and its central tension: Sexual intermediacy appeared more frequently than sexual totality, and being of "male type" didn't make one not a girl or woman.[ii] Most people, he would claim again and again in his many writings on the nature of sex, were not fully male or female but instead fell somewhere between those extremes. Yet, over the course of his fifty-year career, spanning across the turn of the twentieth century, Dickinson continued to categorize his patients as women. Women who underwent hysterectomies and ovariotomies and mastectomies, women who never menstruated, women who were sterile, and women whose genitals transformed as a result of their sexual experiences to become more masculine (you read that right—more on this later) all crossed the threshold of his office. They all had bodies that matched Dickinson's descriptions of intersex. They all remained women.

The story of Dickinson's research and clinical career is a story of how the pathological was not kicked out of the normal but instead brought back into it, and how, rather than constricting, the definition of bodily womanhood expanded. The category *female* could be maintained even when the very organs that were supposed to constitute femaleness went awry or were removed entirely. The body changed, but sex did not. The "female body," then, was more than a collection of female body parts; it was a category out of joint with the material. It tells us something, I argue, that in this particular time and place, a system of binary sex could so willfully refuse the material body. It tells us that binary sex is perhaps not as based on the body as one might think.[iii] His insistence on the womanhood

---

in the notes, the bulk of them were probably written in the late 1920s and early 1930s.

ii   I use the word *woman* in this chapter because it's how the gynecologists I write about by and large categorized their patients and objects of study, but it's a self-conscious shorthand that very much doesn't assume that all people with ovaries or uteruses or vaginas or breasts or high estrogen levels are women. As will become clear, gynecologists ran into trouble when attempting to base womanhood on the possession of a certain bodily configuration. I also use *woman* and *female* interchangeably, following Dickinson's example of not distinguishing between the body and identity/social role, given that the distinction didn't emerge until midcentury.

iii   TERFs ("trans-exclusionary radical feminists," rarely used in expanded form since their feminism is garbage—think the intellectual descendants of Janice Raymond, still arguing that trans women are actually "biological men" and trans men are tragic dupes; see Lewis, *Enemy Feminisms*, and Heaney, *Feminism Against Cisness*) complain all the time about trans people not acknowledging the material

of his patients, despite their bodies, exemplifies the labor of producing what we now call cisness; his categorization practices expose the unnaturalness of a near-universal fit between body, identity, and social position.

This foray into Dickinson's gynecological theory and practice sits squarely at the nexus of two conversations on the historical construction and definition of sex and, by bringing them together, makes an intervention in each. First, I build on scholarship that considers the various ways that doctors and legal figures have attempted to determine the "true" sex of individuals with sexually ambiguous bodies, many of whom would probably be designated intersex today. This work has tracked the shifting debates over which body parts should define sex and how at some moments the purported truth of bodily sex paled in comparison to social roles, with the body sometimes being overruled as long as the person in question picked a role and stuck to it.[1] Those histories, however, primarily take as their focus bodies that were obviously troublesome and explicitly on the edge or outside of male and female categories—usually people with genitals that were revealed in carceral or medical settings to fall between what some viewer imagined as a male or female type, or with unexpected gonads suddenly discovered.

But some bodies—like those of Dickinson's patients—skewed beyond norms and did not cause any particular trouble. They were made normal again, their oddities and anomalies rendered invisible. Rather than taking as its starting point the bodies that found themselves ejected from the binary, this chapter looks to those whose nonnormative traits were worked back into normative womanhood. I examine cases of women with so-called male hair patterns, with labia that supposedly became more masculine due to excessive sexual activity, who lacked ovaries and uteruses, or who were plagued by menstrual irregularities, and I ask why and how those women stayed women.

This chapter also intervenes in a conversation about the ever-narrowing definition of maleness and femaleness. In this historiographical tradition,

---

reality of sex. As this chapter shows, though, the construction of a stable binary model of sex that they so desperately cling to fundamentally depended on ignoring material bodies. While I'm utterly disinterested in attempting to fight TERF rhetoric on the basis of what is scientifically "true" about bodies, I find a broad-strokes rejection of their claims to materiality satisfying in a whoever-smelt-it-dealt-it kind of way. On problems with trying to fight TERFism with science, see the conclusion.

sex itself became more precisely defined over time and people increasingly found themselves outside the bounds of the normal.[2] Dickinson, along with his collaborator Lura Beam, provides a stark counterexample. They did not impose a narrow definition of normal womanhood and then kick anyone who did not fit it out of the category. Rather, they expanded the definition of womanhood. They multiplied the possible contents of the category and made space within it for traits that were otherwise coded as masculine, or at least deficiently feminine, whether in external appearance, physiological function, or social and sexual behavior.

Categorical expansion, though, didn't challenge sex and gender norms. On the contrary, it made a male/female binary work, despite evidence that there was far more variation within sexes than between them. If nothing could push a woman outside of womanhood, then there was no need to rethink the system. This chapter, like much of this book, is thus an interrogation of ignorance. Instead of being reclassified, Dickinson's patients' divergence from ideal forms was *un*known; they made it into the archive not as intersex but as women. I argue, implicitly, that the anxiety that accumulated around bodies named as intersex or otherwise of "doubtful sex," as the common historical phrase put it, had to be affectively constructed in contrast to an acceptance of bodies that were just as threatening to binary sex classifications but *not* deemed doubtful at all.

All of this, of course, happened in a particular historical context. As Joanne Meyerowitz has argued, in order for the possibility that a person could "change sex" to emerge, the right social and scientific conditions had to materialize, and despite earlier interest in modifying sexual traits with hormones, that did not culminate until the middle of the twentieth century with the development of "transsexuality" as a named object.[3] Dickinson was of an earlier age, working with the tools and imagination of science in the first decades of the twentieth century. In this moment, scientists did not define sex by amount of body hair or precise genital configuration or menstrual function, but rather by the presence of testes or ovaries or, in the latter part of the time under study, endocrine profile.[4] As we know from the preceding two chapters, though, sex in this time was not so simple. Multiple versions existed, sometimes in tension and sometimes in relatively cozy coexistence. This chapter examines the awkward prehistory of the moment that, as Meyerowitz puts it, sex changed: a moment when scientists and doctors imagined sex to be malleable and expansive, but could not envision a world in which most people did not still belong to a static binary sex category. They had generated a taxonomy of sexual deviance that lurked everywhere; they had not yet adapted to the ways that their deviants were not only living but exceeding those categories.[5]

Let me say from the beginning that there is no smoking gun as to why, exactly, sex classifications shook out the way they did in Dickinson's work. While I have closely attended to Dickinson's informal notes and unpublished writings for clues about his thinking, they remain clues rather than conclusive evidence of change over time or an indication of fully formed theories. The lack of dates on the majority of Dickinson's notes doesn't help. The overall goal here, however, is more expansive than a neat, certain narrative. This chapter attempts to make strange the unchanging nature of sex in gynecological practice, with Dickinson as microcosm. I am also, to be abundantly clear, not trying to prove what kinds of bodies or identities Dickinson's patients "really" had.

Historians have provided ample evidence that the precise meanings and criteria of sex have changed over time. I am not as confident that we've spent enough time considering how weird it is that those changes didn't result in a wholesale reconfiguration of binary sex categories. So here's the takeaway: The preservation of womanhood when actual bodies don't match normative forms at all is really quite bizarre. It is even more so when considering historians' arguments about the increasing pathologization of sexual deviance and fragility of sexual normality in this period, alongside the well-documented scientific legitimacy of universal bisexuality and similar nonbinary forms of sex.[6] Something strange was afoot when there were so many ways to do womanhood wrong, and yet it remained difficult to end up outside of it. Recognizing this disjuncture destabilizes the naturalness of the category *woman*, and, I hope, the desperate linkage between womanhood and a specific assembly of bodily traits that undergirds so much contemporary anti-trans thought.

If the proliferation of sexes at Cold Spring Harbor demonstrated alternate enactments depending on the differing research aims of individual researchers and institutions, the incongruities of Dickinson's sortings indicate that incoherent sex can exist—perhaps in an even more contradictory mode—within the work of a single scientist. In the immediacy of the clinic, the incompatibility of scientific knowledge and social possibility grew more conspicuous. While his forebears in animal research and contemporaries in eugenics cared about human applications, they were further removed from the people who would experience those effects. For Dickinson, though, the "normal" body was not an abstract one birthed several generations in the future, but something he sought to produce using scientific mastery in each patient interaction. Generalizing from individual cases to population-level claims lent itself to cutting-edge assertions; confronting what those ideas meant for an individual on his examination table apparently proved more difficult. As Dickinson contended with a schism between theory and practice, he united two incompatible models of sex and produced another form of incoherence.

White Supremacist Gynecology

The question of how gynecology would address the category of womanhood predated Dickinson's entrance into the field, and it posed significant problems for the construction and preservation of white supremacy. As early as the mid-nineteenth century, medical men with an interest in women's reproductive systems found themselves navigating tensions around whose bodies could serve as a medical model of femaleness and what surgical interventions threatened womanhood itself. In particular, gynecologists grappled with the contradiction inherent to using Black women as experimental subjects for white women's benefit while seeing Black women as inherently sexually different, and with efforts to build their own expertise through ovariotomy while fearing the masculinizing effects of the surgery that they had performed.[iv] Though gynecology did not imagine itself as interested in questions of sex difference or the meaning of sex, these debates laid the groundwork for the coexistence of multiple versions of sex in twentieth-century gynecology.

In the United States, the development of gynecology as a medical, and especially surgical, specialty was rooted in experimentation on Black women. Throughout the 1840s and 1850s, J. Marion Sims, the so-called father of gynecology, routinely operated on enslaved women—frequently on three women named Anarcha, Lucy, and Betsey—to invent surgical techniques for treating vesico-vaginal fistulae. As Deirdre Cooper Owens has argued, Sims's experiments raised the question of whether Black and white women's bodies were anatomically the same, given that doctors intended the treatments invented on Black women's bodies to be transferrable across racial lines.[7] This forced them to hold two things in their minds

---

iv   The removal of ovaries went by various terms in the period under study: *Ovariotomy, ovariectomy,* and *oophorectomy* all circulated, with some debate about which was most accurate. I've used *ovariotomy,* which was the more common usage in my sources.

simultaneously: It was ethically acceptable to exploit Black women for experimentation because they were so fundamentally different from white women, and also Black women were close enough to white women that the same surgical techniques could be exported from their bodies to white women's. Sims's experiments also contributed to the separation of gender from the body at the site of Blackness. C. Riley Snorton has built on Hortense Spillers's concept of Black "female flesh 'ungendered'" to argue that enslavement laid the groundwork for science to think about sex as malleable.[8] Snorton points specifically to Sims's experiments on Anarcha, Betsey, and Lucy as the materialization of this "ungendering," as these women were "rendered as raw material for making the field of 'women's medicine,' from which they were excluded as women."[9] In other words, early American gynecological practice built its knowledge through the exploitation of Black women whom it did not see as women for the purpose of healing white women, whose bodies were medically the same as Black women's even as they were purportedly different enough to exclude Black women from full womanhood. American gynecology, born of a desire to protect white women's health, held contradiction at the junction of race and sex within itself from the beginning.[10]

Dickinson also practiced gynecology in the same milieu that so troubled Mabel Earle in chapter 2: one in which the masculinization of women as a result of gynecological surgery was a constant specter. The nineteenth-century popularization of ovariotomy produced significant anxiety about the role of the ovary in creating and maintaining femaleness. While some doctors touted the procedure as a cure-all for women's ills with no deleterious effect on femininity, others considered it a dangerous overreach with disastrous consequences that could render women sexless.[v] The 1870s, in particular, saw a boom in the procedure after Robert Battey proposed

v   See chapter 2 for a discussion of how ovariotomy was discussed in relation to eugenic sterilization. As most historians tell it, the age of the ovariotomy began when Ephraim McDowell, a Virginia-born physician, performed the first removal of an ovarian tumor via removal of the entire ovary in 1809. As Sally Frampton has pointed out in *Belly-Rippers*, McDowell's "novel" surgery was predated by decades of conversations among doctors as to both the possibility and ethics of removing ovaries. While McDowell seems to have performed the first intentional removal of an entire ovary, partial and accidental removals had been reported before 1809. In many respects, the historiographic repetition of McDowell's heroic first speaks more to physicians' success in building a historical mythos around the ovariotomy than to the clarity of ovariotomy's origin.

the "normal ovariotomy."[vi] Previously, surgeons had reserved the removal of the ovaries for cases of obvious tumors; now, they began turning to ovariotomies to treat menstrual woes and psychiatric problems even when there was no physical indication of ovarian pathology.[11]

Removing ovaries somewhat indiscriminately came with advantages in that it promoted a vision of male gynecological expertise and potentially relieved some symptoms.[vii] Moreover, ovariotomy had its uses in maintaining gendered hierarchies, both in terms of male doctors over their female patients and husbands and fathers over their wives and daughters; as Barbara Ehrenreich and Deirdre English have argued, ovariotomy was sometimes employed as a tool of social control intended to pacify "unruly" women.[12] With greater numbers of ovariotomies, however, came greater concerns about the negative effects that such operations would have on white women's femininity. In these conversations, the ovaries came to be seen as the absolute seat and symbol of womanhood.[13] Removing them "unsexed" women and, as in the context of eugenic sterilization, threatened to produce both physically and socially masculinizing effects, from the growth of a beard to disinterest in motherly caretaking. It was in large part as a result of these fears, as well as criticism of overly enthusiastic surgeons failing to exercise due caution, that the popularity of ovariotomy as

vi   Mabel Earle, if you recall, sent to Charles Davenport a summary of the article in which Battey first articulated the potential utility of the "normal ovariotomy."

vii   It's entirely possible that for some patients ovariotomy really did do something useful physiologically speaking, creating additional motivation for the continuance of the procedure. Studd, in "Ovariotomy for Menstrual Madness," and Komagamine and colleagues, in "Battey's Operation," have both offered recent medical viewpoints on this phenomenon. Studd suggests that removal of the ovaries would have helped mitigate premenstrual dysphoric disorder by suppressing gonadotropin-releasing hormone. Komagamine and colleagues reviewed nineteenth-century ovariotomy cases and argued that at least three of the cases with detailed descriptions indicated that patients had ovarian teratomas that might have caused a form of encephalitis. While many assumptions must be made to link nineteenth-century menstrual madness to the contemporary diagnoses of premenstrual dysphoric disorder or teratoma specifically, these authors' points stand that some patients may indeed have seen improvement due to the physical changes that would have resulted from ovariotomy. As Warner has pointed out, just because bloodletting is no longer legitimate medical practice does not mean it didn't have an observable effect in the nineteenth century. On nineteenth-century medical identity and interventionist practices, see Warner, *Therapeutic Perspective*.

a cure for insanity, menstrual troubles, and other forms of women's un-wellness waned by the late nineteenth century.[viii]

Ovariotomy debates demonstrate that gynecologists were thinking about sex characteristics as being modifiable through medical interven-tion decades earlier than the advent of hormone theory. As scientists and clinicians witnessed the effects of ovariotomy on the body, they linked sex development and menstruation to ovarian function, which, recall-ing chapter 1, substantially precedes Frank Lillie's innovative work on the freemartin that made a similar argument.[14] Gynecological debates, along-side the zoological and eugenic thinking I've already discussed, helped create the conditions for thinking about sexual characteristics as mallea-ble, and they suggested that it was possible to surgically create a class of women lacking both the reproductive capacity and the organs that sup-posedly made them women without changing their sex. These conditions would shape the way that Dickinson's work proceeded.

Dickinson was not an obscure figure with fringe ideas, but was firmly situated in the medical theories of his time. After graduating medical school in 1882, Dickinson trained in and began practicing medicine amid the ovariotomy debates, and he published much of his writing at a mo-ment when theories of universal bisexuality were ascendant. Dickinson's writings reflect a shared sense among his colleagues that, as exemplified by the work of scientists like Frank Lillie, Oscar Riddle, and Eugen Stein-ach, all bodies contained varying degrees of maleness and femaleness.[15] Though some of his work was certainly controversial, especially that on birth control and female sexual pleasure, Dickinson commanded respect and wielded significant influence throughout his career. His interests were wide-ranging, and his writing prolific: He ran a private gynecology practice in Brooklyn, was an early advocate of birth control and founder of the Na-tional Committee on Maternal Health, and wrote on topics ranging from the instruction of new doctors to the dangers of corset-wearing. Dickin-son recorded more than five thousand patient case histories, which Lura Beam, his research associate and the executive secretary of the National Committee on Maternal Health, used as the basis for their "co-written"

viii   As Mitchinson argues in "Medical Debate," ovariotomy debates revealed deep fissures in the medical profession, and disagreements between surgeons and general practitioners as to the role of the doctor and the importance of con-sidering women's social roles and sexuality when making treatment decisions.

books *A Thousand Marriages* (1931) and *The Single Woman* (1934).[ix] The Dickinson and Beam books were among the most foundational American sexological publications of the first half of the twentieth century, and they influenced the shift toward the aggregate studies of sex that are the subject of chapter 4.[16] He held leadership positions in some of the most important medical organizations of the time and, drawing on his influence and reach, served as a linking force between medicine, the birth control movement, and eugenic research.[17]

In the same decades that Charles Davenport and the Cold Spring Harbor researchers (see chapter 2) enacted sex as simultaneously malleable and inflexible, Dickinson, too, deployed incoherent sex in the service of eugenics. Dickinson was a committed eugenicist. He worked closely with Margaret Sanger in the effort to legalize and provide contraception in the United States, making eugenic arguments for the utility of birth control in maintaining the "quality" of whiteness alongside her.[18] He was thrilled by the development of sterilization techniques, and he attempted to get the American Medical Association to endorse the forced sterilization of women through eugenic state laws after conducting a study—eventually titled "Sterilization Without Unsexing"—on the results of eugenic sterilization in California. Dickinson also saw himself as instrumental to a global eugenic sterilization movement; he even complained that the Nazi surgeon Felix von Mikulicz-Radecki did not cite him in the 1936 guide *Die Praxis der Sterilisierungsoperationen* (The practice of sterilization operations). It was a considerable slight, apparently, because Dickinson had

ix   According to *A Thousand Marriages*, "R. L. Dickinson contributed the material in the form of his carefully recorded observations, and he wrote Chapter IV. L. E. Beam contributed the analysis of the material; its structural organization, the writing of all but the single chapter; and an interpretation influenced by professional study and experience in the field of education" (xx). Which is to say, Beam basically wrote the book that Dickinson got credit for, in a long tradition of men getting to put their names on books that women wrote. Julian Carter, for the record, emphasizes Beam's role in the writing of the two books, and treats her as essentially sole author. There is a lot of merit in Carter's approach; in terms of citational politics, it's probably the most ethical one. At the same time, the books were taken so seriously *because* they had Dickinson's name on them, and they are remembered as firmly within his body of work. I see them as, if not quite a collaboration in the final product, utterly entangled with Dickinson's earlier research and enactments of sex, so I am therefore hesitant to try to separate Beam's and Dickinson's contributions out entirely. Carter, "On Mother-Love," 118.

"got[ten] him invitations and the funds to come to this country" and "all the German work" was based on "all the California and American science."[19] In other words, sexual health in the service of eugenics was quite important to him and informed much of his work.

Dickinson also came of age when the model of racialized sex I articulated in chapter 1, in which full dimorphism was attainable only at the evolutionary pinnacle of white humanity, dominated. That model manifested in subtle ways in his work, in the language that he used to describe intersex characteristics, and in how he represented ideal white forms of binary sex. The vast majority of Dickinson's patients—the women who came to him with small vulvas and scanty periods, the male-type girls— were white. To recategorize them as anything other than obviously women would have defied a racial logic dependent on degrees of sexual dimorphism and potentially threatened Dickinson's own mission to improve "the race" through sexual health. Expanding the bounds of womanhood to make room for these imperfect white bodies provided a way out of this conundrum.

Sex, More or Less

Like many gynecologists before him, Dickinson possessed an understanding of sex that contained multitudes. In both published writings and private research notes, Dickinson articulated sex not as a stark, immovable binary but in terms of degrees of intensity. Vast numbers of people did not naturally manifest the extremes of full maleness or femaleness and instead were less than fully male or female by dint of their unique anatomy and physiology; both medical interventions and sexual behavior could affect a body's amount of sexedness. While this chapter is primarily concerned with the bodies of women who stayed women, Dickinson's musings on the category *intersex* show cracks in his use of a binary scheme to preserve white sexual normality—and therefore eugenic ideals—in his clinical practice.

Intersex intrigued Dickinson as worth pursuing in order to better understand sex as a whole. In a file of Dickinson's notes consisting of over seventy pages—perhaps preparations for his 1933 textbook *Human Sex Anatomy*—one can see him puzzling through the wide possibilities for lesser sex intensity. These research files consist of hastily jotted notes and early drafts of more formal writing, as well as sketches of individual bodies and rough graphs of aggregate analyses. The folders are filled with clippings from journal and newspaper articles on cases of both human and animal hermaphroditism, circus oddities like bearded ladies, and so-called primitive peoples around the world whose sexual characteristics purportedly deviated from the expectations assigned to white bodies. Dickinson carefully trimmed these fragments of scholarship and reporting from their original publications and pasted them onto individual backings. He took extensive notes on the latest work about intersex and sexual inversion, and he collected scraps of paper and index cards on which he noted citations for future reference.[20] It's clear from sifting through his papers that even though he didn't publish extensively on the limits of sexual categories, their indistinct borders nonetheless captivated his interest over a significant span of time.

Dickinson was not convinced "true" hermaphroditism—the presence of functioning, gamete-producing ovaries and testes in one body—ever occurred. "There is no such thing as a hermaphrodite[,] a truly double sexed individual with a functioning ovary and a functioning testicle," Dickinson noted.[21] This was a relatively common position for researchers to take in this moment, aligned with what contemporary scholars have argued was largely an effort to classify hermaphroditism out of existence by making its definition incredibly narrow.[x] He then derisively added "our name is stupid . . . pseudohermaphrodite," before summarizing the work that had been done on intersex so far as "a whole literature on what does not exist."[22] Nor did he believe in the possibility of "complete sexlessness."[23] No one was double-sexed, and no one was not sexed—however, many if not most people fell somewhere between full maleness and full femaleness. Dickinson instead

x   This perspective largely emerged in the early nineteenth century as doctors and scientists, particularly Isidore Geoffroy Saint-Hilaire and James Young Simpson, developed new taxonomic systems for hermaphrodites that split confusingly sexed bodies into multiple types, either according to excess or deficiency in Saint-Hilaire's case, or whether they actually possessed both male and female organs or only appeared to in Simpson's. In the latter part of the nineteenth century, Theodor Albrecht Edwin Klebs used developments in gonadal science to update those taxonomies: Now, "true" hermaphroditism, in contrast to "pseudohermaphroditism," would only apply with the confirmation of both ovarian and testicular tissue in one body. Dreger, *Hermaphrodites*, 141–45. Elizabeth Reis points to the "possibility of fraud," as well as scientists' desires for definitive answers, as reasons that Klebs's taxonomy took off. *Bodies in Doubt*, 30–32.

turned his focus to the common occurrence of vaguely intermediate sex characteristics less extreme than "true" hermaphroditism.

These forms of intermediacy were incredibly common. In an undated sheet of typed notes, Dickinson summarized his thinking on the matter as "marked sex-defect is rare; minor sex defects are frequent."[24] He did not specify *how* frequent, but he included in the realm of "minor sex defects" women with big muscles and men with wide hips, regardless of the perceived normalcy of their genitals or other markers of sex difference. Essentially, anything other than the most perfect manifestation of one's sex could exclude a person from full maleness or femaleness and into a less-sexed state. To illustrate this conception of sex, Dickinson drew a set of two diagrams, one in pencil and one typed and inked (figures 3.1 and 3.2).[25] With minor variations, both depicted a roughly V-shaped curve, reminiscent of a chasm one might fall into, with "full feminine endowment" at the highest peak on one end and "complete virility" at the highest point on the other.[xi] The diagrams did not convey frequency (i.e., the peaks did not represent quantity of people) but rather intensity of sex. Between full sex intensities—in "the valley [of] the descending scale of endowment"— were several other stages: the aforementioned "mannish woman" and "effeminate man," then a step down to "female organs, male secondaries [i.e., secondary sex characteristics]" and "male organs, female secondaries." Finally, "neuters" were at the bottom, in "the canyon," referred to as "sexless" in contrast to his rejection of sexlessness elsewhere in his notes.[xii] In both diagrams, Dickinson penciled in a "bridge" across the canyon at the level of the mannish woman and effeminate man, which he noted was for "uniting the narrow gap between the next to the lowest classes" under the heading "intersex."[xiii] Sex was not taxonomic but scalar.

The diagrams open up all kinds of questions of representation—are the flat areas supposed to represent distinct types? what happens between complete virility and the effeminate man?—but what is striking is how little

---

xi   In the pencil version of the diagram, two higher peaks exist on the ends for the "greatly oversexed," suggesting the possibility of too much of a good thing.
xii   Because none of the notes in this file are dated, it's hard to say if Dickinson changed his opinion on the possibility of sexlessness over time, or if the "neuter" was hypothetical. In either case, he noted beneath figure 3.2 that the diagram positioned the neuter as sexless in a "reversal of the old concepyion [*sic*]" of "full powered completeness of both sexes in one individual."
xiii   *Intersex* presumably being a less stupid term than *pseudohermaphrodite*.

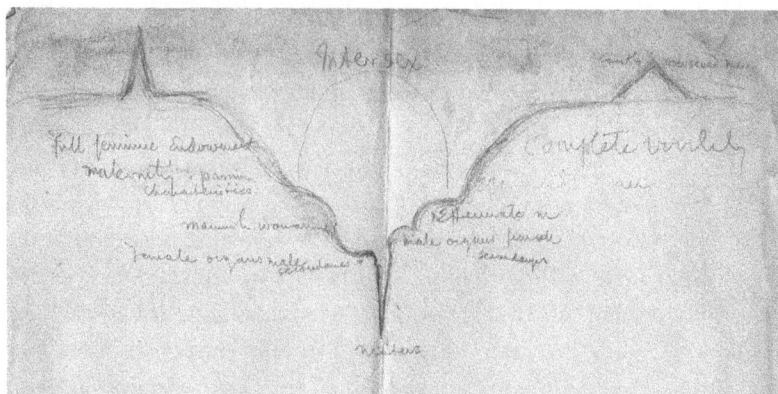

FIG 3.1. Visualization of scale of sexual endowment, drawn in pencil by Robert Latou Dickinson. It represents not a spectrum of sex but rather degrees of intensity of maleness or femaleness, with greater intensity at either end. Unlike the ink version (figure 3.2), this diagram has peaks at each end for the "greatly oversexed."

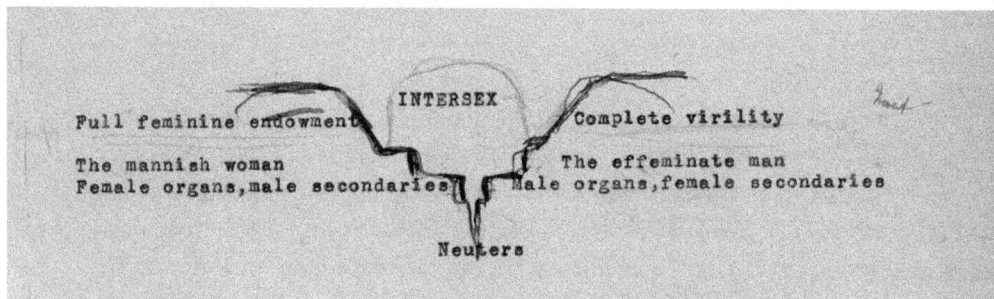

FIG 3.2. An ink version of the scale of sexual endowment in figure 3.1, also drawn by Robert Latou Dickinson. This iteration does not provide for the possibility of being "oversexed." Dickinson drew it as part of a typewritten document titled "Intersex."

one needs to deviate in order to start their descent toward the neuter.[xiv] These two diagrams represented a world in which full sex dimorphism was merely one possibility among many. The fully endowed people were the most "strongly sexed," and presumably, as in discussions about white sexual dimorphism in chapter 1, were the beneficiaries of complete development in the evolutionarily inflected model of sex that proliferated

xiv What is perhaps more interesting about the representation in the pencil drawing are the marks of erasing and trying to get the curves right, which suggests an attempt at precision, even though he doesn't appear to be graphing any actual data here.

at this time. All the other bodies, the less fully endowed, came in a variety of shapes. As Dickinson jotted in his notes on gynecologist Franz von Neugebauer's study of hermaphroditism, "There are infinite gradations both in bodily build, [and] also in the psychic sexual realm."[26]

Dickinson's interest in gradations emerged in another diagram (figure 3.3), repeated several times in different media in his papers and eventually published in his textbook *Human Sex Anatomy*. Sometimes in pencil, sometimes in paint, "Copulator, female, male & intersex" shows the outline and average measurements of a spectrum of organs from clitoris to penis, arranged from smallest to largest, in both "flaccid" and "erect" states, with "intersex" in the middle.[27] The diagram renders the intersex organ as intermediary in both size and shape: Dickinson drew it using the same general outline as his average penis drawing and at the same angle with which he represented the "maximum erect" clitoris. In his preparatory materials, he noted, "On the dimensions of the copulatory [*sic*] of the intersex group— This is almost always of the intermediate dimension between the average male and female."[28] For Dickinson, the intersex body was not outside the realm of male and female but located within that system; it was not only a mix of two sexes, but quite similar to each of them. In the diagram, the composite intersex organ (based, Dickinson specified, on forty-five cases for the flaccid and eight cases for the erect) does not look substantially different from the normal male or female average. Like the "mannish woman" in the previous diagram, the "intersex copulator" is closely adjacent to normal.

In Dickinson's view, intersex was always lurking nearby, and "full sex endowment" was absolutely not a given. For the most part, Dickinson thought, doctors were unlikely to find themselves treating "pseudohermaphroditic" patients with genitals that looked like the middle of the comparative chart.[29] But a "border-line group" appeared in medical offices "not infrequently," and these were the ones that doctors needed to better understand.[30] "The very masculine woman, flat breasted, mustached, narrow hipped, big boned, muscular, deep voiced, aggressive, may belong among the intersexed," Dickinson wrote in an explanation of the sex endowment graph. "So may the timid male with full breasts and hips and smooth face be suspected of defective gonads."[31] It might not take until adulthood for a doctor to begin considering whether a patient's sex strayed too close to the intersex chasm. "Any girl baby or child with a large clitoris is to be suspected of being a defective male," he scrawled in his notes.[32] Suspicion was the watchword: "Suspect every huge clitoris every undescended testicle," he penned, underline included.[33]

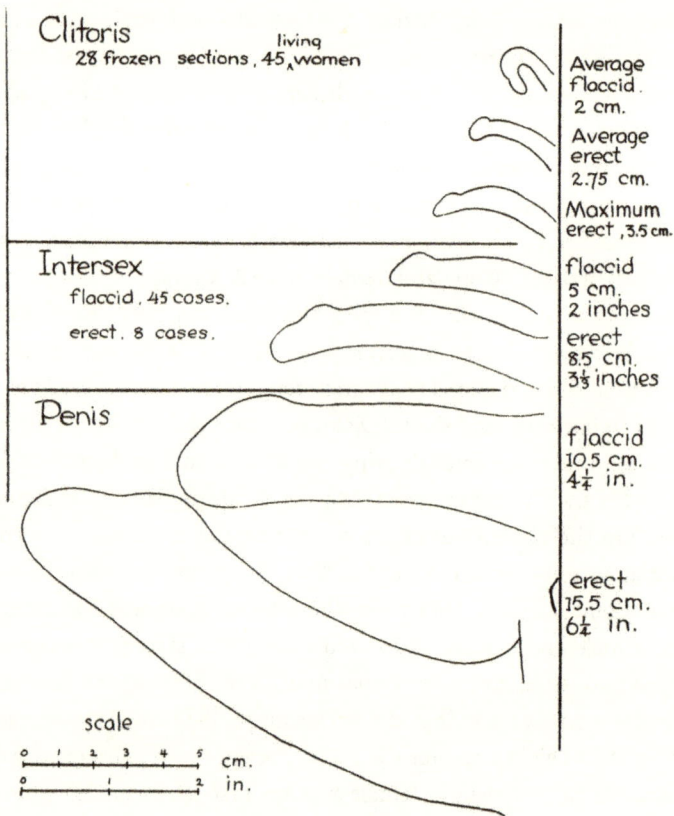

Clitoris
living
28 frozen sections, 45 women

Average
flaccid.
2 cm.

Average
erect
2.75 cm.

Maximum
erect, 3.5 cm.

Intersex
flaccid, 45 cases.
erect, 8 cases.

flaccid
5 cm.
2 inches

erect
8.5 cm.
3½ inches

Penis

flaccid
10.5 cm.
4¼ in.

erect
15.5 cm.
6¼ in.

scale

0 1 2 3 4 5 cm.
0 1 2 in.

Copulator, female, male & intersex

FIG 3.3. Diagram drawn by Robert Latou Dickinson showing the size and shape of a homologous organ that manifested as the clitoris at its smallest and the penis at its largest. Between those extremes is an "intersex copulator," shown here as intermediate in both size and shape. The diagram can be found in multiple forms in Dickinson's papers; this printed version is taken from his published textbook, *Human Sex Anatomy* (1933).

Dickinson worried, though, that doctors were not particularly good at turning their suspicions into treatment decisions. Throughout his career, Dickinson engaged on a quest to identify what was "normal" in order to better identify and treat the pathological, because he had concluded that while clinicians assumed they could tell the abnormal from the normal, they didn't actually know what most bodies looked like and did. As Dickinson put it in the first line of his 1933 book, *Human Sex Anatomy*, "Many of our present beliefs concerning average sex experience and normal sex life have the status of surmises standing on foundations no more secure than general impressions and scattering personal histories."[34] For example, as he wrote in a draft of a letter he planned to send to several men with a request for penis measurements, "We hear much of dyspareunia due to large penis," but there is "a remarkable absence of any data of the normal [here he adds in 'or average'] male organ."[35] In another set of notes, he critiqued what he saw as the "constant use of term hypertrophy of clitoris" when no one knew what the measurements of an average clitoris were.[36]

Identifying a sexually abnormal body, then, was impossible without first knowing the norm.[xv] Unfortunately for Dickinson, a conclusive way to ascertain someone's sex had eluded his colleagues in the field. "Only autopsy with microscopic findings can surely determine the sex," he noted resignedly, adding with an almost audible sigh "(and not always then)."[37] Ultimately, he concluded, "sex may not be determinable during life."[38] Dickinson cautioned that in two out of seven cases he knew of where a large clitoris was amputated so that a patient could live a more normal life as a woman, testes later descended to create a now-conflicting body. "Four chances to one a man guesses wrong," he wrote.[39] This had considerable implications for the caution doctors needed to exercise when deciding how to deal with people in the border zone that he envisioned. "If it takes an autopsy to determine which sex gland is in charge, and if there are cases in which even the microscope cannot decide," he wrote, "the medical man must take care."[40]

Dickinson certainly did so. Though he had delineated a world in which large swaths of people occupied a precarious position on the edge of full sex endowment, and many might even be considered intersex by his standards, Dickinson seems to have been unwilling or unable to imagine his patients as anything other than fully endowed women. "Minor sex defects are frequent," he wrote, and with the stroke of a pen, deviations from the norm became, themselves, the norm. Failure to attain full sex intensity did not disqualify a body from its nearest binary category, even as Dickinson's own reasoning would indicate a quick slippage away from the precipices of male and female. When faced with the prevalence of apparently incomplete sex differentiation, Dickinson widened the meaning

xv A tension, or at least a comfort with contradiction, emerges: Dickinson had developed a classification system that included intersex, even as he admitted to a lack of knowledge of sexual norms. I guess he knew "full sex endowment" when he saw it.

of womanhood to include and encompass a wide variety of "defects" and ejected from the category only the most extreme cases. This refusal to kick actual patients out of womanhood shows the process through which the notion of the exceptional protects the idea of the normal, even as the exception seems more common than the proverbial rule.

The Normal Woman

In the abstract, Dickinson seemed quite confident that sex came in a range of intensities; he knew the difference between "full sex endowment" and intersex forms of embodiment. In practice, the model crumbled. Dickinson's papers include hundreds of typed case histories, based on decades' worth of patient files kept meticulously over the course of his career. These case histories, likely prepared in their edited and typewritten form by Lura Beam in the early 1930s, formed the basis of the co-credited books *A Thousand Marriages* and *The Single Woman*.[xvi] They distilled into digestible form Dickinson's experiences with patients who displayed an expanse of gynecological problems. At no point is there any indication either that these patients fell into the intersex canyon that so interested him or that he considered them anything other than normal women. Gynecological disease was firmly situated as something that happened within the category *woman*, not something that challenged its bounds.

The gendered language of the case histories indicates this stasis.[xvii] They never refer to patients as intersex or anything approaching it; on the contrary, the descriptions refer with overtly feminine terms to women whose bodies would have fit closer to the low-intensity space of Dickinson's sex endowment schema. Dickinson encountered, for example, a "bride" with irregular menstruation and "very small and insensitive" ova-

xvi   I was not able to ascertain with certainty who composed and typed them (if you figure it out, please let me know so I can stop losing sleep over it) but would warrant a guess that Beam did so as part of her work on *Thousand Marriages* and *Single Woman*, with the caveat that many of them are marked with the letters "BS," which might indicate the initials of a different researcher, or something else entirely.

xvii   Could an argument be made that Dickinson just didn't have other language? Sure. But sexologists famously loved to make up words to describe all kinds of sexual deviance, so I'm not sure Dickinson or any of his colleagues would have felt particularly limited by the insufficiency of language.

ries.[41] A "school girl" presented in Dickinson's office with the "male type of pubic hair."[42] A "fine girl" possessed "undeveloped sex organs."[43] A "delicate, reserved, girl," whose menstruation suddenly stopped at the age of twenty-two, had "small breasts and nipples, underdeveloped uterus (an inch and a half cavity), very tiny ovaries, the size of currants; infantile vulva."[44] A "well to do young woman" had a "funnel shape male pelvis with narrow outlet."[45] One summary describes an "honest, hardworking, self-respecting, quiet-mannered girl" with the line "clitoris is erectile or suspicious."[46] These patients were not described as masculine, pathological, or otherwise negatively. Rather, the histories emphasize their womanly qualities: fine, delicate, quiet-mannered. These specimens of womanhood were rendered merely unfortunate, not pushed down the intersex slope, regardless of how "suspicious" their bodies might be (remember: *suspect every huge clitoris*").

Descriptions of patients who had undergone surgical removal of the supposed biological seats of womanhood frame them in similar terms. A "wealthy wife" had no reproductive organs left after undergoing "hysterectomy, oophorectomy, salpingectomy and appendectomy."[47] The aforementioned "quiet-mannered" girl had had "tubes removed, leaving as usual some ovary, hanging the uterus between the four ligaments" early in her medical relationship with Dickinson. When she was twenty-five, Dickinson attempted to perform a hysterectomy, which he abandoned due to abdominal adhesions; instead, he completed a perineal neurectomy because the woman masturbated too much.[48] For the nymphomaniac insanity of a "bank clerk's daughter," a "prominent nerve disease specialist suggest[ed] castration; complete vaginal hysterectomy, both ovaries and tubes removed."[49] Throughout the case histories, one finds the removal of ovaries, uteruses, cervices, and Fallopian tubes. Nothing was particularly remarkable about this, apparently; earlier anxieties about the masculinizing effects of gynecological surgery cast no shadow here.[50] The possibility of destroying womanhood had been swept under the rug of legitimate medical intervention.

Even when terminology used to describe a patient occasionally skewed away from the explicitly gendered, sex designation did not come into question. For a "high school graduate," one case history noted the "clitoris is big and like the male glans."[51] Another patient, referred to only as a "child," had been "rather a tomboy and her mother brought her [to a doctor] because of very big labia minora" when she was young. The case was otherwise unexceptional; as an adult, she still had hypertrophic labia, but this and trouble conceiving and suspected frigidity apparently amounted to nothing worth documenting.[52] Some sets of case histories containing fewer details only referred to subjects as "the patient," but they, too, show no signs of questioning said patients' womanhood. One noted a woman who had had a hysterectomy followed by a "nervous breakdown" and remarked that "clitoris projects beyond the labia; could possibly touch chair when sitting."[53] Pairing physical signs of pathology with a failure of heterosexual feeling, another said of a woman whose labial atrophy had resulted after a period of hypertrophy, "She never has had the least flicker of desire or feeling for any man."[54] Nor was there an impact on pronoun use; the existence of a "clitoris . . . like the male glans," for example, did not preclude the use of "she" and "her." No matter the specificities of a patient's body, Dickinson and Beam's commitment to her womanhood never wavered.

Moving beyond the individual patient into the aggregate allowed Dickinson and Beam to further expand what "normal" womanhood might be. In addition to including a wide spectrum of anatomy, they emphasized that pain and pathology were merely part of being a woman.[55] In her composite analysis of the case histories, Beam made visible the intense frequency of dysmenorrhea, irregular menstruation, sterility, dislike of sex with one's husband, and neuroses of all kinds.[56] Menstruation was especially problematic. "Some form of menstrual disorder was reported by 490, or nearly half of the patients," Beam wrote in A Thousand Marriages, "who complained of dysmenorrhea, menorrhagia, amenorrhea, or irregular periods, alone or in the various combinations."[57] As Julie-Marie Strange has argued, narratives around menstruation essentially made being a woman a biologically lose-lose situation in the late nineteenth and early twentieth centuries. On one hand, menstruation itself came to be seen as pathological, bringing with it pain, weakness, and possible insanity. On the other, to not menstruate meant something was wrong and that a woman's femininity had somehow failed to manifest.[58] In Beam's quantifications, a significant portion of women's bodies struggled to do the thing that ostensibly confirmed their femaleness, but a less than fully functioning reproductive system was not indicative of a failure to be fully sexed. It's not entirely clear what Beam thought about these findings (even less so what the patients thought)—the text does not explicitly discuss her reactions to them. Havelock Ellis's foreword to A Thousand Marriages, however, provides one perspective. Ellis, though enthusiastic about the project, posited that the study of a gynecologist's records is "necessarily limited to women who come to him in the first place as patients," and therefore "overloaded on the side of trouble" rather than a true representation of average experience.[59] Perhaps Beam shared this opinion as she presented a detailed analysis of the many common problems that affected patients, without commenting on what their prevalence meant. Still, *A Thousand Marriages*'s conclusion emphasized that the women studied were "socially normal" and a "representative cross section of one type of American society . . . in good general health, but needing the advice of an obstetrician or gynecologist."[60] It doesn't seem as though Beam thought of these women as tremendously unlike the general population.

When discussing sexual behavior, Beam took a more interrogative posture. Julian Carter has argued that in both *A Thousand Marriages* and *The Single Woman*, Dickinson and Beam "problematize the conceptual division of the world into normal people and perverts"—a different set of possibilities than male and female, and normally sexed ("fully endowed" in Dickinson's language) and insufficiently so, but adjacent nonetheless.[61] Carter reads *A Thousand Marriages*, especially, as a "queer rejection of sexual or indeed any typological classification as scientifically or socially meaningful."[62] And indeed, as Carter so convincingly elucidates, Beam's own rejection of the pathologized category of the lesbian for herself translated, in her analysis of Dickinson's patient files, to a willingness to separate sex between women from homosexuality.[63] Beam instead focused on the tendency of sexual desire and behavior to change over time and across circumstance, whether adjacent to homosexuality or not. "Frigidity," she wrote, for example, "should not be elevated into the position of an organic disease or disorder whereby women may be classified as those who have and those who have not."[64] It instead resulted from a mix of factors, some inherent and some environmental or relational, and could wax and wane with time. Beam noted in her assessment that "there are no proofs that . . . the frigid woman inclines toward the masculine or any other type" and that patients presenting with frigidity "presumably met the accepted social tests for womanhood before marriage."[65] This resistance to typification extended further: Beam questioned, too, whether celibacy, impotence, men's dependence on their mothers, or coitus interruptus were perversions, and where one would draw a line between exhibitionism and wanting to be admired for one's fashion choices, leading her to ultimately ask, "How many of these thousand cases observed for a long time can be considered as without a sexual perversion in some form?"[66]

Beam did not say whether questioning the barrier between full womanhood and something else, like that between normality and perversion, was also on her mind. If, as Carter posits, Beam was reluctant to claim a lesbian identity for herself and for her research subjects given its patholo-gized state, then articulating any distance from full womanhood would likely have been even less imaginable.[67] Though Beam didn't frame her analysis in terms of questions of womanhood, she implicitly portrayed a world in which no women were normal without being made so.[xviii] "The typical bride is unable to feel ardently responsive and seems to herself in regression from the glow of engagement," wrote Beam of the sexual trou-bles frequently experienced by newly married women.[68] A distinct lack of postnuptial sexual pleasure derived, Beam argued, from a widespread lack of information. Women—and their husbands—did not know enough about their own anatomy and physiology, or the mechanics of sex, with-out extensive education. This led to poor "sexual adjustment," manifest-ing as painful rather than pleasurable sexual encounters, paralyzing fears of pregnancy, general feelings of revulsion toward sex, and, finally, giving up on the whole thing and being doomed to a sexless marriage.[69] Beam does not make such an argument outright; still, the unnaturalness of het-erosexuality haunts her discussion of bridal sexuality. I find some intellec-tual kinship in this move, in my own use of Beam's writings to propose the unnaturalness of cis womanhood.

Beam doubted researchers' ability to fully abstract individual experi-ences into theoretical constructs and reflected on this in her methodolog-ical discussion in *A Thousand Marriages*. "Figures are not necessarily true," she cautioned. "Neither are classifications."[70] Anticipating feminist meth-odologies that followed, truth was closest at hand when a single patient recounted their lived experience. As the researcher distilled those stories into composites and groups, they became more and more distorted as they came to look increasingly objective. The Dickinson and Beam books, like the eugenic heredity studies before them, set a course for the turn to ag-gregate studies of sex that would culminate in the Kinsey studies of the 1940s and early 1950s (the topic of chapter 4). For Beam, though, this tran-sition was an incomplete and uncomfortable one.

Nonetheless, Beam's resistance to neat classification only brushed against sex categories. In *The Single Woman*, she explicitly acknowledged that "the authors incline towards acceptance of the theory of bisexuality."[xix]

xviii   Calling Simone de Beauvoir.
xix   Dickinson and Beam, *Single Woman*, 214. Bisexuality here refers to the the-ory of universal bisexuality discussed above, in which all bodies contain aspects

However, an analysis driven by a questioning of fixity of sex categories themselves is much harder to find within either text than the rejection of identarian homosexuality that Carter addresses. While Beam allows for the possibility that the world is not split into the normal and the perverse, her blurring of categories does not extend to sex itself in any practical manner. No one profiled in *A Thousand Marriages* or *The Single Woman* seems to exist in the unsettled interstices of sex. Immediately after Beam notes that shared acceptance of universal bisexuality as a sexual paradigm, she claims that although her research located some "male qualities of character as adapted to environment in work and power" among the unmarried women she studied, "so far as these data tell anything no transposition to male feeling or manifestation of male sexuality is recorded."[71] Beam and Dickinson both subscribed to a vision of sex as more than binary, yet insisted that they saw no evidence of this in practice.

Again, this is not an argument that these patients were anything other than women, identified as anything other than what Dickinson and Beam assumed of them, or would have been at all suspected by anyone at the time to have not fit squarely within femaleness. It is an effort to make strange the ease with which bodies and behaviors that Dickinson and Beam saw as pushing against or even exceeding the boundaries of womanhood were regarded as having no bearing on sex classification. Dickinson and Beam did not sort patients' bodies according to a bisexual enactment of sex or Dickinson's specific sex endowment account; all patients remained women, no matter what form their bodies took. Both *A Thousand Marriages* and *The Single Woman* expanded what normal womanhood could contain, but they could not break free of the category *woman*. This coexistence of possibility and foreclosure encapsulates the fraught position sex science found itself in at this moment: believing in an unfixed, malleable form of sex, and unwilling or incapable of reimagining the implications of that knowledge for social life.

Changing Bodies, Stable Categories

That coexistence is visible in another area of Dickinson's research, in which he investigated supposed changes to the vulva as a result of sexual behavior. From the first years of the twentieth century onward, Dickinson

---

of maleness and femaleness, not bisexuality in its contemporary meaning as a sexual identity.

developed and espoused a theory that excessive sexual excitement, particularly masturbation and lesbian sex but also too much heterosexual sex and sexual violence, produced changes of the genitals that could be decoded by a skilled doctor. In Dickinson's writings on these changes and how to interpret them, the body produced stable meanings that could be objectively read while simultaneously consisting of unstable material molded by touch. Once again, even as an essential marker of womanhood transformed to become more masculine, sex designations did not change. Sexual experience modified women's genitals, but there was no clear physical boundary that, once crossed, required movement out of a category or from one category to another. Sex classification wasn't entirely about the body.

Scholarship on histories of the malleable vulva, clitoral hypertrophy, and the impressionable vagina has primarily approached the history of women's deviant genital anatomy through the lens of sexuality and racial difference. Deviant genitals, in these histories, might be the result of a masculine attraction to other women made visible on the body; occur either naturally or as a result of sexual behavior among white lesbians and Black women (who are not named as lesbians, but always already assumed to be sexually deviant in the relevant primary sources); or physically signal prostitution and nymphomaniacism.[72] Scholars have also thoroughly documented the nineteenth-century fascination with what European travel writers and scientists called the "Hottentot apron," by which they referred to the labial hypertrophy claimed to be common among African women and unmistakable evidence of racial difference, popularized by Georges Cuvier.[73] These nineteenth-century descriptions of African women's bodies detail traits such as, in one representative example, a clitoris "with a well-developed prepuce, and far more conspicuously situated than in the European female," and labia minora that were "largely developed, lax, pendulous triangular lobes."[74] As explained in chapter 1, descriptions such as these served to emphasize a supposed failure of sexual dimorphism.

This scholarship makes a compelling case for the tendency of sexologists and allied scientists to make claims about visual signs of sexual deviance, particularly homosexuality and racial difference, written on the genitals. I argue that in addition to their assertions about sexuality and race, these sexologists were saying something about sex itself. The body was not simply male or female. "Female" did not mean one kind of body but an assemblage of dissimilar ones crowded under the same signifier.

Even if not reclassified as male specifically, the lesbian body was not the same "female"—and therefore not the same sex—as the nonlesbian body, nor was the Black woman's body the same as the white woman's body; at the same time, these bodies could exist outside of normative sex without necessitating a categorical shift in sex classification. In the historiography, the woman with excessive genitals is positioned as an outlier: the white lesbian, the Black woman, the victim of sexual violence, the sex worker, the nymphomaniac. This outlier status has emphasized how certain kinds of women were pushed out of normative womanhood based on the shape of their genitals. Certainly, this was a way that scientists rooted articulations of natural racial hierarchies in purported bodily difference. But the close measurement and comparison of genitals was not reserved for the already-deviant. It was the normal white woman who was the primary focus of Dickinson's investigations in this regard.

In 1902, Dickinson published an article entitled "Hypertrophies of the Labia Minora and Their Significance" that outlined his genital change theory for the first time. He wrote that he had been "surprised to discover" that out of 1,000 cases, 361 women showed at least two of "various vulvar hypertrophies."[xx] This collection of changes were caused, Dickinson proposed, by "prolonged sexual excitation." They included protruding labia that were "thickened, elongated, curled in on themselves, thrown into tiny, close-set, irregular folds that cross at all angles, as in a cock's comb"; pigment deposits and white spots; unequal size of labia; increased vascularity; "flap-like" structures; and increased clitoral size.[xxi] Patients might also show "enlargement and changes in the areolae or in the breasts, resembling those of pregnancy," as well as particularly strong pelvic floor muscles.[75]

xx   Dickinson, "Hypertrophies," 227. The article repeatedly conflates the meaning and relevance of the labia and clitoris; as a result, this discussion does likewise.

xxi   Dickinson, "Hypertrophies," 226. While Dickinson's use of "cock's comb" as a descriptor here likely speaks only to visual similarity, it should be noted that the presence or absence of a literal avian "cock's comb" was a key marker of changes in sex characteristics in experiments carried out on birds at places like Cold Spring Harbor. These experiments came later than the hypertrophy article, so it would be incorrect to make a causal link, but the comb's role in later writings on sex transformation suggests that it was a particularly sexed bodily structure and should be read as such in Dickinson's writing.

Labial and clitoral hypertrophy not only rendered women's genitals pathologically female but made them specifically more masculine. While Dickinson emphasized that only 10 percent of the women studied had "well-marked and distinctive changes" in clitoral size, he did so after remarking that "a slight permanent increase in size is probably present in most victims of repeated excitation." Given this apparently small percentage, clitoral enlargement "should not have undue stress laid upon it." After all, on only one occasion had he observed "penis-like elongation."[76] Still, Dickinson took the time to sketch and include in the article two images of an enlarged clitoris "to show the approximation to the male type of organ."[77] The drawing's caption again pointed out "the male form of the glans."[78] These visuals would, he hoped, help explain "the ease with which these strange cases may claim to be hermaphrodites."[79]

The bodies that had been so changed were the bodies of otherwise "normal" women. The very same bodily changes that could indicate a history of prostitution or lesbianism—that would ostensibly appear in bodies further removed from "full sex endowment"—could happen in married, childbearing white women. At no point did Dickinson propose that these women, with what could only be described by his own reasoning as effectively intersex genitals, were anything other than women. He did not even call their heterosexuality into question. He merely mentioned that he "very rarely and never among cultivated individuals" heard reference to "mutual masturbation" or learning to masturbate "from others," and then moved on.[xxii] The point of the article, after all, was how common these vulvar changes were: One in three patients exhibited them.[xxiii]

That's not to say that vulvar changes had nothing to do with homosexuality. Dickinson contributed his sketches and words to George Henry's 1941 *Sex Variants* in the form of an appendix entitled "The Gynecology of Homosexuality." In it, he maintained his position that sexual stimulation could result in changes to the vulva. This time, however, he focused on women with a confirmed homosexual past and, for some of them, present. Among the thirty-one women examined for the *Sex Variants* study

xxii   Dickinson, "Hypertrophies," 250. These would have been part of a contemporaneous understanding of lesbian sex, particularly in settings like women's colleges, boarding schools, prisons, and reformatories. See, e.g., Otis, "Perversion Not Commonly Noted."
xxiii   Dickinson, "Hypertrophies," 227. Dickinson does not speculate that perhaps one in three patients was a lesbian. Alas.

by Dr. L. Mary Moench and further analyzed by Dickinson, twenty-six apparently displayed varying degrees of labial protrusion, with nineteen out of those exhibiting other changes like "bedded grain-like follicles" and "the curtain-fold atrophy."[80] As in the 1902 study, Dickinson noted several times the "male type" or otherwise masculine characteristics that they had developed.[xxiv]

For all this bodily masculinity, Dickinson again backed away from sex reclassification. In the entire set of examinations performed in preparation for writing the appendix, he said, "no intersex genitals were discovered."[81] This was, again, despite evidence that would have otherwise fit into Dickinson's imagining of sex as a range of intensities. One woman written about in the appendix, her name given as Alberta I., apparently had a "very unusual" clitoris that was "twice as wide, and over three times as long as Dickinson's average, and longer than any among the 1,087 vulvas he has drawn from measurements, except those four which he has called 'intersex.'"[82] Dickinson drew her genitals to show "clitoris, largest in series," and likened her measurements to those of intersex bodies, but Alberta, with her "female build" and "female type pelvis," is called a woman and never imagined to be intersex. Her case history ends with the line "the intersex organ is not here considered."[83] Dickinson again came close to deeming someone not female but stopped short at casting actual classificatory doubt.

In the context of genital change, Dickinson avoided making claims about inherent and discrete sexual types of person. Dickinson reported that in comparison to the genital measurements he had taken that served as the basis for *A Thousand Marriages* and *The Single Woman*, the sex-variant women showed much more clitoris erectility and longer clitoral length.[84] Yet these were not innate distinctions. Homosexuals did not have different bodies; rather, sexual practices modified bodies. Dickinson had more to say about the similarities between women who had homosexual experience and those who did not than he did about their differences. He structured much of the appendix around juxtapositions of homosexuals' exam results and those of patients from his clinical practice—which he distinguishes in a chart as "sex variants" and "office patients"—because "certain striking *likenesses in genital anatomical morphology in autoeroticism*

xxiv   Dickinson, "Gynecology of Homosexuality," for example, refers to "pelvis of a moderately male type" (1097), "woman of the masculine type" (1091), "pelvis with male elements" (1092), and a "boyish individual" (1086), among others.

and homosexual practice [emphasis in the original] make a comparison of the two desirable."[xxv] Masturbation could result in the same changes as "oral-genital techniques among homosexuals, and their use of rhythms of symphysis to symphysis pressures."[xxvi] And while certain traits were worth a doctor inquiring as to a patient's potentially queer sexual habits, "no definite local findings could be classified as peculiar to homosexual practices."[85] Jennifer Terry astutely observes that by Dickinson's own reasoning, it would have been impossible to tell what specific kinds of sexual friction had led to hypertrophy, because the same enlargements could have resulted from lesbian sex, heterosexual sex, or masturbation.[86] Any woman's genitals could change as a result of sexual experience. This would not force their expulsion from womanhood; the category *woman* could contain genital multitudes.

Dickinson was particularly concerned with the genital transformations of white women. Indeed, while many of his contemporaries maintained more interest in the inherent difference of the genitals of Black women, Dickinson rejected the oft-proclaimed idea that African women tended to have larger labia (the aforementioned "Hottentot apron").[87] At least one of Dickinson's most celebrated colleagues, Havelock Ellis, questioned whether Dickinson had adequately factored race into his analysis of vulvar change. Ellis critiqued Dickinson's 1902 article in his own *Studies in the Psychology of Sex* precisely because Dickinson had proposed that masturbation was the cause of labial hypertrophy without acknowledging possible racial influences. Ellis agreed that masturbation was probably the

xxv   Dickinson, "Gynecology of Homosexuality," 1072. The chart is on page 1079. The distinction of "sex variants" and "office patients," rather than "homosexuals" and "normal women" or something similar, suggests again that Dickinson was not particularly invested in making categorical distinctions between types of women.
xxvi   Dickinson, "Gynecology of Homosexuality," 1074. "Symphysis to symphysis pressures," or in queer dialect, scissoring, seems to have been of particular interest to Dickinson, who wrote a three-paragraph footnote detailing exactly how two people with vulvas might engage in "imitation of coitus" (1076–77). For the *Sex Variants* appendix, Dickinson supplied diagrams captioned "clitoris to clitoris?" and "clitoris pressure between women?" (1129); there is also a particularly artful sketch in his papers (Box 7, Folder 46, RLD). His interest may have been prurient, but I find it notable that he painstakingly drew detailed comparisons between his imagined version of lesbian sex and heterosexual sex, perhaps indicating a commensurability that throws another wrench into ontologically distinct sexes.

cause of genital alterations in "civilized European women," but he found it "too absolute a statement" to apply in all cases of labial hypertrophy.[88] After all, Ellis wrote, "it is highly probable that the nymphae [labia minora], like the clitoris, are congenitally more prominent in some of the lower human races, as they are also in the apes." In African cases, "there is not the slightest reason to suppose that these women practice any manipulations," and likewise, at least in some European women, prominent labia were probably the result of a woman being "organically of somewhat infantile type."[89]

Reading Dickinson and Ellis together pairs a malleable white body with the inherent hermaphroditism of the "primitive" body. Kyla Schuller has described an anxious tension in the work of late nineteenth-century physicians Elizabeth Blackwell and Mary Walker that offers one possibility: The white woman's vagina was the epitome of the civilized but oversensitive nervous system, which enabled superior white evolutionary development through its receptivity to stimulation but therefore risked damage.[90] While Dickinson doesn't cite Blackwell or Walker in his hypertrophy work, traces of such a model persist, albeit in a confused manner. Normal white women's bodies react to stimulation, which changes them; the changes, however, bring their bodies closer toward the racialized hermaphroditism discussed in chapter 1. Dickinson's attempts to rescue whiteness as genitally normal—white women's genitals start out small and neat and only become adjacent to primitivity through improper stimulation—end up emphasizing the fragility of the white body and its tenuous grasp on sexual dimorphism.

Dickinson did not articulate a specific racial theory of genital change, and in fact his discussion of race in the context of labial hypertrophy radiates contradiction as much as his approach to sex. Dickinson had drawn Ellis's ire with his assertion that labial hypertrophy "is no racial or tribal peculiarity" and that accounts of African women's thigh-length labia were clearly exaggerations based on the measurements he had found in the relevant literature.[91] Among white women he had examined, he had found "at least 29 cases among [his] notes . . . as long as those of most of the carefully recorded cases among the Hottentots."[92] Differing racial implications of labial hypertrophy emerge, though, in his various accounts of the phenomenon: It was not a racial feature but simply the result of sexual manipulation, while at the same time, Dickinson's accounts suggest a residual link between genital excess and Blackness. The 1902 article had a specific paragraph on "changes in the blonde," in which Dickinson noted that "enlargement of the labia occurs in the blonde to the same degree but perhaps less commonly than in the bru[nette]."[93] The whitest white women, then, were perhaps less likely to experience labial enlargement than those of darker complexion. While increased pigmentation was unlikely to occur in blonde women, in an exceptional case, the labia of one blonde woman had turned "the deepest blue-black tint."[xxvii] In "northern peoples" in general, the archetypical case to the fullest extent of change would have "brownish-black skin folds" protruding from between the labia majora, which themselves would have a "strong growth of pubic hair" (which he associates elsewhere with masculinity). Labial hypertrophy was more common among women with darker skin, and it resulted in a darkening of white women's flesh. Labial hypertrophy was, crucially, a racial transformation.

Though white women were apparently the ones masturbating and, in Dickinson's words, *soixante-neuf*-ing toward "male-type" genitals, Dickinson reinforced the connection between Blackness and genital excess in his infrequent direct engagement with Black women's bodies. One of his go-to examples of intersex genitals was a sketch of the clitoris of a "negro porter," which, while not explicitly described as indicative of underlying racial truths, is one of the only sketches in his 1933 textbook *Human Sex Anatomy* to be labeled by race, and one of only three examples of individual intersex cases in that text.[94] One of the few other racially marked images, labeled "Negress," is unsurprisingly located in the section on labial hypertrophy.[95] In the *Sex Variants* appendix, while only two of the twenty-four subjects described were specifically named as a "negro singer" and a "negress," their descriptions were the first and the last of the appendix, bookending the rest.[xxviii] The latter, named Myrtle K., reported that she was able to vaginally penetrate partners with her clitoris.[96] The only other such case "claiming a clitoris so large as to permit entry into the passages of a lover" is a racially unmarked person named Susan N.[97] One diagram

---

xxvii Dickinson, "Hypertrophies," 244. Pigment deposition, though not otherwise discussed here, was another common change to the vulva.

xxviii Dickinson, "Gynecology of Homosexuality," 1082, 1096. Dickinson remarks in the appendix that "many [subjects] were of mixed race, there being fourteen different nationalities in the heredities, with the four British backgrounds predominating" (1074). None of the "mixed race" subjects are described as such; moreover, it's not clear what Dickinson means here by "race." Given the quick shift to "nationalities" and "British backgrounds," I suspect he was referring to people who were, for example, of both Italian and Irish ancestry.

of Susan, however, includes the note "above, pigment in negress"—again, it's not clear precisely what Dickinson means here.[98] Perhaps he is describing Susan as Black in the caption, without doing so in the written description. Alternatively, he may have been making a cross-racial comparison. Either way, the two cases that involve a woman using her clitoris to take a penetrative sexual role are associated with Blackness. The same "negro porter" sketch that appears in *Human Sex Anatomy*, meanwhile, is also included in the *Sex Variants* appendix.[99] Dickinson, consciously or not, imbued both intersex status and labial hypertrophy with racial meaning.

A few years after the *Sex Variants* appendix, Dickinson began to explore the possibility that men's bodies, too, might document past masturbatory or homosexual experience. He outlined in 1946 what new data would be required for a study on male genitals in relation to sexual behavior.[100] His accompanying description of the project suggested that "when we can develop male case records specifying erection and ejaculation induced by friction or traction applied to limited areas or surfaces—as in these women [who have been studied]—it may be that such surfaces will show the same characters as on the vulva."[101] The full study never came to fruition, but Dickinson completed examinations of five men that Alfred Kinsey had interviewed for *Sexual Behavior in the Human Male* as proof of concept, and the descriptions of Dickinson's findings are highly suggestive. One man's masturbation had apparently resulted in "a wrinkled soft projection, just like . . . the edge of a thickened labium of the vulva."[102] Another man apparently exhibited corrugation "exactly of the *pattern of the labia minora* [emphasis in the original] that is hypertrophied."[103] The fifth subject had a "glans hidden completely behind a radiating series of rugae or wrinkles," with "the sameness of pattern with the labia minora."[104] Dickinson concluded that in three of the five men, masturbatory habits were clearly betrayed by the body. While the numbers here are admittedly small and the descriptions brief, I find it intriguing that Dickinson referred to these men's genitals as coming to more closely resemble women's in both shape and surface patterning. These men certainly did not become women. The partial transformation of their genitals to more closely resemble labia did not interrupt their classification as male. Bodily metamorphosis without expulsion from binary sex categories apparently went two ways.

Dickinson's unwillingness to adhere to his own expansive model of sex thus manifested again in his writing on labial hypertrophy and clitoris size. Some women became less female in what was (and remains) the most visible indicator of sex—the genitals—and they still remained women. At least in these texts, there was no angst about whether or not they should be women, regardless of how thoroughly their bodies were described as masculine. Again, this is not to say *these women were actually intersex*; it is to say, *what an impressive ability these scientists had to accept incoherence*. The shape of the genitals and the malleability of the body could be proof of deviance, or an accepted fact of normalcy. They could exclude one from full womanhood or be shrugged away. The body could be malleable without a change in its meaning. This tolerance of contradiction enabled a static sex binary to persist even as bodies themselves changed.

Solidifying Sex

When Dickinson retired from clinical practice in 1924, he turned his attention to the refinement of chaos into order. He took the voluminous records of his four decades of treating patients and transformed them, often with the help of Beam and other collaborators, into a streamlined, schematic version of sex. I won't pretend to know Dickinson's motivation for his later projects. It seems telling, though, that two of his major late projects purported to show, in singular fashion, what the body looked like. In addition to developing *A Thousand Marriages* and *The Single Woman* with Beam, Dickinson undertook the creation of both an atlas of human sexual anatomy and plaster statues embodying the average American man and woman. These efforts produced simplified and idealized forms of normal womanhood and may have contributed to the squaring of Dickinson's conflicting enactments of sex: on one hand, an understanding of sex as a matter of degree more than kind, and, on the other, a certainty that all of his patients were uncomplicatedly female. The atlas and the statues refined variation among individuals—all the mess of the clinic—into representations of norms and ideals. In the process of this work, exceptions, even frequent ones, were smoothed away as Dickinson once more abstracted sex.

So purified, male and female became their most paradigmatic forms, no longer ranges in an expanse of intensities. The 1933 textbook *Human Sex Anatomy* consolidated the massive diversity Dickinson encountered in his practice into two-dimensional representations intended to teach clinicians what bodies really looked like. The statues, named Norma and Normman and first displayed in 1944, were intended to represent the average American woman and man. For this project, Dickinson and his collaborator, sculptor Abram Belskie, determined their contours by compositing thousands of measurements to show in three dimensions what the only two kinds of body looked like. If Dickinson's earlier work, including the Beam-led studies, expanded the contents of "woman" and created space between maleness and femaleness, *Human Sex Anatomy* and the Norma/Normman statues imposed strict limits, presenting a narrow set of options with a built-in performance of certainty as to what women's bodies were.

Human Sex Anatomy was a response to what Dickinson felt was a tragic series of "gaps in the knowledge of normal anatomy of sex."[105] While doctors had devoted considerable time to documenting pathology and learning how to treat it, considerably less effort had been made, he lamented, to establish a solid understanding of the everyday functioning of the body in its sexual processes.[106] Identifying the normal was crucial for teaching young doctors, who would struggle to know if something was wrong with a patient if they didn't have a standard to assess deviation from.[107] The book consists of descriptive overviews of anatomy and physiology relevant to the study and medical treatment of the sexed body, ranging from the skeletal anatomy of the pelvis to the physical processes underway in heterosexual intercourse, supported by 175 illustrations. Most of the images had been drawn by Dickinson himself, and at least half of them were based on Dickinson's measurements of his own patients.[xxix]

The title page refers to the book as "A Topographical Hand Atlas, "a moniker that epitomizes Dickinson's mission to describe and illustrate sexual anatomy in a format that could be easily consulted in an instructional or clinical setting. Glancing at *Human Sex Anatomy* would be as straightforward as referring to a map. Atlases, though, in the words of Lorraine

xxix   Dickinson, *Human Sex Anatomy*, viii. On the same page, Dickinson thanked Louise Stevens Bryant for her "editorial work." Bryant was the executive secretary of the Committee on Maternal Health and was likely involved in the project for that reason; she also contributed as an editor to *A Thousand Marriages* and *The Single Woman*. She was also Lura Beam's partner for more than three decades. It's hard to say if Beam influenced the book in any way, or the extent of Bryant's "editorial work" (though, given how much of her collaborations with Dickinson Beam wrote, I suspect Bryant's contributions went beyond proofreading), and Bryant and Beam may have differed entirely in their views on classification. Nonetheless, I point this out as a reminder of the personal entanglements that shaped sex research in this moment, with curiosity about how that particular relationship might have impacted the text.

Daston and Peter Galison, "[train] the eye to pick out certain kinds of objects as exemplary . . . and to regard them in a certain way."[108] Anatomy texts, meanwhile, promulgate prototypical visions: As Lisa Jean Moore and Adele E. Clarke have argued, they circulate "shared images" of a simplified, universal body, which then become lodged into understandings of what the body is.[109] In Dickinson's atlas, the combination of an overwhelmingly diagrammatic representation of the binarily sexed body and an invitation for direct comparison between the body and the page taught, implicitly, a method of seeing the body that favored obvious certainty. Though there might be some variation among individual patients, sex was actually quite simple, simple enough even to be represented in line drawings, simple enough that with *Human Sex Anatomy* in hand any medical student would soon come to visually recognize and understand its intricacies. The book cleaned up the multiplicity of the clinical encounter and presented, instead, a neater vision of the sexed body, packaged in crisp lines and clearly labeled body parts.

This outcome was not, however, a mere reflection of anatomy as a genre. Dickinson was expressly interested in visual minimalism as a teaching tool. To that end, Dickinson emphasized the importance of life-size but uncomplicated diagrammatic drawing. This was, in part, a practical concern. Sparser drawings were cheaper to produce and could be printed on lighter, unglazed paper that made the book easier to transport.[110] *Human Sex Anatomy* aimed to provide doctors with a set of images and descriptions of normal bodies such that they could easily assess their own patients by looking back and forth between page and body on the exam table; it had to be as manageable as possible.[111] To this end, Dickinson advocated for life-size drawings in medical records, because they allowed for better record-keeping of individuals by enabling direct comparisons and measurements via calipers and tracing.[112] A lucidity of images was also necessary for teaching purposes. Dickinson found that shading was occasionally required in some of the illustrations for clarity, but he preferred "a restrained form of more detailed picturing" and argued that "for that directness which resides in simplification the diagram is usually better suited."[113] The thousands of fleshy, varied bodies that entered Dickinson's office thus became a handful of black-and-white line drawings. While in some of the illustrations Dickinson provided a range of averages or multiple sketches based on different doctors' models, others gave a single paradigmatic example: Here is the position of the ovaries seen from the front, here are the nerves of the pelvic floor. The variations to which

Dickinson was so attuned in his other writing and in his collaborations with Beam flattened as it transformed from useful knowledge into distracting noise.

This did not mean that Dickinson was fully comfortable with his stripped-down depictions. Throughout the textual commentary in *Human Sex Anatomy*, Dickinson repeatedly pointed to the difficulties inherent to drawing a guide to the sexed body. "There are some anatomical points that seem to run rather true to form," he wrote in the introduction. The distance between the opening of the vagina and the subpubic arch, he noted, was basically the same for most women—over 80 percent of his patients shared the same small range of measurements. He added, though, that "there are others in which a wide variation appears to be the rule."[114] The thickness of the labia majora, for example, demonstrated a "a consistent inconsistency" among patients.[xxx] Reflecting his interest in less-endowed sexual forms, Dickinson included a few examples of intersex genitals, including the "intersex copulator" diagram mentioned above. He noted the homologies between male and female genitals and described the clitoris as "a miniature penis."[115] The book gestured toward the complexity of real live flesh that it otherwise hid.

Just under a decade later, Dickinson gave the world another refinement of sex, this time without acknowledging other possibilities. Norma and Normman, the pair of statues purported to represent the average female and male body, respectively, solidified the eugenic ideals of binary white sex (figure 3.4). In 1942, Dickinson and Belskie produced the set of matching half-life-size statues based on an analysis of the body measurements of approximately fifteen thousand white, "native" American men and women between the ages of eighteen and twenty-four. The statues, beyond visualizing a numerical mean, made a normative statement about eugenically fit bodies.[116] After forty years of intimate encounters with diverse manifestations of femaleness, Dickinson had finally created the perfect woman. She was a eugenic dream: white, young, heterosexual, able-bodied.[117] She was also, tacitly, a defense against incoherence. Norma and Normman scoured out the anomalies of the clinic and replaced an abstract understanding of sex as defying easy categorization with the unyielding materiality of two obviously male and female statues.

---

xxx   Dickinson, *Human Sex Anatomy*, 6. Unsurprising, given all of the masturbating and such.

Norma and Normman furnished a reassuring contrast to the parade of variation among the mostly white bodies that came through Dickinson's exam room.[118] Dickinson, as discussed, directed much of his gynecological expertise toward expanding eugenic sterilization and birth control, and his work on improving marital relations was as much invested in racial improvement as improving the sex lives of the married masses.[119] An uncomfortable situation thus emerged. Obstetrics and gynecology had coalesced precisely because of concerns about the reproductive frailty of white women.[120] Sexual dimorphism as a sign of evolutionary advancement crashed into the fact that white women seemed to overwhelmingly fail at being women. Had Dickinson's model of "minor sex defects" being "frequent" been fully applied to his patients, hundreds of white women would have become less than perfectly dimorphic. On some level, they did, if we take seriously the labial hypertrophy studies. Such an anxiety is not explicit in Dickinson's research notes or his published writings. But the existence of Norma and Normman suggests that a delicate balance of racial thinking, hinged on whiteness as success in sexual dimorphism, needed some concrete proof of concept. Surely, fifteen thousand measurements didn't lie; here was the sum of white American humanity in its most distilled form, and it manifested in a mere two kinds of body.

Alas, it was not so simple. The sculptures certainly did important work of popularizing the notion that the ideal female body was white, American, muscular but not too muscular, able, and paired with a similarly perfect male opposite. Articles in popular magazines like *Time* and *American Weekly* reported on the stony couple's implications for how American bodies had apparently improved over time.[121] A Norma or Normman of one's own could be had: Reproductions of the sculptures sold at rates of $75 (Norma) and $85 (Normman),[xxxi] available in a range of finishes and with six different display pedestal options.[122] Most notably, a 1945 "Search for Norma" contest invited young women to submit their own measurements and imagine themselves as part of the mass of female bodies whose numbers had been transformed into monumental figures, painted purest, snowy white. Prizes ranged from $10 in war stamps to $100 in war bonds for the entrants whose bodies were most similar to Norma's.[123] Clearly, the statues took hold as a means of portraying to the public what their bodies should look like—white, fit, and binarily sexed. Unfortunately for the

xxxi   Thank you to Joanne Meyerowitz for pointing out the insidiousness of the wage gap.

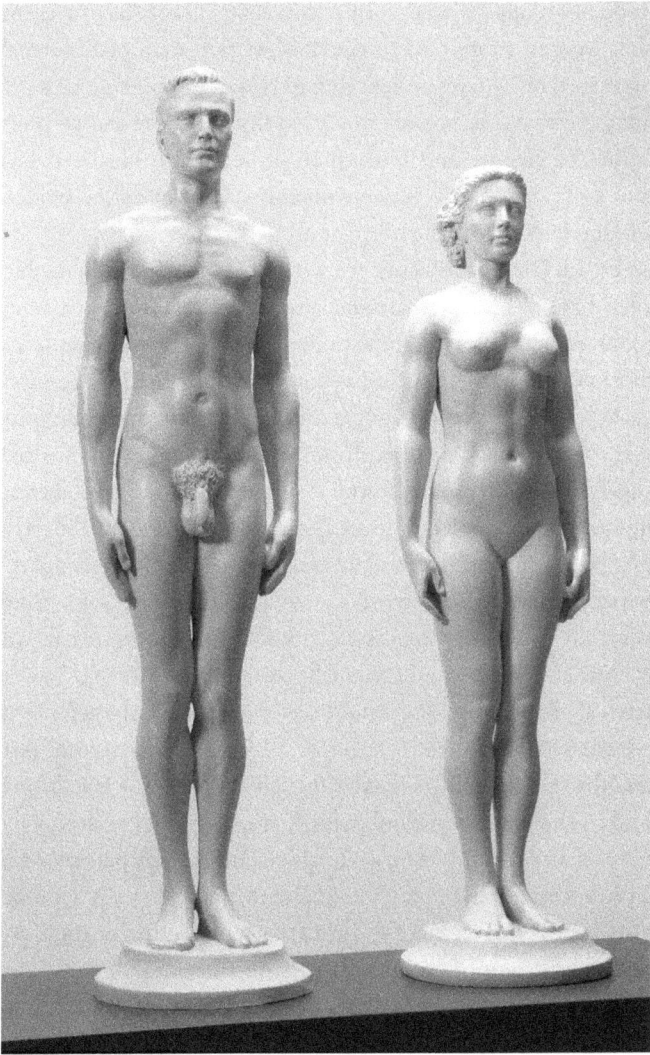

FIG 5.4. Photo of the statues Normman (left) and Norma (right), created by Robert Latou Dickinson and Abram Belskie. Each is half life-size and made of plaster; both are nude, standing with arms at their sides, and are intended to represent the average American male and female, respectively. The photographed set (many were produced) is held by the Dickinson-Belskie collection, Center for the History of Medicine in the Francis A. Countway Library of Medicine. Photograph by Samantha van Gerbig, Collection of Historical Scientific Instruments, Harvard University.

hope of an appreciable norm, only about 1 percent of the measurements that women submitted to the competition matched Norma's.[124] I found no record of Dickinson reacting to this disappointing news. This discrepancy, however, is less an ironic oddity than the entire point. The rigid bodies of Norma and Normman protected an idealized form of binary sex, and particularly white womanhood, from actual evidence that few women looked like the norm at all.

For all Dickinson's concern with the condition of labia, Norma did not even have genitals. Normman proudly displays an uncircumcised penis (covered with a leaf in some publications), framed by sculpted pubic hair; Norma's modesty, however, is maintained by her legs being cemented together. A hairless mons pubis arcs tidily between them, while, from the rear, only a shadow beneath her dubiously spherical buttocks gives any suggestion of further anatomical detail. This sans genital condition seems unique to Norma. Dickinson and Belskie's other collaborative works display vulvas on infant models and as disembodied examples. Their comparative model of adult vulvas wrought in clay features prominently and centrally positioned representations of the "homosexual" and the "masturbator" and their attendant hypertrophies. Norma, however, was safe from such changes. She would never have the enlarged clitoris of the inconclusively racialized lesbian or a masculine pattern of pubic hair. She could shore up white sexual dimorphism precisely because she was idealized to the point of nothingness. So much for the anatomy of sex; lacking vulva and vagina, Norma, like all of Dickinson's patients, was unequivocally a woman. The incoherence within Dickinson's approach to sex begins not to resolve, exactly, but to reach some kind of equilibrium through the replacement of fleshy bodies with stone ones. Bodies transformed into schematics offered clean lines, the perfection that an individual woman could never hope to achieve.

Sexual Knowledge and Social Possibility

Not all women could be Norma. Dickinson had developed a theory of not-uncommon intermediate manifestations of sex; still, he did not find it within himself to imagine a matching social realm of possibility. Part of this inability to reckon with a full reconsideration of binary sex was, as I've suggested, an outgrowth of eugenic and white supremacist investment in sex as a measure of racial difference and hierarchy. This, I think, explains the disjuncture at the highest, systemic level. In the granular, everyday practice of treating patients, of assessing their bodies without reassessing their sex, these deeply engrained logics of sex and race probably reared their heads as well. Dickinson's reasoning here, however, did not make it into the archival record—I wouldn't expect it to, given how much of it probably manifested in implicit assumptions rather than carefully weighed mental treatises on preserving eugenic notions of sexual dimorphism in the assessment of every patient's body and gynecological needs. There are, though, some hints from historical scholarship and Dickinson's own records about what happens when theoretical science meets a social world that relies on different enactments of sex.

Geertje Mak has illuminated how some turn-of-the-century physicians privileged social ease over what they saw as scientific truth when it came to identifying sex. Some doctors would surgically modify—or not surgically modify—the bodies of people of so-called doubtful sex based on the sex they felt themselves to be or already lived as, rather than according to scientific theories of which gonads meant what.[125] Sometimes physicians would not even tell patients that they had found within their bodies gonads other than what they had expected, so as not to cause psychological and social harm.[126] As Mak puts it, "Most physicians were very cautious not to unnecessarily disturb the female self-perception of their patients."[127] While there is no direct evidence to that end in Dickinson's work—I, at least, did not find in his records any cases where he found gonads that conflicted with a patient's original sex assignment and then kept that information from them—the discrepancy between his theoretical understandings of sex and his practical treatment of patients fits this narrative of physicians attempting to protect patients from a need to rethink their sex designation. Dickinson was perhaps more interested in helping patients adjust to the sexual lives they had, and less so in making his patients' lives more difficult in the service of scientific accuracy.

In this way, Dickinson's thinking on more-than-binary manifestations of sex hit the limits of what he imagined to be socially possible. "We [doctors and scientists] were interested in diagnoses and classifications, now we are deep in social service and moral adjustment and vocational problems," Dickinson wrote in the context of intersex patients, describing the slippage of sexual science into social issues over the course of his career. "Shall the intersex individual decide to be a boy or a girl, a man or a woman? Shall the borderline people pretend to be normal? Where does our responsibility come in?"[128] While there absolutely was a thriving social world of sexual deviants in the early decades of the twentieth century—especially in New York City, Dickinson's home—Dickinson could not envision entrance into that world as a positive outcome for his patients.[xxxii]

Consider the case of B. K., a sixteen-year-old who came to Dickinson's attention in 1922 after they made an appearance in the Children's Court of Brooklyn. The presiding judge, Robert J. Wilkin, had referred B. K. to a Dr. Herbert Chase for assessment because their sex was not immediately apparent. Chase subsequently passed B. K. along to Dickinson, who took what would become a yearslong interest in them.[xxxiii] In more than fifteen years of correspondence with colleagues about B. K.'s case, Dickinson seems primarily concerned with figuring out how to help intersex people live within a social binary, rather than changing minds about sex. Though Dickinson was clearly fascinated by B. K. based on his unnervingly thorough case history, detailed drawings of B. K.'s body, and collection of photographs of B. K., Dickinson resisted making claims as to whether B. K. was male or female. Dickinson's undated handwritten notes on B. K. include the line "penis nearly 2 ½ inches" alongside "what does she want to be."[129] He further added on the same page, apparently considering B. K.'s options, "to change to boy wd [sic] have to move—mother's folks in East."[xxxiv] These sentence fragments, informal though they are, indicate the crux of Dickinson's thinking: Even when anatomy might be iden-

xxxii   On queer social worlds around the turn of the twentieth century, see Chauncey, *Gay New York*, and Heaney, *New Woman*, among many others. The upshot is that there were, in fact, plenty of people in this period who had a much more flexible understanding of sex than medico-legal "experts" did, often involving third-sex and otherwise nonbinary possibilities, who probably would not have seen expulsion from binary sex categories as an outright tragedy.

xxxiii   B. K.'s name has been crossed out of Dickinson's records. While it is, on some pages, possible to read through the redaction, I decided to maintain their anonymity, given the horrifyingly invasive photographs that accompany the written documents. B. K. was put through enough for Dickinson's curiosity— I'm not going to make them satisfy anyone else's. A moment of archival honesty: Last time I tried to go back to my scans of the B. K. sources, I had a panic attack. My fellow transes doing trans scholarship, this shit is hard and I hope you're doing okay. You have my permission to put this book down and do whatever you need to do to get it out of your head. Dickinson to Robert T. Frank, April 30, 1927, Box 4, Folder 25, RLD.

xxxiv   Handwritten notes from examination of BK, Box 4, Folder 25, RLD. This imagined need to start life over to hide a transition from one sex to another foretells demands for transsexuals to do so a few decades later—see Stone, "*Empire* Strikes Back."

tifiable within the confines of binary sex, social outcomes must still be a factor in determining how a person should live.

Lest we consider Dickinson a kindly grandfather physician on the lookout for his patients, this approach to B. K.'s welfare did not last. Dickinson's attachment to expertise and scientific progress wrenched the outcome of the case away from B. K.'s interests. Dickinson desperately wanted to publish on the case even as he seemed reluctant to draw a conclusion about B. K. within a paradigm of binary sex and, on some level, to want to privilege B. K.'s desires. Dickinson repeatedly wrote that he wanted to publish something about B. K. but was thwarted by Chase's reluctance to do so: "I have full drawings and history. I urged Chase to publish," he wrote in a letter to gynecologist Robert T. Frank. "Year after year I have pushed Chase."[130] To Wilkin, he expressed a similar frustration: "[It] has been a matter of much chagrin to me to be unable to push Dr. Chase into action on the [B. K.] child."[131] Science, and presumably Dickinson's publication record, was missing out, thanks to Chase's failure to publicly share a report on B. K.

This annoyance with Chase emerged with the development of new endocrinological diagnostic technologies, which led Dickinson to consider that B. K.'s sex might now be determinable within a binary framework. "If that intergrade individual is still within your control or reach," Dickinson wrote to Wilkin in 1927, "you will be interested to hear that we have had a recent development which determines sex by the presence in the blood, or absence from it of the ovarian hormone."[132] Perhaps this test could be conducted on B. K. and finally supply some clarity. On the same day, Dickinson wrote to Chase, "If [B. K.] is still your patient, why not get Dr. Robert Frank to determine whether that individual has the ovarian hormone in the blood or not and settle the matter, unless it has been settled by operation. I hear you found spermatozoa. Is this the case? Have you ever published? My drawings and study and my three cases have awaited your decision."[133] Dickinson's initial willingness to allow B. K. to choose their own fate found its limit at his inability to imagine a livable life outside of a sex binary, and a desire to make claims of expertise in the pages of a scientific journal.

B. K., classified as intersex, was more explicitly excluded from the normative category of woman than Dickinson's other patients, but they were not, in other respects, an anomalous edge case. Their story corresponds to those of the women who served as research material for *A Thousand Marriages* and the labial hypertrophy studies, whose often-contradictory bodies Dickinson deployed to expand the bounds of white womanhood in the service of protecting binary sex, all the while benefiting from a production of sexual complexity that aided his own professional growth. B. K.'s body supported Dickinson's theory of sex in which intermediate forms were not only possible but common, even as B. K. was denied the possibility of living a more- (or less-)than binary life. Possibilities were opened with one hand and closed with the other. Multiple versions of sex existed at once, bound together by their incoherence.

Dickinson's career contained within it a seeming disjuncture between his abstract understanding of sex as a matter of degrees and his consistent placing of women with deviant bodily characteristics within the category of female. But this was not a disjuncture at all. It was merely how sex worked, deriving its power from incoherence and coexisting, contradictory sexes. Dickinson's work encapsulates the preservation of binary categories through the understanding that, despite the evidence, and despite the availability of more complex medical and scientific knowledge of sex, only a few extreme cases would actually be declared ineligible for inclusion in them. These bodies were anomalies, not a comment on the system itself, and therefore nothing needed to be reimagined. In this way, more bodies could be brought, even if grudgingly, into the norm, so that the wheels of white supremacy and scientific control over sex could keep turning. It is a stark reminder that the expansion of categories is not inherently liberatory.

4

Variable Sex in Statistical Research

"The world is not to be divided into sheep and goats," remarked Alfred Kinsey, Wardell Pomeroy, and Clyde Martin, authors of the 1948 best-seller *Sexual Behavior in the Human Male*. "Not all things are black nor all things white. It is a fundamental of taxonomy that nature rarely deals with discrete categories. Only the human mind invents categories and tries to force facts into separated pigeon-holes."[1] There were not, they argued, homosexuals and heterosexuals; on the contrary, vast swaths of the American population reported having had same- and different-sex sexual encounters in a variety of proportions. The authors' revelation of the prevalence of homosexual experience solidified the Kinsey studies' iconoclasm in popular consciousness and the history of sexuality.[2] Their rejection of either/or categories birthed the "Kinsey scale," a seven-point rating system that enabled individuals to place themselves on a spectrum of sexual desire from exclusive heterosexuality (0) to exclusive homosexuality (6). As the authors emphasized, "The living world is a continuum in each and every one of its aspects."[3]

As for the people spanning that continuum of sexual behavior, they were one of two types: male and female. Sex in binary form structured the research project as the first stage of sorting study participants and as a settled variable used for statistical comparison. Though the Kinsey authors professed a commitment to capturing complexity and breaking down sexual typologies, in practice they reified binary sex and its categories "male" and "female" as natural. Their homosexual-heterosexual continuum relied on self-evident sameness and difference of sex; the researchers published their findings in two books split by sex, the aforementioned *Sexual Behavior in the Human Male* and, in 1953, *Sexual Behavior in the Human Female*, known collectively as the "Kinsey Reports." In these two volumes, Kinsey and his colleagues insisted that they were leaving the past of sex research behind for modern statistical analysis and an appreciation for the wide array of human sexual possibility. Instead, they engaged in a time-honored tradition: articulating sex as more nuanced than static male and female categories, and then ignoring that knowledge in favor of binary simplicity.

The Kinsey studies' methods and findings have been well documented.[4] Kinsey, Pomeroy, Martin, and, for the second volume, Paul Gebhard used intensive one-on-one interviews of more than sixteen thousand Americans to collect data on "what people do sexually."[5] Interviews consisted of hundreds of questions that sought information ranging from age at first knowledge of condoms, to preference for having sex in the light or in darkness, to frequency of dreams involving sex with animals.[6] The Kinsey team began interviewing in 1938 and used sex histories taken through the end of 1949 for the two books; additional research continued through the early 1950s. Though, as we'll soon see, they did not interview quite as wide a swath of the American population as their reviewers would have liked, the researchers nonetheless carried out a broader survey than any of their predecessors had. Interviewees came from urban and rural places (primarily in the Northeast and Midwest) and from a variety of class backgrounds and occupational statuses. They ranged from the respectable— physicians, college students, housewives—to the marginalized: sex workers, the incarcerated, career thieves. Most interviewees and all of those whose data were presented in the books were white, though Black interviewees numbered around two thousand, and hundreds of interviewees were classified into a general "nonwhite" category. Kinsey, Martin, and research staff transferred each interview's answers onto punched cards for mechanized sorting and tabulation, which analyzed the results for frequency, incidence, and correlation. Many findings were relatively mundane (frequency of nocturnal emissions, for example, were equivalent for men of both rural and urban groups), but several were shocking. Premarital and extramarital activity were wildly common, as was homosexual experience.

The studies found that most Americans had at some point or another broken either a law or a moral standard in their sexual practices.[7]

To be sure, the scale of these studies and the attention paid to them were unprecedented. *Human Male* and *Human Female* epitomized the turn to vast collections of data that the Eugenics Record Office (ERO) heredity studies and Robert Latou Dickinson and Lura Beam's *A Thousand Marriages*, among others, had heralded (see chapters 2 and 3). The number of people interviewed for *Human Male* and *Human Female* dwarfed the participant numbers of any prior study. More massive still was the public and scientific reaction to the books. They were covered in *Life* and *Newsweek* and countless other magazines; representations of sex surveys started to appear in films and novels; radio broadcasts offered commentary on the Reports.[8] Experts held symposia to discuss the methods and potential impacts of the work. One could buy a set of cocktail napkins—with thirty-six different designs—that featured cartoons and satirical "sex-tistics" and came in a box that read "Sexual Misbehavior in the Human Male" in the same font and layout as the cover of *Human Male*.[9] Despite both books being unwieldy tomes packed with sums and graphs, publishers could barely keep up with sales—*Human Male* went into its sixth printing only ten days following its release, and it had sold more than two hundred thousand copies by two months later. *Human Female* performed just as well. Their reach was global: The Reports were translated into French, Spanish, Italian, Japanese, and several other languages.[10]

Thanks to the wide publicization of their dramatic discoveries and ensuing virulent discussion, the Kinsey Reports have been remembered as an instigator of enormous change in thinking about sexuality in the mid-twentieth century.[11] In the moment, readers predicted that would be their effect; the books' scientific and medical audiences and the broader public understood a revolution to be at hand.[12] The editors of a collection of essays on *Human Male* give a clear picture of what they thought the book would achieve: They thought "that the world will be a happier, more peaceful place, that there will be more justice in it, and that individuals properly educated in terms of sexuality will make better children, more law-abiding citizens, happier, better—much better—parents," thanks to the improvement of scientific knowledge about sex.[13] Charles Walter Clarke, the executive director of the American Social Hygiene Association, said that *Human Male* "could have great influence on the programs of social hygiene societies," and he hoped that it would improve "human welfare" and "assist the process of evolution toward 'the good life.'"[14] Robert Latou Dickinson, a staunch supporter, claimed that the findings would improve marital relations across America.[i] Though many of the authors' contemporaries critiqued their methods, they also made clear their expectation that the Reports would transform American views on sex and American society as a whole. Anthropologist Ashley Montagu, for example, criticized *Human Male*'s implication that just because homosexuality was frequently found in American men didn't mean it was natural, but nonetheless called the book a "contribution of unparalleled importance."[15] Margaret Mead was not a fan of Kinsey's turning sex into what she called a "meaningless act," yet she referred to *Human Male* as a "cultural phenomenon" that "has upset the balance in our society between ignorance and knowledge."[16] The publication of *Human Male* was regularly compared to the dropping of an atomic bomb: It would irrevocably change the world.[17]

Historians, too, have credited the Kinsey Reports as launching dramatic shifts in how Americans and American sex science thought about sex. Certainly, sexual mores were already changing, thanks to previous sex studies, increasing demands for birth control and abortion access, and the gender role disruptions of World War II.[18] Kinsey took advantage of those existing cracks and broke open the possibilities for sex in America. The work contributed to a shift from religion to science as the arbiter of sexual behavior.[19] The findings inspired psychiatrists to rethink their pathologization of homosexuality, laying the groundwork for the removal of homosexuality from the *Diagnostic and Statistical Manual of Mental Disorders* (DSM) in 1973.[20] The Reports introduced new discussions of sex into American life and culture and shaped concepts of national identity, masculinity, and femininity.[21] They contributed to the development of gay identity and community.[22] Altogether, they advanced, Howard Chiang has argued, an "epistemic turning point" in twentieth-century sexology.[23]

The Kinsey team and their contemporaries, along with those who have written their history, have largely framed the Reports as a break with tradition. This chapter argues that as much as the Kinsey Reports may have changed, they reinscribed. Continuities with older approaches tied

i   Dickinson, in Geddes and Curie, *About the Kinsey Report*, 159–66. The cover of this particular collection of essays about *Human Male* tells us something about the importance of new statistical methods. Rather than feature some sort of tastefully embracing couple, the cover is decorated with a graph line superimposed on a grid, complemented by an abstract, shadowy man's head and shoulders.

the bombshell research to the past, and even apparent disruptions often contained existing assumptions. New statistical methods treated sex as a static variable, while a focus on frequency of incidence rendered trans and intersex people anomalous and not worth counting.[ii] The new character of "deviant" sexual behavior *as* behavior rather than marked on the body may have helped normalize a range of acts and desires, but that behavior was assigned to one of two normally sexed bodies, thus reclaiming deviance into binary sex. Kinsey and colleagues claimed to undo categories, but their work imagined a world in which social lines and hierarchies could not easily be crossed. They presented a posteugenic conception of race as cultural rather than biological, but continued to affix sexual difference to Blackness. If the Kinsey Reports mark a turning point, then, it is one at which sex scientists split sexual behavior from the body, enabling new perspectives on sexuality while binarizing bodies into female and male. In other words, the processes of recapturing multiplicitous manifestations of sex into a stable binary seen in previous chapters was not only a feature of nineteenth- and early twentieth-century sex science. With innovations in mathematical tools and a greater interest in population-level data rather than individual findings, the Kinsey researchers did so even more effectively and invisibly than their predecessors. The Reports thus provide another stark example of how knowledge production practices can render the anomalous ignorable.

In contrast to earlier sex researchers who had puzzled over sex as potentially malleable, indistinct, and not entirely binary, Kinsey and his associates pivoted away from questions of what constituted sex and what belonged in—and fell outside of—male and female categories. For the Kinsey team, studying sex as in the biologized typification of the body took a back seat to studying sexual behavior—who was doing what with whom, and where, and when, and how. They included in *Human Male* a discussion of sexual inversion before dismissing it, and in *Human Female* they devoted several chapters to comparisons of male and female anatomy in the hope of finding an explanation for differences in behavior.[iii] For the most part, though, they cared about topics like masturbation techniques;

---

ii   In the late 1940s, what was then called *transvestism* was still considered by scientists like Kinsey to be a sexual practice akin to fetishism. The word *transsexual* had only recently been coined in 1949. I've opted to anachronistically use the word *trans* in this chapter; see the introduction for why.

iii   They didn't.

age of first sexual experience; sex before, within, and outside of marriage; and whether any of this had to do with religious background, education level, and rural or urban living. Their extensive questioning, though it involved some inquiries about the body, was ultimately quite different than the research of most of their forebears. The Kinsey team were far more interested in using *male* and *female* as variables to measure frequency and incidence of sexual behavior. As a result, those hundreds of thousands of copies of *Human Male* and *Human Female* handed to Americans expert proof that sex was a binary of male and female.

Modern Statistical Methods

In the decades that preceded the Kinsey studies, sex researchers had been shifting away from the singular case study and toward quantitative analyses of hundreds of massed individuals. The data-gathering work of the ERO (discussed in chapter 2) was part of this changing methodological landscape, as was the work of Robert Latou Dickinson and Lura Beam (see chapter 3). From the late 1920s onward, several sexologists aimed to find the truth of sex in greater numbers and their hidden patterns. Katherine Bement Davis, Lewis Terman, and George Henry, among others, surveyed, interviewed, and examined hundreds and sometimes thousands of people about their sexual behaviors, their bodies, and their desires.[24] The Kinsey researchers expanded on these early efforts and transformed them into the research enterprise that would become the high-water mark of twentieth-century sex research in the United States.[iv]

The Kinsey team's methodological approach was also part of a much broader adoption of statistics, specifically, as a crucial way to knowledge.[25] In the life sciences, the latter decades of the nineteenth century and early decades of the twentieth saw the advancement of statistical theory in Europe, particularly in the work of Francis Galton, Karl Pearson, and R. A. Fisher, as interest grew in a quantitative study of biology to more rigorously understand heredity and eugenic possibilities.[26] New mathematical techniques and concepts like *standard deviation* and *normal distribution* and *regression* emerged to analyze variation, correlation, and other relationships. Research institutes and a journal, *Biometrika*, supported the invention and dissemination of these tools. By the turn of the twentieth

iv   Kinsey's interest in large collections stemmed from his entomological career. Drucker, *Classification of Sex*, chap. 1.

century, biostatistics and biometry had traveled across the Atlantic. With his 1899 textbook *Statistical Methods with Special Reference to Biological Variation*, among the first books to be published in the United States on the "newer statistical methods in their application to biology," Charles Davenport (see chapter 2) stoked American fascination with wedding these mathematical innovations to the study of biology.[27] Soon after, biologist Raymond Pearl returned to the United States after a year spent studying in London with Pearson and began his own proselytization of statistical methods around agricultural (re)production and population research.[28] As the methods that had been brewing for several decades reached their peak, they defined a new norm of research methodology in the life sciences. Statistics became both the way to turn the chaos of large data sets into ordered knowledge and a marker of rigorous science.[29]

When the Kinsey team began their work in the late 1930s, statistical methods had recently evolved again. The Great Depression and the collapse of American agriculture, Emmanuel Didier suggests, prompted the need for methods to collect and analyze massive quantities of information for New Deal policy and programs.[30] Representative sampling, initially proposed in 1895 but soon forgotten, was reborn in the crucible of economic collapse.[31] The cutting edge of statistical research sharpened as statisticians showed that data could not be collected willy-nilly. Samples— the segment of a population under study, used to make claims about a broader population—must be "representative" of the whole, so that the small group from which data were collected could stand as a microcosm of a larger entity.[32] The reliability of conclusions drawn about the total population based on the samples could be checked with new "probable error" calculations.[33] Research that did not conform to these new standards of rigor became suspect. In other words, just as Alfred Kinsey was beginning to shift his sights from wing morphology in gall wasps to sexual behavior in humans, increasingly refined statistical methods became the best way to achieve precision and accuracy in scientific research.[34] These developments in statistics shaped both the team's approach and scientific reactions to the Reports.

For the Kinsey researchers, the use of modern statistics was, itself, their primary methodological intervention into sex science. Both books include a lengthy methods section—around a hundred pages in each— that explain their sampling approach and interview technique, and why they offer an improvement over older studies. In part, this was likely a preemptive girding against suspicion of the seriousness of their work, given its frank sexual content. These sections weren't, however, merely symptomatic of a defensive posture. The researchers' extensive discussion of method indicated to their readers that they were able to access such revelatory data and make such stunning statements about American sex life because of their novel methodological approach.

In both *Human Male* and *Human Female*, Kinsey and colleagues articulated the novelty of their research by comparing it to the foolish efforts of their predecessors. Recall how in chapter 1 of this book, researchers writing about hyenas' mysterious sexual morphology constantly referred to the "absurd" belief of "the ancients" that hyenas could change sex, which served as a foil for their modern methods. The Kinsey authors deployed a similar strategy—so much for novelty—to situate their work as uniquely modern. In *Human Female*, Kinsey placed the origins of the entire project in his inability to find answers to his students' questions about sex in existing scholarship. In attempting to track down scientific information about human sexuality, he lamented, "We discovered that scientific understanding of human sexual behavior was more poorly established than the understanding of almost any other function of the human body."[35] What existed, alas, was "little better than the ancient belief . . . that sexual responses originate in the heart."[36] Like their hyena-studying brethren, the Kinsey team's research was far more evolved than that of yesteryear.[v]

Kinsey and colleagues were guided by the assumption that taxonomy is the way to knowledge. As Donna Drucker has elucidated, Kinsey's background in gall wasp research and his identity and training as a taxonomist dramatically shaped the form that his sex research would take.[37] The team rooted their work in the tradition of natural historical taxonomy but emphasized that they were aligned with "modern taxonomy." Instead of "naming, describing, and classifying," they were interested in "the measurement of variation in a series of individuals which stand as representatives of the species" under study.[38] This modern approach required large sample sizes; hence, what would follow in *Human Male* was based on data collected from 5,300 individuals. The authors were doing something wholly different than the medical and psychological researchers who came before, who were not able to meaningfully study variation because they used singular case studies. The "old" method was "dangerous" and

---

v   Kinsey himself did not have a reputation for modern scientific leanings among his collaborators. Paul Gebhard and Alan Johnson referred to Kinsey as "basically an early twentieth-century naturalist." *Kinsey Data*, 25.

"the antithesis of analyses based on large and statistically well selected samples of the sort the modern taxonomist employs."[39] Their strong language (danger! antithesis!) made it unmistakable that the Kinsey team were not doing that shoddy kind of "science."

The authors acknowledged the "important contributions" of scholarship based on case studies, like the work of Havelock Ellis, Sigmund Freud, Richard von Krafft-Ebing, and Magnus Hirschfeld, even as they distanced themselves from the past by criticizing a collective lack of statistical soundness.[40] These great men made it possible to conceive of a science of sex, but none of them had gotten beyond "accumulat[ing] great bodies of sexual facts about particular people."[41] They had no sense of frequencies or averages. Only a sparse few sex studies made the authors' cut for being "in any sense taxonomic," and all nineteen of those had major flaws.[42] The Kinsey team criticized their overreliance on college students as data sources, excessive concentration of subjects in New York City, and use of questionnaires rather than direct interviews or physical exams. Most importantly, their poorly-thought-out, unrepresentative samples made it impossible to generalize from study populations to a broader public. Of Dickinson and Beam's work, they noted, "there is no selection of the sample," with all study participants being drawn from one region, and mostly from the middle and upper classes. As a result "the findings . . . cannot be transferred to other segments of the population."[43] Pearl may have helped bring statistics to biology, but even he wasn't immune from such errors: In his 1925 book, *The Biology of Population Growth*, the samples were too small, and the comparisons of rural and urban subjects were dubious because of his failure to use educational levels as a variable.[44] In the hands of the Kinsey team, sex research had changed for the better.

These discussions of method made implicit normative claims about rigorous sex research. As the Kinsey team set an agenda for the future of sex science—*Human Male* was only a "first step" toward "sounder generalizations about . . . sexual behavior"—they erected new epistemological scaffolding for what constitutes knowledge about sex. The modern way to sexual knowledge was through mathematical precision, and the most interesting problems were to be solved at the level of population, not the individual.[vi] *How many* became the driving problematic, which centered

vi   The Kinsey studies coincided with the beginnings of a shift from eugenics to population control, in which the latter continued to hinge on male/female dyadic reproduction as a cause of social problems. "Population" here was not

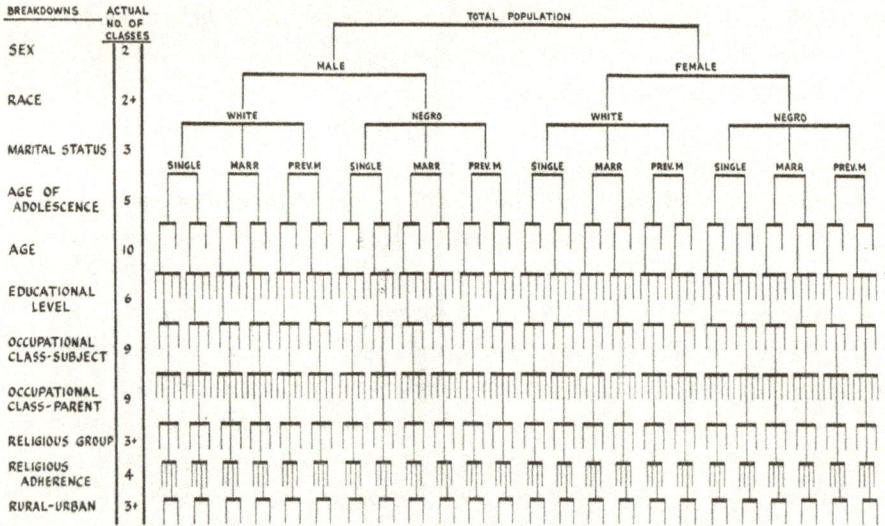

FIG 4.1. A schematic of the "twelve-way breakdown" used in the Kinsey studies to categorize interviewees and compare data across groups, from *Human Male*. The breakdown starts with the "total population" in the study that is then split into "male" and "female," which are then further narrowed according to other specifications.

frequency and incidence as the information of greatest concern.[vii] Debates about sex itself were subsumed under questions about data collection and math.

The Kinsey researchers turned their massive collection of individuals into comparable subgroups by sorting them into the categories "male" and "female." Ten thousand individual interview sheets were meaningless, incomprehensible fragments, just like the ERO's pedigree charts in chapter 2; carefully sorted, they would be made to reveal their secret patterns. Thus, the creation of an analytical apparatus in which easily demarcated

---

only a large, governable group of people in a biopolitical sense, but also specifically linked to anxiety about overpopulation and to new forms of quantifying aggregate life. See Murphy, *Economization of Life*, and Allen, "Old Wine."
vii   Howard Chiang has argued that the Kinsey studies shifted the conversation about sexual normality and abnormality from *clinical* to *statistical* evaluation—the individual body mattered less than how its behavior compared to others. "Liberating Sex," 52.

and usable categories offered the path of least resistance.[viii] The team used what they called a "twelve-way breakdown" to compare incidence and frequency across groups. These twelve variables included race (though the authors declined to include data for anyone who wasn't white), marital status, age, educational level, and occupational class, as well as several others.[45] The first and most important of these was a two-way "breakdown into male and female populations."[46] In a graphical representation that bears a striking resemblance to a pedigree chart, the total population forks into two possibilities: "male" and "female" (figure 4.1).[47] Only after that split did further breakdowns occur—"female" and "male" each branched into "white" and "Negro," each of which further disaggregates into "single," "married," and "previously married." Sex, conceived as a static binary, was the primary difference from which all others would stem. The authors could have made their first split according to race or marital status (recall Dickinson and Beam's *A Thousand Marriages* and *The Single Woman*). Instead, they naturalized maleness and femaleness as the most fundamental human distinction. Given their attention to methods, the underlying message was that binary sex was the right way to structure research. There might be a terrific variety of sexual behaviors engaged in by Americans, but there were only two sexes, so stable they could carry the weight of the largest and most revolutionary sex study ever undertaken.

Statistical Problems of the Kinsey Reports

Upon its publication, *Human Male* garnered praise and critique among scientists in equal measure, with commentators expounding on the authors' methods, the findings of the research themselves, and the potential moral impact that the Reports might have on American society, for better or for worse.[ix] The methodological aspect of these discussions tended

---

viii Recall that the committee reviewing the ERO in 1935 suggested that they use Hollerith punched cards to mechanically tabulate their data. This was precisely the method that the Kinsey team used, which enabled them to process far more data far more quickly than earlier methods of hand tabulation did.

ix Drucker, "'A Most Interesting Chapter.'" Miriam Reumann has discussed other reactions beyond science in *American Sexual Character*, including hope that the research would lead to greater sexual freedom and alarm at the evidence of apparently rampant sexual immorality. Criticisms voiced at a 1948 American Social Hygiene Association Kinsey symposium referenced above, collected in a volume titled *Problems of Sexual Behavior*, included John M. Cooper's theological

to linger on the Kinsey team's use of statistics. Were their findings, with their potentially tremendous significance for understanding the sex lives of (white) Americans, based in good science, or had they done something wrong to produce such shocking numbers? This feedback had surprisingly little to do with sex or sexual behavior. The main concerns revolved around how they had counted. Just as the Kinsey team's use of statistical methods was their innovation, math became a focal point in assessments of the studies' validity, with sex categories disregarded as a subject of interrogation even as the use of other categories came into question.

One of the foremost critiques of the Kinsey Reports involved their sampling method. Rather than deploy representative sampling, they had used what they referred to as "100% sampling"—they picked particular populations, like the members of a school district's parent-teacher association (PTA), the residents of a college dorm, or the congregants of a particular church, and then over time took sex histories from every, or almost every, person in that group. They sometimes built in a snowball technique, beginning with one member of a community and gaining access to others through word of mouth. This enabled them to use positive social pressures to obtain participation in a study that was likely to generate raised eyebrows, if not outright refusals. The authors maintained that "loyalty to a group project" and hearing from friends about the experience of contributing to the study would allow them to gather a greater number of sex histories than asking random people to divulge intimate secrets.[48]

This approach violated up-to-date statistical practices that favored "probability sampling" (a form of representative sampling that uses random participant selection, such that members of a given population have an equal chance of being chosen for the sample). As Paul Gebhard and

---

take that the Reports absolved everyone from sexual responsibility and self-improvement, and psychiatrist Jule Eisenbud's issue with the Kinsey team's lack of engagement with unconscious motivations. At the same symposium, Margaret Mead lampshaded the Kinsey team's lack of attention to the social importance and physical pleasure of sex. "The major abstraction which I think any anthropologist from Mars would get out of the Kinsey report as to what are the beliefs about sex in the United States," Mead reflected, "is that it is an impersonal, meaningless act which men have to perform fairly often—but oftener if they haven't been to school much. . . . This is one of the must Puritanical documents I have ever read in my life" (65, 67). It's true—*Human Male* is a spectacularly unsexy book. Specifically methodological critiques, though, were primarily statistical.

Alan Johnson put it in their 1979 recollection of the study, the 100 percent sampling method that the Kinsey team preferred "was regarded as an alien or anachronistic idea completely at variance with modern sampling theory and method."[49] With modern probability sampling, participants would not all be part of a common group, like a PTA, nor would they form a proportion of the sample in excess of their proportion within the larger population. The Kinsey team's 100 percent method might have skewed results: Sampling all members of a specific church might have introduced an unusual number of people less likely to engage in premarital sex; a group of incarcerated men might be unusually likely to have homosexual experience. Exhibited behaviors might be unique to, say, college students, who were disproportionately represented. Representative probability sampling made generalizations more accurate.[x]

One of the highest profile and most intensive critiques of *Human Male* came from a committee appointed by the American Statistical Association (ASA) and sponsored by the National Research Council (NRC) and the Rockefeller Foundation. Its primary consideration was the study's statistical validity. The NRC and the Rockefeller Foundation had directed a large portion—around half—of the budget for the Committee for Research on Problems of Sex (CRPS) toward the Kinsey studies.[xi] With the results of *Human Male* now available, they wanted to determine whether additional support was a good use of their money. In 1950, the NRC requested that the Commission on Statistical Standards of the ASA lend a hand in figuring this out. The way to verify whether the team had done good work was to consult scholars who did not work on sex, but rather on math.[xii]

x   This problem had to be invented and only came to the fore first between 1895 and 1903, when statisticians and scientists debated whether "one could legitimately replace the whole by a part," and again in 1925 and 1934, when they refined that question into one about accuracy in random versus purposive sampling. Desrosières, *Politics of Large Numbers*, 211.
xi   Yes, the same CRPS that funded a significant amount of Cold Spring Harbor research.
xii   Cochran, Mosteller, and Tukey (CMT) specify at the outset that "the committee wishes to emphasize that this report is confined to statistical methodology." They were not interested in digging into the sexual content itself. Part of this was likely strategic. As Tabea Cornel has argued, CMT de-sexed the Kinsey team's work to heighten the objectivity of their assessment, studiously minimizing sex acts with "abstract terms like 'behavior.'" Cochran, Mosteller, and Tukey, *Statistical Problems*, 1; Cornel, "Contested Numbers," 13.

Statisticians William Cochran, Frederick Mosteller, and John Tukey (hereafter CMT) were tasked with assessing the methodological rigor of *Human Male* and providing feedback in anticipation of the forthcoming *Human Female* volume. While the process of conducting the review was decidedly not without conflict and outright criticism, CMT's "overall impression of [the Kinsey team's] work to date is favorable," and they found that in contrast to preceding work "KPM's work is outstandingly good."[50] At the same time, they had substantial critiques of the authors' approach to data collection, with particular misgivings about "two main logical gaps" that left them concerned about whether the arguments of *Human Male* were "justified in terms of the usual standards of scientific accuracy."[xiii] They published their assessment first as a lengthy article in the *Journal of the American Statistical Association*, then as a book-length version, both with the title "Statistical Problems of the Kinsey Report" (it's telling that they were not called "Statistical Successes of the Kinsey Report").

CMT worried about an interview process based on subjects' accounts of experiences, which may or may not conform to actual past behavior.[51] Relying on after-the-fact recollections left the researchers—and their data— open to both the failures of human memory and outright dishonesty.[xiv] The team's data might have been skewed from the beginning. CMT's main concern, though, had to do with the 100 percent sampling decision.[xv] They cautioned that "groups might be biased with respect to some factor whose importance was not realized."[52] Worse, studies that attempted to make claims about a larger population based on a nonrepresentative sample resulted in inferences that were "often tortuous and weak."[53] The Kinsey Reports were better than what had come before, but their exact findings were questionable.

These disagreements about methodology dovetailed with disagreements about sexual normality. In correspondence to the NRC marked

---

xiii   Cochran, Mosteller, and Tukey, *Statistical Problems*, 263, 305. CMT did not ask if, given that the usual standards of scientific accuracy couldn't effectively study sexual behavior, they were very useful standards. A salient example of the method stubbornly limiting what it's possible to know rather than researchers adapting their method.

xiv   Cochran, Mosteller, and Tukey, *Statistical Problems*, 83. CMT noted that some reviewers steeped in psychoanalytic approaches brought up unintentional repression and distortion as well.

xv   Peter Hegarty has called the CMT report "a strong statement of the ideals of sampling theory." *Gentlemen's Disagreement*, 146.

"very private," CMT voiced their speculation that the authors had fallen prey to "the fallacy which can be phrased 'The more socially unacceptable behavior a sex study finds, the better it is.'" Their suspicions stemmed from the way that the team were "very careful in stressing objectivity" with regard to interview practice but not selection of participants. This, CMT worried, might indicate "a subconscious recognition that objective selection might, unfortunately, <u>lower</u> [underlined in the original] the apparent incidence of socially unacceptable behavior."[54] CMT felt that finding such high incidence of deviation from expected sexual norms meant something was awry; the Kinsey authors felt assured that their method was correct because they were finding so much deviation from expected sexual norms. In other words, methodological debates were not only about statistics per se, but what kinds of findings justified the use of a particular methodology.[55]

In the end, CMT granted that it would "not have been feasible" for the team to conduct the entire study with probability sampling.[56] Kinsey, Pomeroy, and Martin had argued vehemently that probability sampling wouldn't work in a sex study—too many people would have refused to participate, and it would have been impractical to determine specific groups to take proportionate samples while requiring much more financial and time investment, all to the point of near-impossibility.[57] CMT concurred, calling a full probability sampling scheme "almost certainly . . . impractical."[58] Yet CMT insisted that sampling was *Human Male*'s largest problem and that the team still should have checked the representativeness of their sample. Difficulty didn't "excuse KPM from the responsibility for choosing geographical disproportion in order to save travel time and expense."[59] The methodological issues plaguing *Human Male* were, however, not unique to that project. The Kinsey team was, CMT wrote, "engaged in a complex program of research involving many problems of measurement and sampling, for some of which there appear at present to be no satisfactory solutions."[60] They also noted in regard to the subject recall problem that "until new methods are found, we believe that no sex study of incidence or frequency in large human populations can hope to measure anything but reported behavior."[61] As in the Kinsey team's own account, the future of sex research lay in the refinement of statistical theory and practice, not integrating a complex version of sex into a new statistical framework or foregrounding male/female similarity. The future of sex research would be found in improving statistical techniques to count differences between settled sex categories.

With the addition of post-1930s statistical methods, the Kinsey Reports did offer a new tool for making exceptions meaningless. While the worker bee that didn't conform to a definition of femininity as passive had been cast aside as insignificant by Geddes and Thomson, and Dickinson had refused to relocate the woman with "male-type" genitals outside woman-hood, earlier studies of sex were less mathematically rigid, enabling re-searchers to place individual data points along a much more expansive spectrum of sex possibilities (while happily making racist and ableist population-based claims based on a small handful of cherry-picked exam-ples, of course). Statistical innovations and the Kinsey studies' use of them legitimated, with quantification, the practice of deciding that anomalous forms of sexed bodies didn't matter when thinking about sex as a whole.[62]

It took more than statistics mathematically, though, to produce the Kinsey Reports' reinforcing of binary sex. In the last several decades, other sociological and economic projects had taken seriously the possibility of a more participatory statistics. As Autumn Womack has elucidated, Black social scientists and reformers, most notably W. E. B. Du Bois, deployed a variety of data-collection and interpretation technologies and strategies that sought to emphasize the complexity of African American existence.[63] During the Depression, unemployed workers served as enumerators for federal studies of unemployment, such that they were able to contribute to potential solutions to their own problems through relief jobs.[64] Eventu-ally, however, the New Deal enumerators were accused of being too biased to do objective statistical work, and the success of the push for random representative sampling placed expertise squarely in the hands of pro-fessional statisticians rather than populations under study.[65] Social and political context narrowed possibilities beyond straightforward bureau-cratic and scientific projects. In their effort to situate themselves at the frontier of sex research, the Kinsey team left expansive accounts of sex behind in favor of categories it was simpler to do math with—or, at least, they tried.

The Homosexual Is Not a Species (Again)

The Kinsey team navigated the still-ambient theory of universal bisexual-ity and its relationship to the cutting edge of sex research with different degrees of success in each book. In *Human Male*, the authors rejected bi-sexuality outright and focused instead on their argument that "the homo-sexual" was not a particular kind of person and certainly not the "sexual invert" of old. In doing so, they did help change the status quo: They split sexual behavior from the body, and analysis structured around comparisons of sexual contact with "same" and "different" sexes cast aside an inversion framework. Yet with their focus on behavior and environment, the team's efforts to stake out a claim in the land of modern sex research invoked and relied on more sexual knowledge of the past than they bargained for. By the end of *Human Female*, they had to reckon with bisexuality once more as they searched for biological differences to explain different patterns of sexual behavior between men and women and found only homologues and similarities. As they attempted to contain sex to two binary variables, its greater complexity surfaced anyway and left the Kinsey team stranded between past and future.

Fundamental to the Kinsey team's place in the sex research avant-garde was their discovery of the high incidence of homosexual behavior and subsequent refusal of innate sexual kinds. The authors found that at least 37 percent of men had "some homosexual experience" after adolescence—even if only once—with homosexual experience defined as "physical contact to the point of orgasm" with another man.[66] They emphasized that "this is more than one male in three of the persons that one may meet as he passes along a city street."[67] The figure jumped to 50 percent for men who didn't marry until they were thirty-five.[68] Only around 10 percent of these men, however, were exclusively homosexual; most interviewees had had both homosexual and heterosexual contacts, in a range of proportions.[69] With these findings, the Kinsey team rebutted both the older sexual inversion model of homosexuality that was rooted in a universal bisexuality framing, in which some kind of constitutional (whether physical or psychological) departure from total maleness or femaleness was imagined as the cause of homosexual desire, and a newer model of homosexuality in which one was normally sexed and merely desired other men.[xvi] Both of those models assumed "the homosexual" was a particular type of person. Based on the Kinsey numbers, anyone could engage in homosexual behavior, not only a tiny subset of the broader population.

This dismissal of an antiquated sexual typology coalesced in the Kinsey team's refutation of the idea that homosexual desire and behavior emerge from a state of intermediate sex. In *Human Male*, the authors critiqued the

xvi   Even if you were gay, you were actually straight! But also kind of trans? Ah, sexual categories. See Chauncey, *Gay New York*, and Clement and Velocci, "Modern Sexuality."

way that sexologists had used terms like *sexual inversion, third sex,* and *psychosexual hermaphroditism* to suggest "that individuals engaging in homosexual activity are neither male nor female."[xvii] They shot down common beliefs that homosexuality manifested physically, like homosexual men having broad pelvises, a feminine gait, and "teeth which are more like those of the female," which they framed as unscientific "impressions."[70] They had not identified cross-sex psychological or social traits among men who engaged in homosexual behavior; the authors critiqued the existing assumption that homosexual men were supposed to enjoy music and the arts and prefer work as hairdressers or bookkeepers.[71] "It should be emphasized," they concluded, "that the reality [of homosexuality] is a continuum."[72] Sex, however, was not.

Universal bisexuality was, by the 1940s, already somewhat moribund (though not, as discussed in chapter 5, dead yet; it provided a foundation for trans medicine). It had never found quite as strong a foothold in the United States as it had in Europe, and many middle-class men had been identifying themselves as "queer" but normatively male and masculine since the 1930s.[73] Moreover, while anthropologists and psychologists had taken up the idea of a spectrum of sex in personality and other psychosocial factors, they still mostly used binary sex categories and did not apply bisexuality theory to the body.[74] Though the shift from bisexuality-inflected sexual inversion to same- but otherwise normally sexed object choice was uneven and incomplete, it was nonetheless marked.[75] Given their affinity for the new, it's unsurprising that the Kinsey team embraced this modern rejection of an old-fashioned sexual inversion model. In this respect, the publication of *Human Male* contributed to an ongoing reconfiguration of homosexuality. At the 1948 American Social Hygiene Association Symposium on *Human Male*, for example, George Washington Corner, then chair of the CRPS, declared that the Kinsey team had "weaken[ed] any biological hypothesis about homosexuality that ascribes it to defective hormone balance or to genetic factors for instance in the bisexuality theory formu-

xvii   Kinsey et al., *Sexual Behavior in the Human Male*, 612. The authors note that intersex bodies do exist in varying levels of extremity, but that was separate from choice of sexual partner. Intriguingly, *Human Male* cites Dickinson's *Human Sex Anatomy* here, noting "sometimes the term intersex has been applied to such females [with large clitorises]; but until more is known about the biological basis of the situation, it is not certain that the term intersex should be applied even in these cases" (659).

lated by Richard Goldschmidt."[76] Historian Joanne Meyerowitz, too, has called *Human Male* a "repudiation of biological bisexuality."[77]

The Kinsey team thus contributed to a consolidation of sex research around a male/female binary. In the inversion model the Kinsey team disavowed, the designation of the invert as a "third sex" had, for a time, intrinsically increased the number of sex categories available: Per Karl Heinrich Ulrichs in 1870, "As a *third sex*, we [Urnings] are on the same level as the male or female sex, but we are independent of the male or female sex, fully separate from both."[78] At the risk of stating the obvious, the existence of three sexes means there were not only two. As bisexuality produced something like trisexuality, it also inextricably linked the body and sexual behavior, whether characteristics like wide hips or enlarged clitorises resulted from pathologies of the nervous system or were produced by deviant sexual activity itself (there were many iterations of inversions among sexologists).[79] The Kinsey Reports took this ongoing shift and ran with it. The "homosexual" no longer indicated a type of sexually intermediate person with physically manifested desires.[80] No matter what kind of sexual behavior someone engaged in, they were still fully male or female.

In the process, the authors, invested in the behavioral much more than the physiological, focused mostly on social factors like educational attainment and religious affiliation. Out of the twelve-way breakdown, only two factors, age and age at adolescence, were "biologic" (the authors characterized race as "race-cultural group" and noted they weren't interested in its "biological aspects").[81] The authors admitted that the body influenced sexual behavior "to a degree" but clarified that "the psychologic bases of behavior [are] even more important than the biologic heritage and acquirements" and "the most important external force is the social environment."[82] So decreed, the body was decentered as a locus of sexological inquiry. Homosexuality was behavioral and bisexuality was over, and, with that, the Kinsey team turned attention away from the possibility that homosexuality might be a site for an investigation of permeable boundaries between male and female.

They could not, however, get rid of the body entirely. It was precisely the existence of bodies with clearly delineated binary sexes that enabled them to articulate their unearthing of the high incidence of homosexual behavior. The authors' aversion to classifying heterosexuality and homosexuality as completely distinct from each other was rooted in the finding that wide swaths of the population had sexual encounters in different proportions with people across or on the same side of a sexed line. This necessitated such a line in the first place. The Kinsey team's formulation of sexual behavior as defined in relation to the sex of one's object choice in comparison to one's own otherwise normal sex *required* coherent, stable sexes that could be different or the same.[83] This compounded their perceived methodological needs of participants staying put in distinct categories that had already funneled them into the use of static male and female groups. In practice, the Kinsey studies gave that binary a resounding stamp of approval, aided by government funding, multiple printings, and a media blitz. It also prevented the total separation of the body from sexual behavior; even if behavior couldn't be read on the body, the sameness and difference of "hetero" and "homo" stemmed from bodies themselves.

All of this said, the Kinsey studies did not fully adopt the postinversion form of homosexuality that rendered "the homosexual" as deviant in sexual object choice but not abnormally sexed. They objected to any sexual type of person at all, conceding only that a small percentage of men were "exclusively homosexual" and emphasizing that those men were so described based on histories of actual homosexual contacts and psychosexual reactions, not inherent typology.[84] This refusal of sexual type and centering of behavior was, ironically, a regression to an even older conception of homosexuality (before the term *homosexuality* had been coined). Prior to the development of sexual inversion, what would come to be known as homosexuality was largely considered, in Europe and the United States, sinful or illegal behavior that anyone could engage in.[85] By insisting that there was no homosexual type of person, only homosexual desires and acts, the Kinsey team managed to circle back to the mid-nineteenth century or earlier.[xviii] Likewise, the emphasis on the influence of external factors rather than innate sexual orientation recalls the eighteenth-century "porous envelope" account of sexuality described by Greta LaFleur. In that framing, environments, whether the torrid tropics or the crowded city, acted on the individual to shape their desires and actions and caused sexual deviance.[86] The Kinsey studies, for all their distancing from the past, thus exhibit a sort of temporal incoherence: a step forward into the incipient parceling out of sex, gender, and sexuality, and an atavistic return to the time before the homosexual "became a species."[87] A few years later, with their turn to research on women and cross-

xviii   Pick your favorite date for when "homosexuality" became a thing; specifics don't matter here beyond "things got weird about sometime in the nineteenth century." Go with Foucault's 1870 for simplicity, if you'd like.

sex comparisons, the Kinsey team would find that as with the body, they could not entirely leave bisexuality behind.

Bisexuality Strikes Back

In 1953, *Sexual Behavior in the Human Female* joined *Human Male* in defining the midcentury sexual zeitgeist. Like its predecessor, *Human Female* consisted primarily of the statistical outcomes of thousands of surveys—5,940 women provided the sexual histories that went into the book's analyses.[88] Its calculations, too, included only white Americans, though 934 women of color had been interviewed; the team also decided to exclude women who had been incarcerated because their data proved so different than that of nonincarcerated women.[xix] *Human Female* again declined to use representative probability sampling. However, because of "even more inadequate samples" than in the male study, the authors were more reserved when extrapolating from the study population, noting that "the generalizations made throughout the present volume have . . . been restricted to the particular samples that we have had available."[89] The questions asked and methods of analysis remained largely the same.

The volume differed, however, in its greater attention to comparison. In addition to the calculations made in *Human Male* that correlated particular experiences to rural-urban background or marital status, *Human Female* offered juxtapositions of male and female data at the end of every chapter. The numbers demonstrated significant differences: 62 percent of women had masturbated, compared to 93 percent of men; about half of the women surveyed had experience with premarital sex, while between 68 percent and 98 percent of men did, depending on educational level; women tended to be more interested in sex later in marriages, while men were more interested earlier on.[90] These and other discrepancies demanded explanation, and the authors set out to locate what physical differences may have been their root cause. The book's third section—over two hundred pages—attempted to explain them with a thorough survey of the extant literature on male and female anatomy and physiology. They were particularly interested in processes of sexual response and orgasm:

---

xix   Kinsey et al., *Sexual Behavior in the Human Female*, 22. Incarceration apparently also provides a little nudge toward nonwomanhood, in the assumption that including incarcerated women's interviews will skew the "female" results rather than meaningfully reflect what women "do sexually."

how sexed differences might result in divergent experiences and interpretations of stimuli, and thus which sexual acts and circumstances different bodies seek out. Except they couldn't find an embodied basis of behavioral difference. While *Human Male* dismissed bisexuality theory, *Human Female* ultimately returned to something quite like it as the authors failed to locate causal differences in male and female bodies.

Human Female stressed the presence of sexual homologies, shrinking the gap between male and female. The authors opened their discussion of anatomy with the assertion that male and female bodies emerged from an embryological "common pattern," which set them up for similarity from the earliest stages of life.[91] Even accounting for differing developmental pathways, the authors noted that "each structure in the one sex is homologous to some structure in the other sex."[92] The clitoris, therefore, was "the phallus of the female," and the labia majora and scrotum were so similar that they could be dealt with under a single chapter heading.[93] In nongenital parts of the body—mouth, buttocks, nerve endings more broadly, senses of smell and taste—responses were identical, regardless of sex.[94]

This framing, to be sure, is not identical to the universal bisexuality model espoused by many early twentieth-century scientists. While, for example, Robert Latou Dickinson (see chapter 3) formulated full sex development as only infrequently attained, with bodies frequently falling somewhere in between poles of male and female, and researchers such as Magnus Hirschfeld and Eugen Steinach envisioned all bodies as a mixture of male and female, the Kinsey Reports' version is more reminiscent of the model adhered to by nineteenth- and early twentieth-century zoologists (see chapter 1), where sex distinction increases along an evolutionary path, from primitive hermaphroditism to advanced dimorphism. Kinsey's own work on evolution rejected nineteenth-century theories of speciation, and the Reports make no explicit evolutionary claims. Still, the Kinsey team's rooting of sex differences and lack thereof in shared embryological origins echoes older evolutionary trajectories: An indistinguishable form develops over time into one of two sexed possibilities, leaving traces of that earlier state in homologous anatomy and physiology.[95] It is one more example of the ways in which the Kinsey studies drew, intentionally or not, on the enactments of sex elaborated in earlier chapters of this book and enshrined them in mid-twentieth-century sex science.

After careful comparison, the Kinsey team found anatomical, physiological, psychological, neurological, and hormonal factors to be unlikely sources of differences between men's and women's sexual behavior. "In brief," they concluded, regarding anatomy, "we conclude that the anatomic structures which are most essential to sexual response and orgasm are nearly identical in the human female and male. The differences are relatively few. They are associated with the different functions of the sexes in reproductive processes, but they are of no great significance in the origins and development of sexual response and orgasm. If females and males differ sexually in any basic way, those differences must originate in some other aspect of the biology or psychology of the two sexes." In other words, the authors spent an entire chapter of *Human Female* parsing out the anatomical differences between male and female only to determine that those differences are unimportant. The "other aspects" they point to in this passage did not pan out either. The same failure to find substantive difference continued in their chapter on the physiology of sexual response and orgasm—in that case, after a review of processes like changes in pulse rate, respiration, and muscle movements during sex, they again concluded, "In spite of the widespread and oft-repeated emphasis on the supposed differences between female and male sexuality, we fail to find any anatomic or physiologic basis for such differences."[96]

Turning to psychology, the authors found some differences, but they weren't particularly confident about their impact. Men, for example, were apparently more "responsive to psychosexual stimuli" than women and thus more psychologically conditioned by sexual experiences.[xx] Men reacted more strongly to things like reading erotic literature, "observing animals in coitus," and looking at their own genitals in a mirror, and were more likely to develop fetishes.[xxi] Underlying reasons were again unclear: The researchers admitted "there is nothing yet known in neurologic or physiologic science which explains what we have found."[97] Neurological studies had made "considerable progress," in that "after three decades of research . . . [they were] able to identify some of the internal mechanisms which account for the similarities between female and male sexual responses," though not for the divergences. It was clear that "cerebral differences" had something to do with sexual response, but "what the nature of such cerebral differences may be, we do not know."[98] Their discussion of the impact of hormones is similarly unimpressive. The most

xx   Kinsey et al., *Sexual Behavior in the Human Female*, 688. The authors cite this as a reason that there are more "transvestites [who] are anatomically male" than vice versa (681).
xxi   So much for autogynephilia.

they could say was that "various hormones may affect the levels of sexual responsiveness in the human female and male."[99] Rather, the authors confessed "there is no demonstrated relationship between any of the hormones and an individual's response" to specific sexual stimuli, partners, or practices.[100] None of these branches of science offered definitive answers.

One way to interpret this final section of *Human Female* is as a rejection of binary sex. Drucker, for example, has argued that the concluding chapters indicate Kinsey's growing suspicion of the utility of male and female as categories of analysis, and his rethinking of sex in its entirety.[101] "Analyzing the third section of [*Human Female*] shows how previously solid forms of classification in an author's thinking, such as binary division between men and women," Drucker concludes, "break down under the weight of contrary evidence, making room for new ways of thinking about how to classify human beings sexually."[102] However, I'm not as sure that a binary division did break down, especially not "under the weight of contrary evidence" given how long that evidence had been around. Like sex researchers who came before them, Kinsey and his colleagues opened up the possibility that sex might be more expansive while simultaneously insisting, theoretically or practically, on a binary. We again see the incoherent coexistence of multiple, conflicting models of sex rather than a reconfiguration. Even if Kinsey personally had started to doubt the utility of binary sex, he used them anyway.

Notwithstanding the nonbinary potential of *Human Female*, such questioning doesn't seem to have been taken up in the book's reception. As already discussed, Kinsey's scientific critics were quiet on the issue of male and female categories as analytic variables. Miriam Reumann has amply demonstrated that *Human Female* launched substantial debate and anxiety about the "normal" or "typical" American woman. Some of this conversation even "dismissed the evidence of the report" and suggested that women had lied or were particularly unusual.[103] Readers seem to have flat-out ignored the third section of the book—Reumann notes that several respondents emphasized apparent physiological sexual differences like ease of and speed to orgasm to link female sexual pleasure to marriage (apparently husbands were committed to their wives' sexual fulfillment).[104] If *Human Female* challenged ideas of normative womanhood, it was in terms of sexual morality, not the body.

Countless popular sources portray the subjects of the Kinsey Reports as normally sexed men and women, without reference to the authors' inability to find any physical basis of divergent behavior. They lean into, rather

24% HAD THEIR EYES OPENED EARLY

" MOTHER TOLD ME THE WHOLE STORY. I CAME FROM A FLOWER AND YOU CAME FROM A BEE."

FIG 4.2. Cartoon from a "Sexual Misbehavior in the Human Male" cocktail napkin. It shows a boy and girl (gender difference identified by hair bow) sitting on a stoop, with the boy saying "Mother told me the whole story. I came from a flower and you came from a bee." This is apparently an example of the "sex-tistic" that "24% had their eyes opened early."

than disrupt, sex stereotypes. In the aforementioned cocktail napkins, for example, marketed as "Sexual Misbehavior in the Human Male," the overall joke is about "sex-tistics" in general, and they emphasize sex differences, not similarities. One napkin avers "93% Like Feminine Femmes" and features a cartoon of a man in pants and jacket and a woman wearing a dress, walking arm in arm. The image is captioned "There's something different about you, Fran, that sends me. Maybe it's because you wear skirts." Another (figure 4.2) reports that "24% Had Their Eyes Opened Early"; it shows a young boy and girl (we know one is a girl because of the bow on her head) sitting together, with the boy saying, "Mother told me the whole story. I came from a flower and you came from a bee." Surprise, surprise, most men like normative women (unclear if that missing 7 percent references the homosexually inclined or just men who like butches), and adults know the truth about sex differences. The cocktail napkins are only one example, but the point remains: The striking similarities between male and female bodies and sexual responses were not the information that filtered from the Kinsey Reports into popular consciousness. In the social context of an attachment to normative masculinity as necessary for cold warriors' victory, anxiety about state-threatening homosexuality and subsequent repression of it, and the nuclear family as protection from nuclear destruction, it comes as no shock that the destabilization of sex categories didn't quite work in the public sphere.[105] The third section of *Human Female* brought the Kinsey team back to bisexuality, and the authors and their readers both thoroughly refused the evidence that similarity and homology might be a better analytic frame than difference.

The Roaring Silence of Race

For all their concern about representative sampling and generalizing from the survey population to Americans writ large, commentators were unworried about the impact of an entirely white data pool on the Kinsey team's findings. Race, demarcated as "white" or "Negro," was second only to sex in the twelve-way breakdown, and the Kinsey researchers interviewed hundreds of Black contributors for each volume.[xxii] Yet neither *Human Male* nor *Human Female* factored data from Black interviewees into their calculations and comparisons, with the explanation that there weren't enough Black people in their sample to make meaningful claims.[xxiii] Despite this conspicuous absence of racial analysis, the Kinsey Reports evince the long legacy of the racialization of sexual pathology, while also maintaining eugenic ideas about unassailable social hierarchies. At the same time, the authors presented their conception of race as aligned with a posteugenic turn to culture over biology. In addition to their incoherent use of sex as both a static binary and marked by incredible physical similarity, the Kinsey studies enacted an incoherent use of

xxii   In several places, that binary is complicated with the addition of "non-white."
xxiii   Gershon Legman said the sample indicated "typical American megalomania" and said the book should have been called "Sexual Behavior in 5300 Northeast American Males"—notably missing the point about the whiteness of it all. "Minority Report," 13, 15.

racial classification. Sex research became whiter on a surface level, but its imbrication with race science remained.

Though many scholars have remarked on the whiteness of the Kinsey studies, few have engaged extensively with Kinsey's own relationship with eugenics.[106] Whether Kinsey was "a eugenicist" is difficult to say with any degree of certainty; nor is it a particularly interesting or useful question. However, it's apparent that Alfred Kinsey was part of the network of sex scientists I've traced in this book and that his 1930s work on biology instruction advanced eugenic views. More importantly, while the Reports distance themselves from biological race and don't comment on eugenics at all, the moments in which they do discuss race and social hierarchies reinscribe race as entangled with sexual deviance. In other words, the circular definitions of race and sex discussed in chapter 1 persisted well into the twentieth century. "New" methods for sex science could not escape this foundational connection.

Kinsey credited several actors in the overlapping network of eugenic and sexological science described in previous chapters as indelibly shaping the studies that bore his name. Raymond Pearl's visit to Indiana University in the fall of 1938 tremendously influenced Kinsey's ideas about normalcy and variation, the utility of large data sets, and approaches to sampling and statistics.[107] Robert Yerkes—the student of Charles Davenport whose tenure as CRPS chair produced significant funding for ERO research— received an "especial mention" in the acknowledgments of *Human Male* for his support.[108] In addition to his role as CRPS chair during the initial funding of the Kinsey studies, he had also encouraged Kinsey to focus his work on "presumptively typical and normal male and female whites of our U.S.A. culture."[xxiv] Dickinson was thanked with an entire paragraph recognizing "the benefit of his accumulated experience, and his constant advice."[109] He and Kinsey had spent a considerable amount of time together, with Dickinson visiting Kinsey in Bloomington several times; Dickinson apparently saw Kinsey as his inheritor, and in 1943 he wrote to the director

xxiv Yerkes to Kinsey, September 25, 1943, Alfred Kinsey Correspondence, Kinsey Institute (hereafter AKC). Kinsey assured Yerkes that he was focusing on "securing the picture of the average and the usual portion of the population" and made clear that he welcomed further suggestions. Previously, Kinsey had written expressing interest in cross-group comparisons, particularly because there were "different racial stocks to compare." Kinsey to Yerkes, September 28, 1943; Kinsey to Yerkes, January 14, 1941, AKC.

of the Rockefeller Foundation Medical Sciences Division, Alan Gregg, "I think you can imagine what it means to an Ancient to find someone [i.e., Kinsey] who would be willing to shoulder a frayed mantle."[110] Kinsey was steeped in the social world of eugenic sex science.[xxv]

Near the end of his career as an entomologist, just before beginning his human sexuality work, Kinsey wrote several successful high school biology textbooks, workbooks, and a teaching guide, and in these, his stance on eugenics sharpens into focus. The 1938 *New Introduction to Biology* demonstrates an understanding of a hierarchy of human life based in inherited abilities. "The importance of heredity must never be forgotten," Kinsey wrote. "We only fool ourselves when we say that all men are born equal. While we may have equal rights legally, we differ greatly in our biologic equipments. There are really very few of us who have the necessary heredities to make good Presidents of the United States."[111] A subsequent student activity idea instructs, "Find (in some reference book on heredity, eugenics, or sociology) the story of the Jukes, the Kallikaks, the Darwin, or the Edwards family, and explain how their heredities helped or hurt society."[xxvi] The book's next chapter, "The Value of Scientific Research," opines, "In the care with which the genes are mated lies the best hope of an improved race of mankind."[112] Kinsey imagined it worthwhile for high school students to understand both the social threat of bad genes and the promise of good ones.

Kinsey's teaching guide *Methods in Biology*, published in 1937, more explicitly emphasizes the importance of eugenics instruction. Though he noted that the subject was controversial, it was "one of the most hopeful signs for the future that young people are becoming interested in problems of human breeding."[113] Kinsey admitted that knowledge about heredity was still growing; still, there were some agreed-upon scientific facts. The facts (and I quote at length because this material has, to my knowledge, not yet made it into the Kinsey historiography) consisted almost entirely of the following:

xxv Janice Irvine has discussed in *Disorders of Desire* that one of the many hopes for the Kinsey studies was an improvement of heterosexual marital relations. While beyond the scope of this chapter, it's worth noting that early twentieth-century sex education to that end, including much of Dickinson's work, was explicitly eugenic; it would be intriguing to consider Kinsey as an outgrowth of this milieu. See Haager, "'Sex Education's Many Sides.'"
xxvi Kinsey, *New Introduction*, 392. Histories of these families were some of the most well-known eugenic case studies of heredity.

More than 2 per cent of our population is hopelessly dependent because of mental defects that are known to be hereditary. There is no record to show that this group ever produces any individuals that are socially valuable.

The percentage of hereditary defectives is constantly increasing, because such defects are spread by apparently normal as well as by abnormal individuals. The normal individuals from hereditarily defective families may carry the defects in recessive genes.

We face an indefinite increase in the number of such defectives unless we can in some manner prevent reproduction in these individuals. Already the care of these defectives costs more than the education of all the college students in the country.

The complete isolation of all such defectives (which no state has yet been able to afford), or their sterilization, are the only two methods available for controlling the spread of hereditary defects.

While it would be very difficult to decide which was the socially most important and which the socially least important half of any population, there would be little difficulty in selecting the 10 per cent which is the greatest drain on the advancement of our social institutions. The limitation of reproduction among this 10 per cent may be necessary before we can expect any decrease in the number of helpless dependents. How the control of the group may be accomplished is a matter of debate.

While the mass of the socially worth-while individuals, and even some leaders have come from the middle classes, the data abundantly prove that most leaders have come from the group which is the best equipped in hereditary capacity and environmental training.

Young people who are hereditarily sound and environmentally privileged may contribute to the quality of society by planning to have as many or more children than the average for the whole population. With less than three children to a family, the hereditary line which they represent will soon become extinct, and society deprived of the qualities carried in that line.[114]

He then suggested the proceedings of the Third International Congress of Eugenics as a teacher reference.[115]

The malevolence of eugenics is dense in these pages. Kinsey's assertions contain a pro-sterilization stance, an attachment to a self-evident distinction between worthy normality and unworthy deviance, and a narrative of economic and social drain. It's possible Kinsey's opinions changed between the publication of *Methods in Biology* in 1937 and when work on *Human Male* began in July 1938, but his textbooks nevertheless complicate the image of Kinsey as a progressive interested in "fostering . . . a tolerance for diversity."[116] While the studies that would bear his name for decades to come were credited with inspiring a wide-open future of sexual revolution, they were spearheaded by a scientist who, by his own words, had advised for the prevention of "defective" people through the "limitation of reproduction."[117] And it wasn't just Kinsey himself—as late as 1978, Paul Gebhard gave the Galton Lecture at a symposium of the Eugenics Society in London, where he concluded that "uncontrolled reproduction" would "deplete our finite resources and pollute our environment," and therefore "reproduction must become a privilege and not a right."[xxvii]

The effects of all of this on the Reports are implicit, but notable, and they illustrate the Kinsey studies' position in the midst of a broader transition in scientific racial discourse. As noted, the Kinsey team had interviewed thousands of Black study participants but declined to include their data in the calculations that made it into the books. Apparently, this was a sampling issue: In *Human Male*, the authors explained that "the story for the Negro male cannot be told now, because the Negro sample, while of some size, is not yet sufficient for making analyses comparable to those made here for the white male."[118] They subsequently used hypothetical "Orthodox Jewish males who were Negro, single, between the ages of eighty-five and ninety, illiterate, living in rural areas, and belonging to the Social Register" to exemplify a group "so rare in the American population that they are unimportant for study," for which "it would never be possible to secure a statistically good sample" anyway.[119] For *Human Female*, the team again excluded the "non-white sample . . . because that sample is not large enough to warrant comparisons of the subgroups in it."[xxviii] This dismissal of small numbers was a choice. The pool of men over the age of sixty (Black and white combined) was, the authors warned, "too small to

xxvii   Gebhard suggested limiting births through "adjustments in taxation policy, the rewarding of minimal reproduction, delaying marriage, and devising ways to appeal to self-interest." Not quite pro-sterilization, but, uh, not great either. Gebhard, "Sexuality in the Post-Kinsey Era," 56.
xxviii   Kinsey et al., *Sexual Behavior in the Human Female*, 22. I guess despite their very thoughtful 100 percent sampling approach they just couldn't find enough people of color to interview?

allow statistical analyses of the sort employed for the other age groups." However, the authors reported on the data anyway. They explained that "there was such interest in the sexual fate of the older male that it seems valuable to summarize the data even for these few cases."[120] It seems that the members of small samples were still worth discussing when they were mostly white or had implications for a white population. In the end, despite the criticism about representative sampling, the authors' colleagues let them get away with the decidedly nonrepresentative, completely white samples used for both books. Race was such an important variable that combining Black and white data would ruin the study, yet so unimportant that nonwhite data could disappear without consequences.

Still, even with the absence of quantitative racial comparisons, the books indicate a move toward a cultural definition that continued to racialize sexual pathology. In their explication of the twelve-way breakdown, the authors highlight another one of their modern touches. They specify that for a "Race-cultural Group," the "question is one of race-cultural backgrounds, rather than racial background in the exclusively biologic sense."[xxix] A race-cultural group would instead be determined by "the subject's place of birth, his place of residence during childhood and adolescent years, and the ancestral home of the parents."[121] Such a definition aligns with the concurrent trend emblematized by the 1950 UNESCO Statement on Race, which emphasized the genetic relatedness of human groups and encouraged attention to cultural difference over biological difference.[122] It stressed, too, that differences within racial groups were more substantial than differences between them.[xxx]

Instructions to interviewers, though, demonstrate some uncertainty about racial categories, drawing attention to "blood" as well as appearance even as they reflect a sense that how a person moves through the world has more bearing on their sexual experience than innate biology. "If a person has Negro blood, but passes as a white, he would be called W, with explanatory note," the instructions specify. "If a person is part Indian and part N, but is thought of as N by his associates, he would be called

---

xxix   Kinsey et al., *Sexual Behavior in the Human Male*, 76. They offer eleven possibilities: American and Canadian White, American and Canadian Negro, British (Great Britain), Western and Northern European, Mediterranean European, Latin American, Slavic, Oriental (Asia), Filipino, Polynesian, and American Indian.
xxx   The Kinsey team did not seem to have gotten that latter memo.

N (with also indication as part Indian). After N a check scale indicates the darkness of skin – = very light, ✓ = very dark."[123] There is an admission, implicitly, that race is not as simple as the researchers' racial categories would have it, yet we see here how that complexity is narrowed into a single variable. Even with an explanatory note, the "part Indian and part N" interviewee would end up in the N category (and thus written out of the study). The Kinsey studies ultimately enacted race as a binary variable: white or not. While clearly interviewees did not always fit into such categories, the nuances of their existence overwhelmed the possibilities of the schema on offer.

There is a seemingly willful disinterest in race in most of the questions asked of interviewees, except regarding sexual deviance of particular concern to reformers and the state. Interviewers only asked about the race of a subject's partner(s) in the context of homosexual experience or heterosexual prostitution.[124] Otherwise, the extensive questions presumed that sexual contacts were same-race—interviewees weren't asked about the race of spouses, partners in preadolescent play, premarital petting, premarital coitus, extramarital relations, postmarital intercourse, intercourse with prostitutes as clients, and intercourse with clients as prostitutes.[xxxi] There wasn't even an imagined possibility that a white interviewee might have racial fantasies. The Kinsey team asked everyone about heterosexual and homosexual erotic thoughts and masturbatory zoo-erotic and sadomasochistic imagery, and they inquired about age preferences for homosexual contacts specifically, but they didn't address racial imagery or preferences.[125] An entire chapter of both *Human Male* and *Human Female* was devoted to cross-species sexual contacts; that was evidently easier to conceptualize than cross-racial contacts.

This disregard for interracial sex reflects the authors' tendency to treat social lines as almost entirely uncrossable, with social hierarchy

---

xxxi   Kinsey et al., *Sexual Behavior in the Human Male*, 66–67. While popular narratives often suggest that interracial marriage was illegal across the United States before *Loving v. Virginia*, interracial marriage very much existed pre-1967, particularly in the Northeast and Midwest, the geographic areas that the Kinsey studies mostly drew participants from. See Roberts, "Crossing Two Color Lines," and Fryer, "Guess Who's Been Coming to Dinner?" As far as nonmarital interracial sexual relations go, white anxieties about those were the raison d'être for antimiscegenation laws, justification for lynching, and other forms of racial violence. See Pascoe, *What Comes Naturally*; Lui, *Chinatown Trunk Mystery*; and Feimster, *Southern Horrors*.

an unquestionable fact of life. The authors allow that "there are, admittedly, a few persons who do move between groups" but specify that "most persons do not in actuality move freely with those who belong to other levels."[126] They also establish themselves as not distracted by dreams of a nonhierarchical society: "Social levels are hierarchies which are not supposed to exist in a democratic society, and many people would, therefore, deny their existence," they note in a section titled "Realities of Social Levels."[127] This emphasis on "reality"—a word repeated in some form six times in this brief section—echoes their claims to a particularly attuned sense of objectivity compared to other researchers. "In this country we make it a point that there should be no physical barriers nor legal codes which forbid people to move with almost any social group," the authors say, but this was a fantasy. "Each adult," "each child," and "each group" invariably knows the "boundaries" to which they are subject, and the previous admission of occasional movement disappears in definitive assertions that "persons in one group do not invite persons from the other to their homes for dinner" and "store clerks and office staffs do not move freely with the business executive group."[128] The authors emphasize the seriousness of these differences by explicitly comparing them to race and, implicitly, civilizational development, positing that "the data show that divergencies in the sexual patterns of such social groups may be as great as those which anthropologists have found between the sexual patterns of different racial groups in remote parts of the world."[129] In this world of intense stratification, even communication was difficult: Interviewers had to be well versed in different vernaculars so that they could reword questions according to what a subject "will understand."[130] Using vernacular language correctly could get a reticent subject to offer more information. Linking miscommunication to racial boundaries, the authors provide a lengthy example of an "older Negro male whose first answers were wary and evasive" until the interviewers employed terms like "hustler" and "common law."[xxxii]

Social groups were so dissimilar that a mismatch in sexual mores and language could cause calamitous trouble. "Whenever people of different social levels come into contact," the Kinsey researchers caution, "conflicts between sexual patterns and failures to understand the patterns of other groups . . . provide considerable impediments to any cooperation between the groups."[131] That conflict went far beyond the potential

---

xxxii   Kinsey et al., *Sexual Behavior in the Human Male*, 60. Their one other extended example is a racially unmarked (and thus presumably white) sex worker.

misunderstandings that could emerge in an interview; such differences could be deadly. "Most of the tragedies that develop out of sexual activities are products of . . . conflict between the attitudes of different social levels," they advise, including "loss of social standing, imprisonment, disgrace, and the loss of life itself."[132] The authors indicate that "while there are no sharp boundaries to social levels, there are obstacles" to crossing them.[133] Some were legal obstacles: The few references to interracial sex generally involve some kind of run-in with the law. When discussing the severity of legal punishments for violating sex laws, the authors note that, other than age differences, courts tend to hand down greater punishments when "the coitus . . . involves persons of different racial groups."[134] Similarly, they emphasize that "if the [sexual] behavior involves a relation between persons belonging to different racial groups . . . then the laws against premarital intercourse become convenient tools for punishing these other activities."[135] As a result, sexual contacts outside of one's own social group, including but not limited to racial group, were relatively rare. It was common belief that "most pre-marital intercourse is had with girls who are below the social status of the male," but that wasn't actually the case. Likewise, a significant age gap didn't appear frequently: "Most males have intercourse with girls of about their own age, or . . . only a few years younger."[136] The exception to this was pre- and extramarital sex between white men and Black women (white women "very rarely" had extramarital sex with Black men). The authors explain in *Human Female* that "a small portion of the discrepancy between [that] female and male data may be accounted for by the fact that these interracial contacts were included in the male volume"—hard for a reader to discern, given that *Human Male* didn't actually contain data about the race of sexual partners.[137]

In essence, then, *Human Male* and *Human Female* naturalize social hierarchy—particularly race—and its impermeability even as they refer to these groups as "social" rather than biological. This slippage recalls a similar one in eugenic science, where some groups were biologically better and of higher social status than others because of a relentlessly blurred etiology of environmental and heritable factors.[138] The authors also attach to race in particular two forms of sexual deviance of considerable concern to early twentieth-century reformers and eugenicists: homosexuality and sex work. Although the authors briefly mention that "Negro and white populations for comparable social levels are close if not identical" in their sexual behavior, as noted above the only interviewees asked about the race of partners were those already known to have homosexual contacts or experience with prostitution.[139] The homosexual history questions inquired about race multiple times, including race of the partner in the subject's first postadolescent homosexual experience, "races involved: white, Negro, others" in cases where the interviewee's partner had not had previous homosexual experience, and the subject's own "estimate of extent of homosexuality . . . among Negroes, whites."[140] The researchers asked those who had engaged in "heterosexual prostitution" about "racial groups" that hired them (though those with "homosexual prostitution" experience were asked only about "frequency," "situations," "amounts involved," and "long-time maintenance as prostitute").[141] Both of these groups also received more specific questions about less-normative sexual activities, collectively including "exhibitionistic activity," "anilinctus," "flagellation" (and in homosexual contacts, whether flagellation was "on back, buttocks, genitalia"), "scatology," "urethral insertions," "interest in transvestism," and positions involved "including 69."[xxxiii] Through the connection between interracial sex and both homosexual contacts and sex work, then, race became linked to these other forms of what the authors call "variant techniques"—not just conceptually, but quantifiably, because the only interviewees asked about race were also the only ones asked about those particular sexual behaviors, such that taboo sex acts were far less likely to come up in an interview with someone who exclusively had heterosexual sex with white partners.

The treatment of race in the Kinsey studies had two profound effects despite their minimal engagement with race per se. First, the choices that the Kinsey researchers made about race and sampling mark a crucial moment in the progression of the place of race within sex science between the mid-nineteenth century, where this book began, and the mid-twentieth century, where it will end. At first, sex science's racial investment was the elaboration of a racial hierarchy rooted in varying degrees of sexual dimorphism. Over time, it got whiter. Eugenic research and gynecology in the service of eugenics began to focus more on white bodies that could, potentially, be refined and salvaged for racial advancement through careful breeding.[142] When the Kinsey studies became the pinnacle of American sex research in the late 1940s and early 1950s, white sexual behavior was cemented in scientific and public imagination as the subject of

---

xxxiii   Kinsey et al., *Sexual Behavior in the Human Male*, 68–70. The other time these questions come up is regarding preadolescent experiences, which seems to link them to sexual immaturity.

sexological research.[xxxiv] Though they may have been intended and touted as a bastion of progressive consideration of benign variation, the studies rendered unimportant anyone who could not or would not be counted in large numbers, and in the process made it seem like most Americans were both binarily sexed and white.[xxxv] Second, even with this focus on white sexual behavior, and with the outright biological eugenicism of Kinsey's earlier work replaced by a cultural account of race, his eponymous studies perpetuated long-standing associations between Blackness and deviant sexuality.[143]

Making Transness Invisible

We saw in the last section the power of frequency to establish not only a norm but the importance of whom to study; the Kinsey team had excused the all-white study population for *Human Male* and *Human Female* on the basis that there were simply more white people and not enough data from Black participants to make valid claims about. Through a similar process, the Kinsey studies made deviations from a nascent precursor of cisness disappear. Kinsey and his team were very much interested in increasingly visible communities of transsexuals and transvestites, but that trans population didn't meaningfully make it into either of the books or call into question the stability of the binary sex categories used to make sense of everyone else. Most people were not, in the team's view, trans, and therefore trans people weren't relevant to questions about most people's sexual behavior. Trans people were rendered anomalous and thereby minimized. The idea had begun to take hold that most people happened to naturally be uncomplicatedly male and female, just as the existence of transness became increasingly well known. Cisness, though it would not be named for several decades, had started to emerge.

xxxiv   At least, in the United States. Simultaneously, anthropology leaned hard into an obsession with "primitive" sexuality (e.g., Mead, *Male and Female*, and Ford and Beach, *Patterns of Sexual Behavior*). See Drager, "Early Gender Clinics," for a discussion of the racial politics of twentieth-century trans medicine, and their relationship to culture of Black poverty discourse. Miriam Reumann has noted that at least Kinsey's lack of attention to Black women meant that a hypersexualized population finally got some relief from scrutiny. *American Sexual Character*, 17.
xxxv   Those things had always gone together—maybe we get the attention to binary sex from the whiteness of it all.

As the team embarked on the analyses that would form the basis of the Reports, they had determined that using groups containing fewer than fifty cases produced substantial probabilities of error.[144] Analyzing such samples created a situation in which "the calculations of various statistics for various types of sexual outlet could not be depended upon in more than 1 in 5 cases."[145] Kinsey wrote to researcher Louise Lawrence in 1949 that his team had "twenty or thirty transvestite cases in our histories"—certainly not enough by the team's standards of sampling adequacy.[xxxvi] Specific numbers aside, this wasn't a population in meaningful enough proportion to the rest of the country to study: In an undated letter, Lawrence recounted that Kinsey "thought the problem [i.e., transvestism] was relatively rare."[xxxvii] Recall the hypothetical illiterate, Orthodox, and so on, Black man, the exemplar of a sample group "so rare in the American population that they are unimportant for study."[146] By extension, trans people, too, were unimportant. In a research program most interested in establishing "what average people do sexually," this perception of rarity meant that transness fell out of view, apparently occurring in such small numbers as to be merely noise.[147]

Kinsey himself had a long-standing interest in transvestism and transsexuality. He was a close friend of Harry Benjamin, who is widely recognized as a foundational figure in American trans medicine and a focus of chapter 5. Benjamin introduced Kinsey (to whom Benjamin referred lovingly as "Prok") to Louise Lawrence, a cornerstone of trans life in California in the 1950s and a foremost researcher of trans experiences. For years, starting at least as early as 1950—after the publication of *Human*

xxxvi   Kinsey to Lawrence, October 10, 1949, Louise Lawrence Collection, Kinsey Institute (hereafter LLC). By Gebhard and Johnson's 1979 reassessment of the original data, the *N* of transvestite and transsexual cases had increased to fifty-six—although they still didn't do anything with those cases except note their ages at time of interview. They did find that among their homosexual sample (*n* = 634), respondents who said they "often" cross-dressed ranged from 7 percent (white nondelinquent female) to 16 percent (nonwhite delinquent male). The complexity of categories and meanings of cross-dressing make it not particularly useful to try to draw any conclusion about who might have counted as something recognizably trans-ish today, but those are not small numbers! *Kinsey Data*, 6, 630, 607.

xxxvii   Lawrence to Benjamin, n.d., Series IB, Box 1, Folder 2, LLC. She went on to tell Benjamin that she was "very sure that it is more common than most of us, even prominent doctors, are willing or able to admit" and that she was "going to try to prove to him that [she] was right."

Male but while *Human Female* was still in progress—Lawrence served as an informant for the Kinsey team, and she collected and shared information about her own contacts (she carefully specified that she was "not working FOR [Kinsey]").[148] She and her partner, Gay, visited Kinsey in Bloomington, toured the team's facilities, and met with five transvestites in the Midwest in the hope of securing interviews with them; while on the West Coast, Kinsey, Gebhard, Pomeroy, and Martin visited Lawrence's home to discuss possibilities for further research.[149] The Kinsey team were explicitly interested in learning about transsexuals and transvestites and the instability of sex categories that they signified.

Drucker has argued that Kinsey's ongoing interactions with transvestites and transsexuals contributed to his attention to homology and difficulty finding physiological differences between men and women to explain divergences in sexual behavior.[150] Gebhard and Pomeroy, too, were interested in this work—after Kinsey's death, they expressed a desire to continue putting together a book on the subject.[151] Yet this shared interest and knowledge did not shift the team's methods or approach to sex. The transvestite and transsexual fell under the existing, implicitly biologized categories of male and female, so they were not treated as evidence that these categories might be inadequate as universal types. For example, in *Human Female*, the authors report that a large proportion of transvestites are "anatomically males," which they suggest is a sign that "males who wish to be identified as females are in reality very masculine in their psychologic capacities to be conditioned" (they had previously argued that male desires tend to be more responsive to habit and experience).[152] The Kinsey team thus used transvestism to reinforce the concept that male and female minds and bodies were substantively different from each other.[153] Instructively, Kinsey was unenthusiastic about transsexual genital surgery; Meyerowitz has characterized it as the place where Kinsey "hit the limits of his sexual liberalism."[154] In the Kinsey studies, transvestism and transsexuality were, if not entirely pathologized, uncommon, unimportant, and somewhat unpleasant subsets of maleness and femaleness.

Following the publication of *Human Female*, the Kinsey team's work was defunded, largely thanks to a 1954 attack by Republican Congressman B. Carroll Reece that sought to protect the American taxpayer from financing such "unsavory" research.[155] In 1956, Kinsey died after several months of cardiac illness. Even with these losses, the Institute for Sex Research, now helmed by Paul Gebhard, carried on the project with support from its longtime home, Indiana University.[156] In 1957, the institute obtained a National Institute of Mental Health grant to revisit the original interviews, recode the data, and transfer it to computer tape. Gebhard, along with Alan Johnson, who led the transfer project, published their results in 1979 as *The Kinsey Data: Marginal Tabulations of the 1938-1963 Interviews Conducted by the Institute for Sex Research.*[xxxviii] Gebhard and Johnson's interest in transness, and their brush with a lack of difference between male and female bodies, again did not inspire a rethinking of sex. *The Kinsey Data*'s nearly six hundred pages of tables are categorized, in clear black and white boxes, by male and female. Aside from a brief note and index entries for "transvestism" and "transsexualism," there is no suggestion in the book that sex is anything other than binary and static, as though that entire section of *Human Female* and the work with Lawrence never happened. Questions about sex categories and edge cases were again drowned out by comparisons across male and female populations. Gebhard and Johnson never wrote the book about transness.

This, then, is one of the perils of treating transness as distinctly uncommon. It becomes an exception to a rule that otherwise, ostensibly, functions naturally without external reinforcement. As American sex science, driven by Kinsey, shifted toward statistical analysis with its investment in frequency and incidence, an existing assumption about sex—that the vast majority of people fit into a binary of male and female—made anything imagined as rare not worth counting. The way that normative categories of male and female contain just as many exceptions within them as outside of them, and the means by which they have to be constantly shored up against the potentially destabilizing force of anomalous bodies and ways of being, became incredibly difficult to see.

It's hard to say precisely what happened in sex science after Kinsey, because the vast majority of scholarship on American sexology—including, as you can surmise by now, most of this book—centers either on the late nineteenth and early twentieth centuries or on the Kinsey era.

xxxviii   Gebhard and Johnson, *Kinsey Data*, 6. The book is quite dry reading with, it must be said, occasional moments of mirth. Their "most serious criticism" was that the interview questions themselves "fossilize[d]" early on, and thus interviewers did not stop asking questions that turned out to be irrelevant or add new questions. "We continued," they note, "to ask males which testis hung lower than the other and whether the scrotum was on the left or right side of the central seam of the pants. Both of these matters would have been better dealt with by a urologist and a tailor." Gebhard and Johnson, *Kinsey Data*, 14.

Histories of sexuality and science in the later twentieth century tend to focus on the HIV/AIDS pandemic, reproductive health, the development of sociobiology and evolutionary psychology, medical interventions on intersex bodies, and, increasingly, trans medicine and "conversion therapy." Much remains to be written about later studies of sexual behavior, searches for biological bases of homosexuality and transness, and subsequent research on what sex is beyond the genetic.[xxxix]

What we do know suggests that the Kinsey studies did not successfully invigorate sex science. Peter Hegarty has noted that sex behavior studies continued to be a "'low' science" in the wake of Kinsey, despite hopes to the contrary.[157] William Masters and Virginia Johnson were, for a time, able to build on Kinsey's success to establish their program of research on sexual response, but they shut down their laboratory in 1970.[xl] In the late 1960s and early 1970s, sexology's focus started to shift from science to business, especially the marketing of pharmaceuticals. Earlier ideals of scientific rationality gave way to what Irvine has called a "humanistic" approach that eschewed empiricism in favor of self-actualizing discoveries of desire, emotion, and fulfillment.[158] Sexology never quite resolved into a unified field. Irvine locates this failure in a lack of a cohesive theory of gender and sexuality, especially as feminist and queer movements pushed back against pathologizing and essentialist framings.[xli]

Sexology also lacked a cohesive theory of sex; it became less interested in sex categories and more so in sexuality specifically. Bisexuality theory largely disappeared except for what formed the basis for trans medicine, which, as in the Kinsey studies, was deemed relevant to only a very small, anomalous population. With those exceptions sequestered off, remaining sexologists could focus on men and women. Sex itself became, effectively,

---

xxxix   Irvine's *Disorders of Desire* is a notable exception, though, because it was originally published in 1990, it only historicizes American sexology through the 1980s.

xl   Irvine, *Disorders of Desire*, 56. Notably, Masters and Johnson included a brief chapter in *Human Sexual Response* on similarities between male and female sexual responses in an attempt to make up for otherwise splitting the book into female and male sections, which, they note, "tends to create a false impression . . . and emphasize the differences in sexual response between the two sexes rather than the similarities" (*Human Sexual Response*, 273). The terms *transsexual* and *transvestite* do not appear in the book.

xli   Perhaps unsurprisingly, given the arguments of this book, profits can be had whether or not one has a watertight theory of sex.

a matter of statistical averages. Non-trans bodies that didn't match an image of sexual normality or distinct maleness and femaleness could be made invisible, too—sex was a bimodal distribution, and, of course, there would be small numbers of people on either end of the curves, but they didn't mean anything.[xlii]

Sex and sexuality research past and present is often appraised as progressive. Sometimes it is, but I suspect this characterization has more to do with the Kinsey studies' framing of sex science as progressive than with any sort of overarching reality. Our inheritance from their authors, alongside any depathologization that occurred, is the cementing both of methods that limit the possibilities of research and of persistent categorical rigidity. The turn to statistical sex research might have happened anyway, given the building interest in large sample sizes, but the Kinsey studies without doubt oriented sex research toward analyses that highlighted what was common and dismissed what was not.[xliii] In so doing, they brought into the mid-twentieth century a long-standing tradition of making anomalous, and therefore unimportant, forms of sex that did not align with a stable binary.

xlii   Thanks to Ahmed Ragab for helping me clarify this point.
xliii   There could, of course, be different ways of using statistics—as noted earlier, the path that the Kinsey studies took was not the only possibility.

5

Immaterial Categories in
Trans Medicine

"Each of us will probably die by getting shot by some patient like E. V.,"
wrote urologist Elmer Belt to Harry Benjamin, an endocrinologist and the
so-called father of transsexuality, in February 1960.[1] He was joking—kind
of. Belt had sent along a citation for a new book on heart disease that
struggled to define maleness and femaleness, and the joke was that the two
of them were more likely to be murdered by an angry transsexual than
to die of heart disease.[i] For almost two years prior to Belt sending the

---

i   Throughout this chapter, I avoid precisely defining *trans* and *transness* in
an effort to refuse the incitement to taxonomize and with the recognition
that sexual classifications, whether in the past or the present, can only ever be
approximations. When I say *transsexuality*, I refer specifically to the midcentury definition articulated by Harry Benjamin: A transsexual is someone who
"want[s] to undergo corrective surgery, a so-called 'conversion operation,' so that
their bodies would at least resemble those of the sex to which they feel they belong." Benjamin, *Transsexual Phenomenon*, 14. I use *trans medicine* as a shorthand
for the clinical apparatus that developed around hormonal and surgical interventions for trans people. While this was not an actor's category—something
like *transsexual medicine* would probably be more historically accurate—I gesture
with this phrase and *trans* or *trans people* toward the many ways that these clinical practices exceeded the highly contingent category "transsexual," including

letter, he and Benjamin had been debating whether Edie V. Hutchens ("E. V.") should be eligible for the removal of her penis and testicles and the construction of a vagina.[ii] The longer they delayed, the more desperate Hutchens became, but they hesitated to allow the surgery for fear that Hutchens would regret her transition and turn on them, whether with gun in hand or by other means.

In previous decades, sex became incoherent as exceptions to ideal forms were worked back into normative categories of male and female. In the mid-twentieth century, with trans medicine added to the practical uses of sex research, the construction of conditions for *leaving* one's normative category built on the existing methods for turning people anomalous discussed in chapter 4. Transness became one container for the failure of a static binary model of sex to encompass what people and bodies actually did.[iii] In the process, an early inkling of cisness (though not yet named as such) enabled the forgiveness of all matter of non-trans people's sexed and gendered sins. People who weren't transsexual had a normal relationship to sex and were therefore unexceptional men and women.

Benjamin's and Belt's clinical practices exemplify the divergence between theory and practice seen in previous chapters and epitomize the coexistence of mutability and fixity that animated twentieth-century sex science. As people in the United States began to seek out medical transition in the 1950s and 1960s, doctors closely guarded the technologies of sexual transformation that were being increasingly repurposed from the animal studies of previous decades.[iv] Although these doctors articulated themselves as having a clear sense of who counted as a transsexual and therefore who should be eligible for hormones and surgery on the basis of the extremity of their transsexuality, they feared potential bad outcomes should the wrong person be allowed to change their body. Patients, they thought, might not pass or be able to find work post-transition, and their ensuing regret might lead them to seek vengeance. What doctors

---

their present legacy even as diagnostic and nonmedical categories have shifted. I refer to *transsexuals* specifically when referring to the actors' category.

ii   Patient names have been changed in accordance with Kinsey Institute policy. I maintained the original initials to facilitate reference and, when possible, used the same pseudonyms as Meyerowitz, *How Sex Changed*.

iii   Another is *intersex*, which by the 1950s and 1960s had been separated from transness. See Gill-Peterson, *Histories of the Transgender Child*, esp. chap. 2.

iv   I focus primarily on genital surgery in this chapter; much work remains to be done on the history of endocrine and nongenital surgical transition.

conceptualized as a simple classification system of transsexual or not crumbled in the face of sorting out who should—according to medical professionals—obtain medical transition.[v] Doctors' anxieties about making the wrong call produced decades' worth of vociferous correspondence that renders visible another mess of categories where they met clinical practice. In the aftermath of the Kinsey studies, the meanings of male and female, man and woman, were no longer particularly contested. Now, the ostensible classification problem was who could be considered a "true" transsexual. As usual, though, coherence was not the source of classification's power. A taxonomy of transness emerged, but the anxieties of a handful of doctors shaped the lives of trans people attempting to access transition-related surgeries.

Since its beginnings, trans scholarship has emphasized the importance of performing normative gender and "proving" that one was "really" a diagnosable transsexual in the history of medical transition.[2] This chapter examines another central force in how clinicians made decisions about access to transition technologies in the early years of American trans medicine: fear of transsexuals ruining doctors' lives. While adherence to normative gender roles certainly played a part in whom doctors allowed to medically transition, fears of being sued or otherwise facing retribution from patients could easily shift the balance in the final decision of who could have surgery. Focusing on this aspect of clinical decision making shows how informal evaluative practices rooted in anticipation of bad outcomes became standards of trans care. More importantly for this book, it illuminates how gender operated as a functional, rather than ontological, designation in doctors' understanding of transsexuality. The question for the preeminent practitioners of trans medicine was not whether someone was *really* a woman but if they could *pass as* a woman, nor whether they were *really* trans but would regret transitioning.[vi] Outcomes mat-

v Today, the "not" would fall into the category *cis* or *cisgender*. However, as noted in this book's introduction, the term *cisgender* did not come into use until the 1990s, and, in and of itself, it tends to naturalize the very binary it seeks to call into question, so I have used the somewhat clumsier "not trans" to point to historical actors' assumptions about the standard against which they constructed transness. For a critique of the category *cisgender*, see Enke, "Education of Little Cis."

vi Benjamin also treated trans men according to the same principle. Most of his initial patients, however, were trans women. For brevity, this chapter contends with Benjamin's treatment of trans women specifically, but a comparison

tered more than one's place in a taxonomy of transsexuality. The answers to these questions, however, were not remotely self-evident. That uncertainty carried significant perceived risk of getting it wrong, even as the same uncertainty made space for doctors to insist that only their expertise could be trusted to get it right. It also emphasizes an affective dimension of this incoherence, in which physicians' feelings of uncertainty led them to assign risk to themselves when it was actually trans people who were more likely to suffer. In the culminating chapter of this book, then, the uneven distribution of the consequences of incoherent sex becomes abundantly clear.[3]

For most trans people living in the United States in the mid-twentieth century, even more so than today, accessing the technologies of medical transition was no simple matter. Even after the widely celebrated story of Christine Jorgensen's transition introduced the American public to transness and inspired many trans people to ask their doctors about accessing hormones and surgery, putting theories of malleable sex into practice remained an often unreachable possibility.[4] Few doctors had heard of transsexuality beyond popular media. Fewer still were willing to prescribe hormones or operate; instead, they outright refused to participate in what they saw as patients' delusions. Nonetheless, over the course of the 1950s and early 1960s, Harry Benjamin and a small network of his colleagues developed a reputation for working with trans patients. Their work helped legitimize trans medicine as a field and laid the groundwork for decades of clinical encounters to come.[5] As Benjamin and his colleagues exchanged flurries of letters to decide which of their potential trans patients they would be willing to treat, they produced yet another logic of sexual classification in which the categories, which mattered far less than the power to wield them, often didn't align in practice with the bodies they claimed to organize.

As discussed in the introduction, historians of transness have engaged in extensive thinking about who should be considered the subject of trans history, particularly before the category *trans* existed, and have largely settled on an idea of someone who "moves away" from the gender they were assigned at birth.[6] The post-mid-twentieth-century trans subject tends to be regarded in contrast as self-evident, with the "transsexual" especially

---

of the precise ways that Benjamin's treatment of trans women and trans men differed is a crucial area of further investigation for someone who, unlike me, wants to spend more time crying over Benjamin's papers.

emerging as a particular construct of mid-twentieth-century medical discourse.[7] Looking solely at the category *transsexual* in Benjamin's published writings would, after all, indicate that the "true transsexual" was merely someone who wanted to change their body, especially their genitals, to "at least resemble those of the sex to which they feel they belong."[8] Diagnosis was fairly simple, he wrote: "The request for a conversion operation is typical only for the transsexual and can actually serve as definition."[9] The way that he interacted with transsexuals, however, indicates that the stated classification system did not align with clinical practice. Not everyone who was diagnosable as a transsexual should, in his measure, have the surgery they wanted. Attending to exactly how Benjamin determined who was eligible for surgery demonstrates that the category of the transsexual formed less around a set of coherent gendered contents and more in terms of the possible effects of trans surgery. The mid-twentieth-century transsexual was not the self-evident cousin of the less categorically delineated nineteenth-century trans subject. Figures like Benjamin did not define transness by establishing solid criteria and then assessing people by how closely they matched them; rather, they constructed transsexuality as a space of uncertainty through their fear-based practices and then used that uncertainty to justify their own clinical control while establishing themselves as risk-taking pioneers.

This chapter takes as its base assumption that the categorical transsexual emerged through a set of practices. As scholars in science and technology studies (STS) have argued, and as has been an undercurrent throughout this book, concrete and coherent things do not simply exist out in the world waiting to be described but are, rather, produced through naming and interaction, often in convoluted and contradictory ways.[10] This chapter is the result of tracing such productive classification in action in the creation of the transsexual.[11] I look primarily to unpublished source materials, especially Benjamin's correspondence with colleagues and patients as well as the subset of his clinical records open to researchers at the Kinsey Institute. Doing so makes visible, in messy and granular detail, how Benjamin and his colleagues negotiated the incoherence that emerged as a result of their attempted sortings, which Benjamin, like the scientists and clinicians in previous chapters, largely smoothed over by the time he published.

Formalized diagnostic criteria for transsexuality eventually solidified, both in Benjamin's thinking and in codifications like the 1979 Harry Benjamin International Gender Dysphoria Association (HBIGDA; later World

Professional Association for Transgender Health, or WPATH) Standards of Care and the 1980 addition of gender identity disorder to the *Diagnostic and Statistical Manual of Mental Disorders, Third Edition* (DSM-III). Those classification schemes, however, came after the practices established by early practitioners of trans medicine as they attempted to minimize risk to themselves when interacting with patients. They accrued through a series of habits produced through caution about legal, professional, and personal consequences that practitioners anticipated would result from performing what they called "conversion operations," at the same time that this focus on risk took precedence over questions of what made someone male or female. Coherence of diagnostic categories, though, was not enough to structure trans medicine. Throughout the period under study in this chapter, to be transsexual was not to be treated as a transsexual. Transsexuals, in Benjamin's model, were transsexuals because they wanted hormones and surgery, yet that desire was not actually enough to convince doctors to prescribe or operate. The transsexual was constructed as someone whose desire for surgery simultaneously defined them and was likely to produce a bad outcome: Whether or not patients got the care they needed depended on whether medical professionals felt that they themselves would not face any unwanted consequences from that decision, regardless of patients' categorization as transsexual. The incoherence of sex, then, is not only about contents *within* categories but also about the authority to decide when they are and aren't used in practice.

Mayhem! at the Clinic

When Benjamin first began to see trans patients, few surgeons performed transition-related genital surgery. Christine Jorgensen had traveled to Denmark in 1953 to obtain the surgical services of Dr. Christian Hamburger, but in the ensuing media bonanza and sudden influx of inquiries for treatment from around the globe, the Danish government opted to prohibit the castration of noncitizens in Denmark.[12] This left few options for trans people in the United States, and in the early years of his trans practice, Benjamin struggled to locate surgeons for his patients. Many of his colleagues thought transsexuals were mentally ill and needed to be protected from their own desires; the prospect of removing healthy tissue just because a patient wanted it gone seemed absurd. One surgeon Benjamin had hoped might be willing to operate on his patient Val Barry refused, for example, on the basis that it would be "no more scientific to accede to the desires of a patient to the point of transformation and removal of his sex organs than it would be scientific to accede to the desires of some other patient who suffers from a fixation involving other organs."[13] Furthermore, in the 1950s, trans surgery was decidedly, as Belt described it, "experimental."[14] While doctors affiliated with Magnus Hirschfeld's Institut für Sexualwissenschaft had performed transsexual vaginoplasties in the 1930s, the development of the surgical technique ground to a halt with the rest of the institut's activities when Nazis destroyed it and burned its records in 1933.[15] Doctors in the 1950s had to develop new surgical methods, which, as Belt reported from his own experience, often resulted in complications.[16] Trans medicine, then, offered doctors a chance to be on the cutting edge of a new field, but with concerns about how individual patients would fare, not to mention how it would be publicly perceived, most surgeons had little interest in it.

Others did view surgery as a positive intervention that could improve transsexuals' lives—but they, too, were reluctant to perform it, primarily on legal grounds. While no law explicitly criminalized genital surgery for trans patients, surgeons and their lawyers interpreted an old statute outlawing "mayhem" as prohibiting the castration of patients with healthy organs who wanted their testicles removed. According to Robert Veit Sherwin, a lawyer and friend of Benjamin, what has come to be known as the "mayhem statute" was a broad overreach of a law that had been intended to prevent harm to men who might become soldiers.[17] There was never explicitly a law that said removal or modification of genitals was illegal, but castration could be framed as one of many kinds of illegal maiming within the definition of mayhem.[18] This immediately ground efforts to find a surgeon for Barry to a halt, and it influenced the conditions of access to trans medicine for the next several decades.

The mayhem statute has been something of a black box in trans history, possessing an outsized influence over how medical transition developed but without much clarity on the exact process by which it did so. Sherwin's explanation of the overreach of an outdated and irrelevant law has been largely accepted by trans historians.[19] Yet the mayhem statute didn't emerge suddenly from the mists of legal history. In the twenty years leading up to Barry's case, there had been two high-profile, nationally reported cases in which San Francisco surgeons had run afoul of the law for coercively using surgeries similar to those that trans people wanted. First, in 1928, Leo Stanley, head physician at San Quentin State Penitentiary, was sued for the "mutilation after death" of Clarence "Buck"

Kelly, an incarcerated man who had been executed that year.[20] Following Kelly's hanging, Stanley had removed Kelly's testicles and implanted them into an older prisoner as part of a larger program of sexual rejuvenation experiments.[21] The case made headlines in the *Los Angeles Times*, as Kelly's mother sued on her son's behalf.[22] Presumably, Benjamin, who had spent much of his early career working on rejuvenation, would have been aware of this case.

Then, in 1936, in an even more widely reported case, surgeons Tilton Tillman and Samuel Boyd were sued under mayhem charges after sterilizing the well-known heiress Ann Cooper Hewitt.[23] The wealthy socialite was allegedly sterilized "for society's sake" because she was "feeble minded" and had "erotic tendencies."[24] Hewitt's case against her mother and the surgeons for sterilizing her without her consent appeared in newspapers across the country. The case was eventually dismissed, but Hewitt remained a not-infrequent subject in tabloid columns in the ensuing decade. In 1947, just two years before Barry was refused surgery on the basis that any surgeon involved might be prosecuted under the mayhem law, reports of Hewitt's fourth marriage brought reminders of her mayhem case back into the news.[25] Her run for the US Senate in 1950 produced another wave of headlines about the "Sterilization Case Heiress," and obituaries announced the death in 1956 of "Ann Hewitt, of Sterilization Case."[26]

The fear of the mayhem statute in trans medicine, it appears, stemmed at least in part from the legal history of eugenic experiments and sterilization. Castration had for so long been either a punishment or an involuntary eugenic practice that it could not be conceptualized as something to be willingly chosen.[27] Legal and medical professionals seemed unable to adapt to the idea that someone might want their gonads removed. Midcentury trans's eugenic legacy went beyond the specific technologies of hormonal and surgical interventions.[28] Rather, while I haven't found any explicit discussion of this in archival sources, it seems that efforts to distance postwar science from the violent overreach of eugenics contributed to a reluctance to allow bodily autonomy in trans medicine.[vii] Certainly

vii   Megan Glick has argued that suspicions of plastic surgery at midcentury were part of a post-Nuremberg anxiety around the use of new technologies to "transform the human body." This deeming of trans surgery as unscientific and dangerous, while not explicitly placed in reference to Nazi experimentation by writers at the time, was likely informed by that context. *Infrahumanisms*, 101.

Benjamin, who had presented a talk on the effects of vasectomy at the tenth annual meeting of the Eugenics Research Association in 1922, knew of the changing tides in explicit engagement with eugenic sterilization research by the late 1940s.[viii] In addition, Benjamin himself possessed a rather litigious history. He had sued Morris Fishbein, head of the *Journal of the American Medical Association*, along with publisher Horace Liveright for libel in 1929, after Fishbein called Benjamin a quack in his popular book *New Medical Follies*. More than six years after the initial filing and multiple appeals, Benjamin finally dropped the suit in 1935.[ix] No surprise, then, that he projected his own dogged relationship with lawyers onto patients.

Benjamin did not have any particular expertise in transsexuality when he became the go-to physician for those seeking medical transition. His eventual power and influence as an obligatory passage point in the mid-twentieth-century world of trans medicine came not from an existing fount of knowledge but through a fairly arbitrary series of events.[x] To be sure, Benjamin's years of experience as an endocrinologist and gerontologist "rejuvenating" men, and occasionally women, who felt they had lost their vitality with age, and his work with Eugen Steinach, whose gonad transplants in guinea pigs had shown that sex characteristics could be changed, gave him a baseline understanding of sexual malleability.[29] Beyond that, though, Benjamin largely benefited from knowing the right people in the

viii "Tenth Annual Meeting," 80. The meeting was held in Cold Spring Harbor, putting Benjamin into contact with Charles Davenport, who likewise presented at the conference, along with other members of the Eugenics Record Office.

ix Li, *Wonderous Transformations*, 96–101, 120–21. Max Thorek, whom Benjamin would later ask to operate on Val Barry, tried to intervene in the three years before the formal suit, during which time Benjamin relentlessly sent letters to Fishbein and other American Medical Association leaders and showed up at Liveright's office to threaten legal action. As he tried to calm Benjamin down, Thorek apparently cautioned that Benjamin risked the ire of the whole New York medical establishment if he lost because while Fishbein was powerful with many friends, "there are a lot of fellows that do not like you" (99).

x An obligatory passage point, a key figure in Actor Network Theory, is in this case best defined as an expert who must be consulted or referenced in order to legitimately approach a given problem. The expert becomes "indispensable" after they convince other actors (or after other actors are otherwise convinced) that they cannot get what they want by themselves, whether a material outcome like surgery or the authority to make particular knowledge claims. See Callon, "Some Elements."

right places at the right time; certainly, he did not initially seek out trans patients. In 1938, Benjamin did treat a patient for arthritis who also happened to be a transvestite; he prescribed the patient estrogen in the hope that it would minimize feelings of what would now be called gender dysphoria.[30]

Ten years passed, however, before Benjamin's trans practice took off, which it did largely thanks to Benjamin's place in the network of sex researchers that I've been tracing in this book. Robert Latou Dickinson (see chapter 3) had introduced Benjamin to Alfred Kinsey (see chapter 4) in 1945, and the two soon became "good friends."[xi] In 1948, while conducting sex history interviews, Kinsey met Val Barry, an apparently baffling individual who wanted to have her penis and testicles removed and a vagina surgically constructed. Kinsey referred her to his friend Harry, with the idea that Benjamin might take an interest.[31] Kinsey was right. Benjamin described Val as "one of the most interesting cases of transvestism that has come to my attention" and began what would end up being a four-year-plus process of working to get her access to surgery.[32] As Benjamin's reputation for being willing to recommend surgery and hormones increased, other doctors referred patients to him that they didn't know what to do with or otherwise couldn't treat.[33] When trans people wrote to Jorgensen, who had very publicly transitioned in the early 1950s and, upon her return from Denmark, had become one of Benjamin's first patients, she suggested they write to him.[34] As a network of trans people who corresponded with each other grew, they, too, told each other to write to Benjamin for a sympathetic doctor who believed in the legitimacy of their experience and could help them.[35] Soon, anyone seeking a "conversion operation" essentially had to go through Benjamin's private practice for a surgical referral. Though Benjamin developed a reputation in the 1950s for caring deeply about his patients—he has been referred to as a "benevolent paternalist," as the "patron saint of transsexuals," and as willing to listen to patients with a nonjudgmental openness—his writings also reflect a

xi   Schaefer and Wheeler, "Harry Benjamin's First Ten Cases," 78. Besides being mutual friends with Kinsey, Benjamin and Dickinson occasionally overlapped in their professional activities. They both participated, for example, in a 1946 panel in New York about impotence, where they had a conversation about racial difference in rates of impotence and how apparently "primitive people have no impotence." "Round Table Discussion on Clinical Approach to Male Sexual Disorders—Impotence," transcript, November 22, 1946, Box 6, Folder 51, Robert Latou Dickinson Papers, Center for the History of Medicine, Francis A. Countway Library of Medicine, Harvard University (hereafter RLD).

tremendous investment in his own expertise, a fear of legal repercussions, and an unwillingness to take responsibility for patients.[36] This is not to say he didn't also care about or help his patients. But the two are not mutually exclusive, and Benjamin's own interests shaped the way trans medicine would proceed.

In this context, at the nexus of protecting himself from legal consequences and trying to help a patient access care, Benjamin immediately encountered a legal barrier as he attempted to find a surgeon for Barry. Barry had been admitted to the State of Wisconsin General Hospital's psychiatric ward after reportedly throwing and breaking objects and finally attacking her parents after her hopes for meeting with a surgeon in Chicago—presumably Max Thorek, for reasons that will become apparent— were "dashed" because the surgeon had "left town."[37] Even though the hospital recommended castration and plastic surgery to manage her psychological distress, the attorney general of Wisconsin intervened and refused to allow the surgery because it would be prohibited under the mayhem statute.[xii] Undeterred, Benjamin pressed on in his efforts to find a surgeon for Barry, but legal concerns continued to dissuade anyone from operating. Thorek, best known as the founder of the International College of Surgeons, had initially agreed to perform Barry's surgery but changed his mind after hearing about the legal risks. "As I had expected," Benjamin wrote to Kinsey, "Thorek has been strictly advised by his attorney not to perform the operation on [Val] as he would open himself up to *criminal* charges."[38] Thorek eventually agreed to provide Barry with estro-

---

xii   Benjamin to Karl Bowman, n.d., attached to H. M. Coon to W. B. Campbell, July 19, 1948, box 3, folder B, VB, Harry Benjamin Collection, Kinsey Institute (hereafter HBC). Folders in this collection containing patient correspondence are labeled in the format "patient's last name, patient's first name"; to preserve anonymity, I have shortened these labels to the patient's initials. Coon informed Campbell that after an extensive battery of tests, at a staff meeting of over thirty hospital affiliates, a majority voted to allow surgical intervention. They reasoned that Barry had lived as a girl throughout her childhood and had left school to do "women's work" when her high school would not allow her to attend in feminine clothing, and thus at the age of twenty-two she was unlikely to respond to any psychological treatment. Brain surgery would have been likely to cause personality problems and potentially make it impossible for Barry to care for herself, and refusing surgery risked turning Barry's present extreme distress into psychosis or suicidality. Barry, according to the report, had wanted to "change physically into a girl" since puberty and in the past few years had read widely about her "condition" and "operative procedures which feminized men."

gen but recommended that she seek out a surgeon elsewhere in the world, like Mexico, where there was no such mayhem law.[39] For him, it was not worth the legal risk.

A solution to the legal conundrum eventually emerged: Prospective patients could be required to undergo a psychiatric evaluation to distribute responsibility. Before declining to operate in 1950, Thorek had apparently told Benjamin that he "would perform the operation only on the advice of a psychiatrist, and then only with the permission of the authorities," as Benjamin put it to Barry.[40] To obtain such a recommendation, Benjamin approached Karl Bowman, a psychiatrist at the Langley Porter Clinic in San Francisco. "In attempting to find an urological surgeon interested in [Val]'s case," Benjamin wrote to Bowman, "I was impressed with the fact that probably only an authoritative advice, as it could come from you, could induce a reputable urologist to agree to operate."[41] A few years later in October 1953, a patient named Caren Ecker reported to Benjamin that she had removed her own testicles and wound up in the hospital after she "almost bled to death." Perhaps luckily for Ecker, she lived in Northern California and was referred to Bowman; he recommended a penectomy because Ecker had already castrated herself. Dr. Frank Hinman Jr., a urologist at the University of California Medical School who had been trained by Hugh Hampton Young of the Johns Hopkins University intersex program and had recently published an article titled "Advisability of Surgical Reversal of Sex in Female Pseudohermaphroditism," was willing to complete Ecker's efforts and remove her penis.[42] According to Hinman's lawyer, there was no legal barrier to performing the surgery now that Ecker had already started the process. "However," Ecker wrote to Benjamin that December, following Hinman's meeting with his lawyer, "as it may become an item of controversy, he wants me to obtain a second opinion of both an urologist and a psychiatrist."[43] This would, Ecker's note implies, protect him from legal fallout if that controversy occurred.

The psychiatric evaluation became firmly enshrined as a de facto requirement when Benjamin began working with Elmer Belt. Belt, a Los Angeles–based urologist and the protégé of Hinman's father, Dr. Frank Hinman Sr., became Benjamin's go-to referral for penectomies and vaginoplasties as early as 1954, but he remained concerned about the possible consequences of performing trans surgeries for years.[xiii] The two quickly

xiii   The earliest mention of Belt operating on one of Benjamin's patients is in a letter from Benjamin to Kinsey on September 22, 1954. Based on a 1960 letter

settled on a psychiatric evaluation as a necessary component of patient approval, and their informal practices soon congealed into a set path toward surgery. First, patients contacted Benjamin, whether self-referred or sent to him by another doctor. If the patient seemed suitable for surgery, Benjamin referred them (although he repeatedly insisted that he was not referring to specific surgeons or recommending surgery, merely offering a name of a sympathetic surgeon and "consenting" to surgery) to Belt. Belt then sent patients to Carroll Carlson, a psychiatrist based in Beverly Hills, for assessment.[xiv] By May 1958, this process was so engrained that Belt wrote saying that he had sent a patient to Carlson "in accordance with our established routine."[44]

Despite this routine, the pair struggled to decide who should be eligible for surgical interventions. Rather than negotiate the ostensible gendered truths of patients' lives and identities, their debate focused on who might cause trouble for them if the patient regretted their surgery. Belt and Benjamin were both particularly concerned about Hutchens, the patient Belt joked was likely to shoot them. Though Benjamin encouraged Hutchens to make the trip to California for a consult with Belt in May 1958, by June he wrote to Belt to advise that while Hutchens would probably psychologically benefit from surgery, he had concerns about the practicality of Hutchens living as a woman.[45] Belt agreed, noting that he was reluctant to operate on Hutchens because she would likely have to "undergo physical examination in the course of [her] work as a teacher." Because the surgical technique Belt used was still in its infancy, the artificiality of her vagina would be obvious to any examiner and result in Hutchens being fired. Belt was sure that when this happened, "[Her] resentment against the man who carried out this work will rise and grow—no matter what [she] thinks of [her] feeling now." A lawsuit, Belt was confident, would follow, spelling his professional demise. "Hardly a jury in the world would condone [the surgery]," he concluded.[46]

---

from Benjamin to Belt, suggesting that Belt experimentally remove only one testicle from a patient and implant the other in the abdomen because "you won't castrate anybody that way," it seems like Belt did not want to perform castrations out of a legal concern. Benjamin to Elmer Belt, July 12, 1960, series IIC, Box 3, Folder Belt, Dr. Elmer (1959–1962), HBC.

xiv   Carlson was by no means an expert on transsexuality. He had, however, treated Belt's daughter-in-law for "puerperal insanity," so Belt trusted him. Belt to Benjamin, August 20, 1956, series VIB, Box 23, Folder 34, HBC.

Over the next six months, Belt continued to obsess over the imagined penalties he would face for operating on Hutchens, and in the meantime, Hutchens continued "grinding through the mill they have set up" at the hospital where he would potentially operate—namely, getting psychiatric approval.[47] By early December, Hutchens remained in surgical limbo. Belt had finally decided he would operate and had secured permission from his friend Percy Riggs, superintendent of Hollywood Presbyterian Hospital, to perform the surgery there. Less than two weeks before the scheduled date, though, Belt wrote to say that Riggs had suddenly died of a heart attack. Because their agreement had not been in writing, Belt was unsure if he would still be able to operate and told Hutchens she would have to wait until the following summer. He would probably be able to get permission, he said, but would have to be cautious in the meantime.[48] While Belt's logistical issues certainly played a part in this decision to delay Hutchens's surgery, his letters to Benjamin also indicate a continued fear of legal action. Hutchens, he wrote to Benjamin that December, "would most certainly get anyone in trouble who dared to operate on [her]. I still have that feeling about [her] regardless of what the psychiatrists say."[49] To assuage Belt's fears, Benjamin emphasized the role of the psychiatric evaluation in warding off legal consequences. "I understand your hesitation to operate," he wrote, "although the psychiatric evaluation would protect you."[50] The psychiatric evaluation was not, then, for Hutchens's mental well-being or even diagnosis but for the benefit of Belt, who was afraid of being sued.[51]

Concern about legal censure and public controversy, rather than any attachment to taxonomic clarity, thus produced the requirement of a psychological evaluation. Benjamin did not task psychiatrists with making any kind of diagnosis of transsexuality or innate femininity. "All I would expect of a psychiatrist," he wrote to Belt, "is to pass judgment as to whether the respective patient has a sufficiently normal mentality, to allow him to make his own decision."[52] The desire for surgery in and of itself was supposed to be enough for a diagnosis of transsexualism, but it was not enough to convince Benjamin and Belt that the risk of "treating" transsexualism was worth it. Instead, the psychiatric evaluation existed to protect surgeons from legal backlash. Certainly the law didn't require it—again, nothing specifically outlawed castration or surgical construction of a vagina, so there were no legal specifications for how to go about obtaining either of those surgeries. But with a psychiatric evaluation, if anyone attempted to sue a surgeon, the surgeon could appeal to the authority of the psychiatrist who had signed off. Another expert's approval would diffuse responsibility for the decision in a way that taxonomy alone could not.

The Ghost of Surgical Future

As is clear from the Hutchens case, regret and ensuing anger toward the surgeon were persistent concerns for Benjamin and Belt, and managing that imagined future of regret became a key aspect of deciding who would get surgery and who wouldn't. Benjamin did his best to weed out patients who seemed like they would not pass as successfully as they hoped, or might change their minds later. The problem with not passing was that people might regret their transition; the problem with regret was that they might blame the people who had facilitated their surgical treatment and sue them or, as in Belt's "joke" about Hutchens, show up to their office and shoot them. As Benjamin wrote to Belt about a patient he declined to approve for surgery, "I am afraid if anything is done now it may backfire."[53] The anticipated backfiring remained a practical concern about patients' regrets more than a potential result of misclassification, illustrating a disconnect between what sexual taxonomies are supposed to mean and how they manifest in life experiences.

Transsexuals were apparently particularly prone to psychological states that would lead them to both make poor choices and lash out. Whether these states were due to an inherent trait or difficult lives—an unsettled question—Benjamin saw trans people as constitutionally unstable and annoying to deal with. "Many of these patients are utterly unreliable," he wrote to surgeon Frederick Hartsuiker to explain a patient's seemingly erratic behavior. "After all, nature has made them misfits."[54] To Kinsey's associate Wardell Pomeroy, he confided, "Most of these people are narcissistic, completely lack judgment, and some of them easily develop ideas of persecution."[55] In a letter to Kinsey himself, Benjamin said simply, "Those T.V. cases of mine (transvestism, not television) are a damned nuisance most of the time."[56] Trans people thus wound up on a circular path: They were transsexual because they wanted surgery, and they could not be trusted to make their own decisions about surgery because they were transsexual. It was the classification as transsexual itself that produced uncertainty and thus created incoherence between theory and practice, rather than uncertainty about whether a given person should be classified as transsexual.

Benjamin tended to reject patients he viewed as emotionally volatile, particularly if they seemed likely to question his authority. "Since this young man does not seem to be too cooperative," he said of one patient, "I would not treat [her]."[57] Anyone Benjamin saw as impatient, pushy, or demanding could find themselves ineligible for surgery or at the very least reprimanded. "You will have to learn to be patient," Benjamin wrote to another in a veiled threat.[58] "Otherwise, you may jeopardize your future chances for the operation."[59] Learning to be patient was required in order to be a patient. Benjamin also sometimes cut off correspondence with patients he felt were challenging him. Early on in their three-year correspondence, Carlotta Dorta shared her experience with doctors who resented patients' attempts to assert their own knowledge of transsexuality. "Slowly but surely I reached the conclusion that it doesn't pay to be honest and open-minded about it, specially with medical-doctors, and even more specially, with psychiatrists," Dorta wrote in March 1965. "The minute they have the slightest to suspect or guess that the so-called 'patient' *knows-too much*, or that he knows more than he is supposed to know . . . they cool off and crawl back into their shells, and run away like scared rabbits."[60] Evidently, this experience repeated with Benjamin: He later ended their correspondence owing to the "tone" of a postcard that Dorta had sent to him, as he explained in an October 1968 farewell.[61] He had also told Margaret Harrison that he could "no longer be [her] doctor or [her] friend" because "it was very inconsiderate [underlined in the original] . . . to disturb me on a Sunday morning just to request that I should write to your brother."[62] When Stephen Wagner asked Benjamin to write to a physician about an estrogen dosage, and whether Benjamin might send before-and-after pictures of people who had had surgery, Benjamin's research associate and secretary Virginia Allen wrote on top of the received letter "read—no reply" and "'TOO DEMANDING'" (capitalized in the original) (see figure 5.1).[63]

Presaging the various crises in trans healthcare access of the 2020s, Belt and Benjamin both emphasized the irreversibility of surgical interventions as justification for limits on eligibility. As Benjamin advised Rhonda Wallace, "The operation you are contemplating is a serious and irrevocable step. Safeguards are required."[64] Belt, too, expressed the importance of scrutinizing and denying surgery to anyone who might have second thoughts. Scrutiny in the present, though, was not enough; it also required a talent for future-telling. Belt wrote of one patient in 1958, "So far, by carefully evaluating [patients], we have not had any disappointed

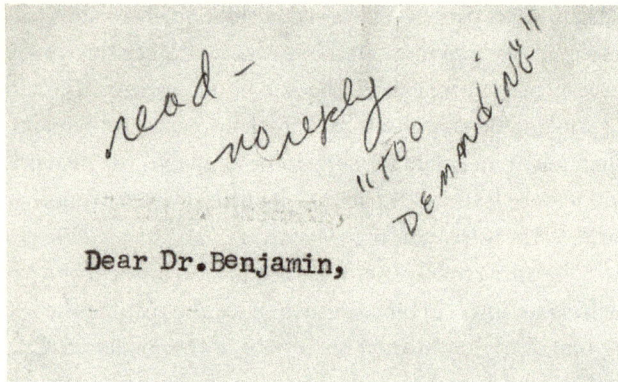

FIG 5.1. Excerpt from a letter to Harry Benjamin, showing a note written in the letter's top margin in the handwriting of Benjamin's research associate and secretary Virginia Allen. It says "read—no reply" and "TOO DEMANDING" (capitalized in the original). The writer had asked Benjamin for advice and to talk to Wagner's doctor about estrogen dosing, and for photos of transsexuals pre- and post-transition. From the collection of the Kinsey Institute, Indiana University. All rights reserved.

patients, but this particular boy seems to be beyond prediction."[65] This unpredictability made Belt want to proceed with extreme caution, and Benjamin concurred: The patient was "one of the most 'dangerous' cases. I say at present, HANDS OFF" (capitalized in the original).[66] In this emotional register, each transsexual was a dangerous, unknown quantity who had to be delicately managed at every step, even as clinical assessors apparently had preternatural powers to see who would cause trouble down the line. Incoherence allowed several conflicting enactments of transsexuality to exist simultaneously, but it also produced anxiety for all involved (some more realistically than others).

Notably, Benjamin's personal criteria for assessing patients had little to do with who was truly a transsexual or truly a woman or a man. Degrees of inherent masculinity and femininity did occasionally come up; Benjamin sometimes used "wrong body" language that implied a true, stable internal gender that could be without doubt identified as masculine or feminine. For example, Benjamin told Val Barry that he "would consider [her] definitely a woman that accidently [sic] possesses the body of a man."[67] When it came to making the final call of whether surgery was appropriate for a given patient, though, practical considerations came to the fore. In a June 1958 letter written amid the back-and-forth over Edie Hutchens's future, Benjamin outlined three prognostic factors to be considered in approving a patient for surgery in order to avoid regret: surgical, psychological, and practical outcomes. Benjamin's faith in Belt's surgical skills, he wrote, meant that he was not concerned about the first. The second factor was increased happiness and decreased fear of being arrested for cross-dressing. Benjamin mentioned that all of his patients whom he had approved for surgery had experienced an improvement in psychological well-being as a result of hormonal treatments and also "due to the realization that they have come as close to the female sex as medicine can provide." The practical outcome mattered most, because only those with the best outcomes in that regard would benefit psychologically. This "practical outcome," as Benjamin put it, "refers to the prospect of producing a reasonably successful 'woman.' In this respect, the physical structure and appearance of the patient is of importance. If this appearance is unchangeably masculine, the outcome is, of course, not only problematical but definitely doubtful, if not unfavorable." This practical outcome, he continued, was likely to be problematic in the case of Hutchens and was why he thought Hutchens should resume living as a man instead of continuing with her transition.[68] In Benjamin's account, Hutchens would come to regret her transition because she was unlikely to successfully pass as a woman, which outweighed any categorical designation.

In other words, what mattered for these outcomes was not who really was a woman, or even really a transsexual, but who would look feminine enough to pass. In a 1957 letter to Pomeroy, for example, Benjamin mentioned a patient who had managed to have surgery even though he had not agreed to it. "The reason that I did not consent was the strong masculine appearance," he wrote, though he admitted that this patient had had a positive surgical outcome, largely because rhinoplasty had given her a more feminine-looking face.[69] Being a "convincing" woman, meanwhile, could be grounds for surgical approval. Benjamin wrote in a 1958 assessment of a patient for Belt, "I know too little about him to pass an opinion, but agree with you that he may look very well as a girl. He probably ought to be one."[70] By focusing on external appearance, Benjamin and Belt made an unshakable internal gender identity and a diagnosis of transsexuality necessary but insufficient criteria for granting a patient access to surgery.

Classifications could also be overshadowed by economic concerns. Benjamin worried about patients' capacity to "do all the things that women do (household duties, etc.)" and especially to find employment as women.[71] Kinsey told Benjamin that all of the patients he knew of who had had surgery and socially transitioned were struggling to find jobs, unless they were willing to earn "their living on the lecture platform or some type of public exhibition."[72] This was also a key point in the Hutchens case: Benjamin and Belt feared that Hutchens would be fired from her job as a result of not passing. Labor issues were so important that while Benjamin made decisions about who could have surgery in part based on who he thought would be able to pass afterward, he sometimes suggested that patients have genital surgery but not attempt to live as women. To a doctor who wanted surgery but was concerned about losing professional status, Benjamin wrote, "How would you feel to have genitals resembling those of a woman, and yet 'Masquerade' as a man in order to preserve your professional work standing?" and confided that he knew of at least two such cases.[73] Benjamin wrote to Kinsey about another patient who had returned to the United States from a successful surgery in Mexico but was "still working as a male," which was true of three other patients as well.[74] Demands of passing, then, could occasionally be overlooked for the sake of economic stability, while an inability to pass—based solely on Benjamin and Belt's narrow conception of normative feminine appearances and beauty standards—foreclosed surgery because patients would be unable to support themselves financially.[xv]

A focus on passing and the consequences of not passing thus took precedence over gender classification. A prospective patient's failure to adhere to gender norms might lead to regret, but Benjamin and Belt did not categorize people according to intrinsic gender (normative or not) so much as how they might be read by others. While Benjamin believed that transsexuals possessed a degree of what he called "constitutional femininity," he took issue with what he felt was a misreading of this statement that suggested he believed that transsexuals were actually female. Benjamin wrote to the editor of the *Journal of the American Medical Association* to correct an article that had described his perspective as such. The "concept that these subjects (transvestites) are 'constitutionally female' . . . is not and has never been my concept," Benjamin specified. "Naturally an assumption of a certain degree of constitutional femininity is not to say that these subjects *are* constitutionally female."[75] In other words, Benjamin did not assess his patients to determine if they were women—he assumed that they were not.

xv   Neither seems to have had anything to say about pay gaps and gendered labor-market segmentation.

True Transsexuals and Real Women

Scholars have discussed the gatekeeping of trans medicine as a defense of stable binary sex, but neither Benjamin nor his colleagues seemed concerned that the binary or womanhood or femininity were under threat.[76] On the contrary, they viewed transition as largely functional and cosmetic. While estrogen and vaginoplasty, along with new clothes, a new job, and a new set of familial and social relationships, might enable someone to live in the world as a woman and be more comfortable with herself, none of those things would make her a real woman. In these doctors' framework, their patients were not women but transsexuals. Even as "transsexual" became a condition of being for which there was an appropriate medical response, it was also a state of rejecting reality that required careful handling.[77] Regret emerged from the imagined mismatch between the postsurgical body and the truth of sex classification. This truth, of course, was not remotely settled (Benjamin himself argued against the idea of one singular sex and instead broke sex down into multiple components), but that didn't keep some enactments of sex from being emphasized over others when trans people sought medical care.

Benjamin stressed that patients must demonstrate a "realistic assessment" of their future to qualify for surgery. They could not think of themselves as real women, only imitations of such; if they believed otherwise, they were deluding themselves. Benjamin was clear with patients that they would never be women, no matter what they did. "You must realize, of course," he responded to Edith Williams's inquiry, "that living as a woman and taking female hormones does not make you a woman."[78] To Winnie Dunning he wrote, "Please remember that no operation can ever make a normal female out of a male. Sex cannot be changed—only the secondary sex characters."[xvi] So framed, medical transition was a long-term façade, an intervention that could treat the symptoms of gender dysphoria but not one that could produce a fundamental change in someone's sex. Medical transition could certainly help a patient feel better, but only the appearance of the body would be transformed, and only partially at that. The sex category that a patient would fit into would not change, just

xvi   Benjamin to W. J. D., August 18, 1955, series IIC, Box 4, Folder D., W.J., HCB. It's not entirely clear what Benjamin means here by "sex." Does he mean chromosomes? Anatomy other than genitals like skeletal proportions? Some abstract ontological concept? Who can say.

as we've already seen bees without reproductive capacity categorized as female, women masculinized by ovariotomies remain women, and women with genitals transformed by sexual experience kept in their sex category. It was disqualifying to imagine otherwise.

Acceptance of one's true sex suggested that a patient would gratefully accept the outcome of their medical transition, further enshrining it as a selection criterion. Belt in particular found it endlessly frustrating that his patients continued to pester him for more surgical interventions that were beyond his capacity, like implanting ovaries and a uterus. "No matter what we do for these patients they will never be satisfied," he wrote to Benjamin in 1956, after he had operated on several trans women. "As each procedure is performed, they come up with further desires and requests which makes the job of dealing with them and handling their problem very difficult."[79] Belt cautioned a patient similarly. "It is in the nature of things that a transvestite will never be wholly satisfied with her appearance," he wrote to Barbie Owens. "In the most successful operation we have had, a young person with so great a tendency toward femininity that her very perineum was constructed by nature wider than the male . . . the patient came in after all was done expressing dissatisfaction because there was not a uterus with tubes and ovaries projecting into it from above and she could therefore not have a baby. 'You have performed a miracle so far for me, Doctor, why can't you do just this one more thing?'"[80] Even with an innate "tendency toward femininity" that suggests gender and sex had not entirely diverged yet, a "realistic" set of expectations—here taking the form of an acceptance of material limits—was apparently lacking.

Prospective trans patients also stood outside the bounds of rationality in their relation to modern science. In 1969, when journalist Burton Wolfe asked why Belt had stopped performing trans surgeries, Belt explained that he was concerned about lawsuits in which amounts of "money demanded by the dissatisfied transsexual who had dreams of becoming a mother and other such nonsense were beyond the wildest imaginings."[81] Likewise, while turning down Lorna Harding for treatment, Belt claimed that he had stopped performing penectomies and vaginoplasties "due to a series of unfortunate experiences with patients who have felt that they wished this type of work done but who expected more than the surgeon can possibly deliver in the way of alteration even though the limitations of the method were most carefully set forth preoperatively."[82] Benjamin, too, framed too much anticipation of a positive surgical outcome as unrealistic. "Do try hard to give the impression of a well-balanced sensible person [underlined in the original] who does not expect miracles," he wrote to Debbie Mayne as she sought a psychiatrist's approval for surgery.[83] His reference to "miracles" placed transsexual hopes in the realm of the fantastic and likely impossible. Benjamin himself had written to Hartsuiker, Mayne's potential surgeon, saying, "It is quite important in my opinion that the patients retain their realistic attitude toward their own status even if they live the life of a woman to which I feel they are entitled."[84] Here, "status" as a transsexual contrasted with the unscientific fiction of reclassification into womanhood. One could "live a life," but never be. Medical transition and indeed transsexuality in general did not, it seems, threaten binary sex, at least not for Benjamin.

Though Benjamin may have coached Mayne on how to convince Hartsuiker that her expectations for her surgical outcome would not be too high, Benjamin expressed doubts about Mayne's grip on reality. "[Debbie] impressed me as so highly emotional as to be almost called psychoneurotic and certainly very unrealistic," Benjamin wrote in the same letter to Hartsuiker. "I think [Debbie] is a more serious problem than many other transsexualists and it is really often difficult to decide which is the lesser of the two evils: to operate or to refuse operation." Throughout the 1950s and 1960s, Benjamin routinely and intentionally avoided outright rejecting patients to prevent the psychological ramifications of taking away someone's hope for surgery, even when he had no intention of helping them obtain surgical care. When Benjamin learned that the mayhem statute would get in the way of surgeries being performed in California, he wrote to the district attorney, "After having read the memorandum, I do not see how any surgeon anywhere in this country could possibly perform such an operation, but I cannot bring myself to inform the patient [Barry] of the true situation."[85] He continued, "I would not want to take the responsibility for possible development of an acute psychosis or for suicide."[xvii] To Kinsey, Benjamin specified that he "told [Barry] that until

---

xvii   The risk of suicide, though, seemed less of a concern for Belt than the possibility of regret. When a patient threatened suicide if surgery continued to be withheld, Belt wrote to Benjamin in a panic, "This [suicide threat] indicates an instability of character to such an extent that the patient would be dangerous if things did not go exactly to his liking. It gives me very cold feet about taking him on as an operative possibility." Benjamin was less immediately concerned. He wrote confidentially to the patient, saying that he had followed up with Belt and tried to talk him down from not operating. "But if you write to him again," Benjamin cautioned, "please write as sensibly and unemotionally as you can."

Thorek will be able to make the arrangements [for surgery] somewhere in the world (which I doubt he ever will or can), he is using a 'chemical knife' toward the same end."[86] When arrangements elsewhere in the world failed to materialize, Benjamin wrote to Hartsuiker, at that point Barry's last hope, that he "dare not tell [patients waiting for surgery] the bad news the way you wrote it. I will simply say that too many patients have gone to Holland who had no business to go and because they were not referred by any doctor that <u>for the time being</u> [underlined in the original] no further operations could be scheduled. I think that will soften the blow a little."[87]

Conveniently, the psychiatric evaluation that had already been so useful in staving off legal anxieties could be repurposed to buy time. After receiving threatening letters from a patient's father, Benjamin asked Belt to lie to the patient and say that further observation by Carlson would be needed before moving forward. Even though Benjamin felt surgery would be of psychological benefit to this particular patient, he was "a bit fearful of possible consequences." He continued, "Naturally, I would consent to the operation even without re-examining 'Marie' if you and Dr. Carlson consider the operation indicated. Of course, I have never peremptorily refused my consent to the operation; I merely held off until such time that I felt [their] emotional balance was improved."[88] Benjamin then advised Belt, "Should you hear from him, just tell him he has to be patient and wait until I am able to consent to the operation in his case."[89] Benjamin continued to recommend this approach well into the 1970s. During a speech given before the New York Academy of Medicine, he told his audience, "An abrupt refusal to operate must be avoided as it could precipitate dangerous emotional decompensation. The recommendation just to postpone surgery for various reasons is usually accepted."[90] Lying to patients was excusable in the service of avoiding "responsibility," as Benjamin had put it decades earlier, for their emotional responses—and avoiding having to deal with those emotional responses as clinicians. So much, then, for reality. A floating concept of real sex could be weaponized against unrealistic patients, whose supposed tendency to lie was representative of psychological instability, but clinicians could and indeed should lie to their patients about what was really happening. Benjamin was afraid of the very uncertainty that he and Belt had created.

Belt to Benjamin, February 10, 1958, and February 18, 1958, series VIB, Box 23, Folder 34, HBC.

Uneven Distribution

Patients, of course, were not always so willing to go along peacefully, and they often pushed back against doctors who threatened to block the progress of their medical transitions. Debbie Mayne, for example, had first brought up the possibility of transitioning with her physician, Guy Tourney, in 1940, but Tourney "refused to encourage the subject."[91] By 1954, Mayne had arranged to have surgery with Hartsuiker in the Netherlands— but psychiatrist Frederic Worden had determined that Debbie would be better off living as a male homosexual than as a woman.[92] She was not put off: "This girl is going to keep on raising hell until I get my operations," was Debbie's response to the rejection.[xviii] Other patients, too, were not oblivious to the way that they were being manipulated by surgeons using psychiatric testing as a stall tactic. Carla Sawyer, for example, who became part of a University of California at Los Angeles (UCLA) research study in the hope of obtaining approval for surgery there, wrote that she felt "they have been trying to tax my endurance and ability to stand up to a test just to find out how much I am able to take before I bust. I feel as if I have been flattened out, and rolled up and pushed through a knot hole and I told them so, too."[93]

When the UCLA Committee on Transvestism decided not to allow surgery for anyone, Sawyer turned to Daniel Lopez Ferrer in Mexico City, who felt that it would be "justifiable" to operate on her.[xix] But after she had an intense emotional experience due to "making a friend and knowing happiness and by the religious atmosphere of the country" and cried an apparently excessive amount, Lopez Ferrer would not operate because of Sawyer's "emotional instability."[94] Benjamin's response to patients like Sawyer who were fighting their way through a medical process filled with disappointment was to tell them to remain calm. After Sawyer wrote him

xviii   Mayne to Benjamin, March 15, 1954, series IIC, Box 6, Folder M, D, HBC. In an example of Benjamin helping patients at the same time as all of the above, Mayne thanked Benjamin and said he had "stuck by her" in the same letter.
xix   UCLA's gender clinic was run by Robert Stoller, who championed a psychological and psychoanalytic approach to transness and saw those modalities, rather than hormones and surgery, as the best "treatment" for transsexuality, as discussed in Gill-Peterson, *Histories of the Transgender Child*. Stoller complained to Benjamin in 1966 that he had "recently gotten a number of calls from people who are under the mistaken impression that they can have surgery done at UCLA." Stoller to Benjamin, October 5, 1966, Folder 21, Box 25, series VI C, HBC.

from Mexico City, where she had just been denied surgery at the last minute, Benjamin replied, "It seems to me you have to learn to keep your emotions under control. Otherwise you will run into more and more difficulties."[95] When Sawyer replied that she had been having "almost unendurable" anxiety attacks because she didn't know when or if she would be able to have surgery, Benjamin said merely that her "emotional stability" would have to "improve a great deal" before she should try to schedule surgery again.[96] "I find that I am dependent upon a total stranger for rendering a decision on my life," Sawyer reflected in her response. "I realize that my whole life is dangling by a thread and there isn't much of anything that I can do about it. I don't think that you could ask for a greater test of emotional fortitude than my present dilemma."[97]

The medical context of midcentury trans medicine was structured in large part on the assumption that it was clinicians, not patients, who had the capacity for logical decisions and should therefore be empowered to decide what made sense. The result was a bizarre reversal of where the risks of medical transition accrued. While Benjamin and Belt understood themselves to be in most danger, midcentury approaches to trans surgery were far riskier for patients than for doctors, and not only psychologically. Clinicians' reality won out over patients' and the bodily perils of surgery became invisible to them even as trans patients were painfully aware. Like most American surgeons working in the 1950s, Belt had no prior experience working with trans patients. After performing his first few surgeries on trans women in 1956, Belt wrote to Benjamin that he felt that it would be "wise to wait a while before undertaking any plastic operations for people of this type until I see how these progress." The surgeries that he had just completed required several revisions, including subsequent procedures to remove scrotal skin out of which he had attempted to shape labia, and to change the placement of one patient's urethra.[98] Belt described his own treatment of trans patients as "experimental."[99]

Benjamin, Belt, and their patients were highly attuned to the on-the-fly nature of vaginoplasties in the early years of trans surgeries. By the late 1960s, Georges Burou of Casablanca would develop a reputation for his "gold standard" penile inversion technique, but, until then, surgeons produced inconsistent results.[100] Though Benjamin seemed satisfied with Belt's surgical technique, he was dubious about some of the other options. Benjamin admitted that Lopez Ferrer's decision to stop performing castrations and penectomies was for the best, as the "results from the operations performed in Mexico have not been too good" and several patients "immediately had to undergo treatment here again, or have had to have another operation performed."[101] A patient wrote to Benjamin saying she'd heard some women who had gone to Jose Jesus Barbosa in Tijuana had good surgical outcomes, but "others have been mutilated."[102] Benjamin would later write to a patient trying to find a surgeon, "With the exception of very few, chiefly Dr. Burou in Casablanca, I feel that all the doctors doing the operation now more or less secretly, are experimenting, trying to find the right technique."[103] Patients knew how little experience these surgeons had. At a meeting of "male transsexuals" held in 1969 by two Purdue sociologists and attorney Robert Sherwin (of mayhem statute fame), several participants discussed their struggles to find surgeons who knew what they were doing. One woman told the group that she believed that, other than Burou, "places are just experimenting" with surgery. When another participant asked her what she thought about Belt, she said, "He's the biggest butcher of all."[104] Even in 1973, Benjamin confessed, "The various surgeons who experiment with [vaginoplasty] use various methods."[105]

While the inexperience of medical professionals put patients at great risk for physical harm, it presented an opportunity for clinicians wishing to position themselves at the forefront of a new field. Benjamin and Belt translated their inexperience into a positive image of themselves as pioneers, true scientists compared to establishment colleagues who privileged their emotional response to transsexuality over an objective view of the potential for discovery.[106] When UCLA's Committee on Transvestism voted to refuse to allow transsexual surgeries by members of its medical department and within the walls of university-affiliated hospitals, Benjamin described it as "a pretty sad commentary on scientific freedom and objectivity" in which "scientific investigation seems to the committee less important than conforming to legal technicalities and religious prejudices."[107] Benjamin and Belt were, of course, the scientific and objective ones who could overlook how the practices of trans medicine emerged to manage their own fears and annoyances.

Classificatory Distress

Benjamin's heightened capacity to see objective reality found its confirmation in the form of a classification system. In the early 1950s, there were no meaningful diagnostic criteria for transsexuality other than Benjamin's own gut feelings. While an October 1964 letter from psychologist Ruth

Rae Doorbar to Benjamin suggests an approach that put potential surgical candidates through a battery of intelligence, personality, and perception tests, these served more of a research purpose in their early incarnation than a diagnostic one.[108] In the mid-1950s, Benjamin was effectively on his own, diagnosing transsexuality according to his own judgment. Complex diagnostic criteria for transsexuality were developed after Benjamin and Belt had already established their routines for assessing patients for surgical approval.

After over a decade of informal categorizing, Benjamin developed what he called the Sex Orientation Scale, or the SOS.[xx] The scale, based on Kinsey's sexual behavior rating system (see chapter 4), described seven categories of relationship to one's sex, from "Type 0," or those with "normal sex orientation and identification" who find the idea of cross-dressing and surgery "foreign and unpleasant" and comprise "the vast majority of most people"—again rendering transsexuals and transvestites anomalous minorities—through another six degrees of "sex and gender role disorientation and indecision" ending with "Type VI," the true transsexual of high intensity.[109] However, Benjamin emphasized that the types "are not and never can be sharply separated" and were "approximations, schematized and idealized."[110] Most people would "fall in between two types and may even have this or that symptom of still another type."[111] Even with this diagnostic tool, then, transsexuality itself and who would benefit from surgery were anything but self-evident. Matching patient to category, and then to treatment plan, required a clinician's interpretation.

Within the SOS, transsexuals were transsexuals because dressing and living as a sex other than what they were assigned at birth was not enough to alleviate their gender dysphoria, because psychotherapy did not work to relieve their symptoms and, most of all, because they wanted genital surgery. This inability to suppress gender dysphoria simultaneously made someone both a good candidate for surgery because it defined transsexuality and a risky candidate for surgery on the basis of their failure to cope with psychic pain. "Normal" people, meanwhile, were normal because transitioning didn't appeal to them.[112] Locating transsexuality in the desire for

xx   While SOS was not, as far as I can tell, intended to summon the image of a sinking ship begging for help, it does have a certain je ne sais quoi about it in both the past and present context of trans medicine. Dot-dot-dot dash-dash-dash dot-dot-dot, we have hit an iceberg and it is Republican-controlled state governments. I'll never let go, Jack, et cetera.

surgery and with the action of cross-dressing—and normality in the lack of that desire—immediately re-normalized vast swaths of people who otherwise displayed various forms of gender deviance but didn't specifically want genital surgery.[xxi] There was no need to worry about edge cases or blurred categories. Someone either wanted surgery or they didn't. By making a desire for surgery the basis of ejection from "normal" sex categories, Benjamin and his colleagues helped plaster over a century's worth of anomalies in sex research, and contributed to the assumption that most people have a "normal" binary sex and gender that match each other.

Diagnostically and practically, little changed with the development of the SOS. Benjamin occasionally responded to inquiries with an SOS diagnosis—in November 1961, for example, he sent a copy of the SOS to patient Joan Sewell with the note, "Judging by your description, you most likely belong to Type III of Transvestism, as I described it in the enclosed reprint."[113] But references to SOS in correspondence were rare, and I found no indication that Benjamin actually used the scale in any decision making. Patients' scale rating served a primarily organizational and research purpose, with little practical clinical relevance. Ratings appear, for example, in a spreadsheet of all Benjamin's trans patients that included a column listing patients' "TV-TS type," and what appears to be an intake form that has a space to note SOS rating.[114] I located no discussion, though, about whether a particular patient was a particular type and what that meant for their surgical potential. The ratings given after a clinical encounter were effectively the conclusion Benjamin would have come to anyway, because they depended on how much someone wanted surgery, which was already Benjamin's defining feature of transsexuality. If there were now types of person according to degree of gender dysphoria and surgical desire, SOS ratings were an afterthought that justified the recommendations Benjamin had been making for the past decade.[115] Nonetheless, as a post hoc attempt to make his practice seem more systematic and scientific, the SOS gave an air of objectivity to Benjamin's decision-making process.

For Benjamin and Belt, larger questions about what sex was and who might be included in what kinds of categories were eclipsed by the question of who should and should not be allowed to access surgery. Despite Benjamin's claims to transsexual diagnostic expertise, and despite the creation of the SOS, it was still not obvious who should qualify for medical

xxi   Benjamin would have fallen solidly in the camp of "female husbands are not trans," in a slightly anachronistic example. Cf. Manion, *Female Husbands.*

transition. This opened up the possibility for terrible mistakes, which in turn required expert regulation of the process. Assumptions about masculinity and femininity absolutely played a role in the process, especially as time went on and the gender clinics of the later 1960s and 1970s developed institutionalized screening practices.[116] But in the early days of trans medicine, sorting masculine from feminine paled in comparison to assessing possible risk. The quotidian clinical practices that Benjamin and Belt established to handle their anxieties about postsurgical futures would continue to shape the experiences of people attempting to access transition care for decades to come.

"Standards" of "Care"

The threats of mayhem and regret proved a nonissue and the anticipated lawsuits never materialized.[117] In response to rumors that he had quit surgery because of a lawsuit, Belt informed a reporter that he stopped as a "matter of [his] own volition" and that it was "not true that the District Attorney or the County Medical Association or any other agency either threatened or suggested that [he] stop doing these operations."[118] Still, the specter of litigation had a tremendous impact, shaping both the availability of surgery for trans patients and the requirement for psychiatric assessment before a surgeon would operate. For much of the 1950s, Belt was the only urologist in the United States and one of few in the world who would perform transsexual surgeries.[119] His methods for soothing concerns about retribution thus became effective requirements, leading to selection criteria based on a general sense of a patient's likeliness of regretting their surgery and turning to Belt for revenge. It may have been Belt's "own volition," but he nonetheless wrote that threat of lawsuits had the potential to drive up malpractice insurance costs and thereby contributed to the end of his transsexual surgical practice. "This subject has never come up for discussion between our insurance company and our group," he wrote, "but because I felt that it might come up I stopped the practice [of operating on trans patients]."[120] Belt's anticipation of legal trouble was enough to scare him away, further reducing options for trans medical care.

Trans medicine gained legitimacy and coalesced into a recognized field of expertise over the course of the 1970s, thanks in large part to the rise of university-supported gender clinics and research projects that gave clinicians and researchers institutional backing. Positive press coverage and legal victories in favor of trans people further supported a sense of optimism about, as well as public and professional acceptance of, medical transition.[121] This did not, however, result in greater trans self-determination or diminished fears about bad outcomes. On the contrary, clinicians doubled down on limiting access to both hormonal and surgical interventions. What had functioned as informal habits at the height of Benjamin's influence in the 1950s and 1960s were codified as official standards in 1979, when a group of doctors, therapists, and researchers formed a professional organization that would provide a forum for discussions on transsexuality and cement their status as experts on transness. In a nod to Benjamin's eminence, they called their organization the Harry Benjamin International Gender Dysphoria Association (HBIGDA).[xxii] One of HBIGDA's first tasks was to develop a set of criteria by which practitioners could assess patients to determine who would benefit from hormones and surgery, and who should not have access to them. HBIGDA framed the Standards of Care, as they came to be called, as existing "for the sake of proper patient care everywhere."[122] However, the standards also provided a means for them to protect themselves—just as Benjamin and Belt's clinical decision making had done for several decades.

The 1979 Standards of Care, consisting of a set of principles to guide clinical practices, foregrounded regret.[xxiii] Principle 1 sounds extremely familiar: "Hormonal and surgical sex reassignment . . . has effects and consequences which are not, or are not readily, reversible, and may be requested by persons experiencing short-termed delusions or beliefs which may later be reversed."[xxiv] Therefore, such interventions should not be performed electively (Principle 2), because some past patients' surgeries have been "psychologically debilitating" (Principle 3). Thus, it is "professionally improper" to provide transition services on demand, and clini-

xxii   The organization changed its name to the World Professional Association for Transgender Health (WPATH) in 2007. I look forward, with tremendous excitement, to Os Keyes's forthcoming work based on newly recovered sources and oral history interviews with founding members of HBIGDA and many other key figures in late twentieth-century trans medicine.
xxiii   The Standards of Care discussed in this section refer to the original 1979 draft. For the sake of citational brevity, all quotes in this and the following four paragraphs are from that document. The standards have continued to evolve, with the eighth version (the most recent as of this writing) being published in 2022.
xxiv   This remains an anti-trans talking point.

cians must engage in "careful evaluation" of prospective patients. As with Benjamin and Belt's concerns, an inability to pass after surgery could fuel regret. Principle 12 claims, "The best indicator for hormonal and surgical sex-reassignment is how successfully the patient has been in living-out, full-time, vocationally and avocationally, in all social situations, the social role of the genetically other sex and how successful the patient has been in being accepted by others as a member of that genetically other sex."[xxv] For this reason, transsexuals should spend a full year before surgery demonstrating that they can pass, so that they do not surgically transition only to find that they don't (which they would presumably regret).

The psychiatric evaluation remained a key aspect of clinical practice, and as in earlier decades, it did more to protect practitioners from facing consequences than to diagnose or support patients.[xxvi] In fact, of the standards' twelve pages, only about one and a half concerned patients themselves. HBIGDA briefly acknowledged that some doctors had unscrupulously overcharged patients, and it emphasized that trans people often face social and economic discrimination and should not be charged exorbitant fees. The standards also protected patients' right to privacy and information about the physical risks of treatment. No attention was given, however, to what might happen *for patients* if they were not allowed to transition. The only bad outcomes were those that might cause problems for doctors. It was for this reason that clinicians required extra safeguards, like the diffusing of risk between several colleagues. To this end, Principle 15 notes, "Peer review is a commonly accepted procedure in most branches of science and is used primarily to ensure maximal efficiency and correctness of scientific decisions and procedures," and therefore should be a key part of patient assessment. This was particularly important because, as per Principle 16, medical providers "must often rely on possibly unreliable or invalid sources of information," including "patients' verbal reports." "Peer review" referenced a tradition of scientific truth-making and thereby added legitimacy to the enterprise while also establishing that clinicians,

xxv   Note the specificity of "genetic" sex, despite the issues at hand being hormones, surgery, and social role.
xxvi   A desire for hormones and surgeries remained diagnostic of "gender dysphoria." The standards did not have anything to say about a person "being" a particular gender.

with their purportedly unbiased perspective, held the ability and authority to assess "reality" much more than lying or incompetent trans people. Decision making, and therefore liability, remained distributed among several clinicians, just as it had when Benjamin and Belt looked to the psychiatric evaluation to protect themselves. This diffusion was unambiguous; the standards specifically contend that "psychiatrists and psychologists, in deciding to make the recommendation in favor of hormonal and/or surgical sex-reassignment share the moral responsibility for that decision with the physician and/or surgeon who accepts that recommendation."

The standards also echoed Benjamin and Belt's characterization of themselves as brave pioneers. Principle 17 encapsulates risks to providers: "Psychiatrists and psychologists, given the burden of deciding who to recommend for hormonal and surgical sex-reassignment and for whom to refuse such recommendations are subject to extreme social pressure and possible manipulation as to create an atmosphere in which charges of laxity, favoritism, sexism, financial gain, etc., may be made." The person recommending surgery, sadly, "does not enjoy the comfort or security of knowing that his decision would be supported by his peers."[xxvii] Once again, the document framed providers as experiencing the real difficulty in transsexual medicine. The standards outline these encumbrances thusly:

> The care and treatment of sex-reassignment applicants or patients often causes special problems for the professionals offering such care and treatment. These special problems include, but are not limited to, the need for the professional to cooperate with education of the public to justify his or her work, the need to document the case history perhaps more completely than is customary in general patient care, the need to respond to multiple, nonpaying, service applicants and the need to be receptive and responsive to the extra demands for services and assistance often made by sex-reassignment applicants as compared to other patient groups.

All this considered, the standards declared that anyone who operated on a trans patient without at least two written recommendations from psychiatrists or psychologists—which it reiterated are for "peer review"—is "guilty of professional misconduct."

xxvii Apparently trans people do? Certainly no experiences of social pressure, there, either.

Doctors' fear of patients' regret persisted despite experiential evidence that transsexuals who had surgery overwhelmingly did not regret it. Even in the earliest days of American trans medicine, Benjamin wrote to a prospective patient, "All the patients that I have seen following an operation are definitely happier than they were before," though, of course, he followed that assertion with the caution, "but that does not guarantee anything for the future."[123] In 1957, Benjamin noted that all fifteen of his post-op patients were happier than before, and that "not one of them has told me that they have regretted the operation but, on the contrary, assured me that under no circumstances would they want to have it undone, even if such would be possible."[124] Though prior experience suggested regret was unlikely, predictions that it could and would happen helped Benjamin and Belt and the clinicians of HBIGDA solidify their own importance and the role of medical expertise in making decisions about trans bodies and lives. Because of the potential for disaster, one needed an expert to make the right choice. To maintain control over who could access surgery, Benjamin and Belt created an anticipated future that only their careful selection of patients could prevent. Their own track record of happy patients could, in their minds, be justifiably ignored in the face of an anticipated future disaster. This divinatory catastrophizing had a material legacy as trans medicine coalesced around a set of clinical standards intended to stave off imminent disaster. Though trans people rarely changed their minds, trans people's changed minds dictated the trajectory of trans medicine in the second half of the twentieth century and into the present. Practitioners of trans medicine could position themselves as experts precisely because they cast transition outcomes as both possibly good and inevitably bad, with their expertise hinging on a regime of anticipation that came to exist through "simultaneous uncertainty *and* inevitability of the future."[125] Trans people couldn't be trusted to make their own decisions; they needed an expert to do it for them. The experts could never quite be sure; this made them very impressive solvers of a complex problem but, despite their prowess, put them consistently in harm's way. The truth of sex, whatever that might be, was subsumed into a more practical question of risk and reward, and a contest for the authority to make truth claims about the future. In the multiple realities of trans medicine, incoherence struck again, this time as the source of clinical power.

Coda: A Once and Present Shitshow

The approach set in motion by Benjamin, Belt, and the founding members of HBIGDA continues to perpetuate an image of the emotionally unstable trans person likely to regret their transition. Regret, rather than trans desire, remains the structuring principle of trans medicine.[xxviii] Since the mid-twentieth century, access to transition care has become more available in some respects, though still tremendously limited by expense, waiting times, and general lack of access to healthcare. Or, at least, it had become more available. Now, we're on the precipice—and in some places in the United States and the United Kingdom, off the cliff—of Belt's hopes and fears coming true. Transition-related care is becoming illegal.

Over the multiple iterations of this project, I've written several endings for this chapter. In 2019, when I drafted the first version, it was a salty story about my own experience needing to obtain a psychological evaluation letter to get top surgery, even though my surgeon assured it me was only a formality (for "insurance reasons"). By 2021, a number of revisions and an article version published in *TSQ* featured a series of "As of this writing . . ." notes about various legislative attempts to restrict transition care for youth. Each iteration felt immediately dated as the current panic about transition, which, if I had to stick an origin date on it, I'd locate with Jesse Singal's 2018 *Atlantic* article, has taken greater hold.[126] They haven't felt dated in a sheepish kind of way where something that was of the zeitgeist no longer is, and now the conclusion is the equivalent of early 2000s bootcut jeans. They've felt dated in a horrific kind of, oh, you thought that was bad, what an innocent time, way.

I won't document the specifics of where we're at now, near the end of 2024, as I write this, because things continue to change on a near-daily basis. In the space of two years, we've gone from initial failed attempts to successful legislation that will remove trans kids from their parents if they're allowed to transition, that makes providing transition care a felony, and that limits the care available to adults. This doesn't include the sports nonsense, the book banning, and the Don't Say Gay garbage, or the overarching attack on reproductive freedom and bodily autonomy marked most emphatically by the overturning of *Roe v. Wade*. It's unclear

---

xxviii   See Andrea Long Chu, "My New Vagina Won't Make Me Happy," *New York Times*, November 24, 2018, for an incisive critique of how this manifests in what Chu terms "compassion-mongering."

to me what fresh hell will be unleashed next. I wish I could end this in some even-handed way where I acknowledge the way that Benjamin and Belt enabled, or at least increased the momentum of, possibilities for medical transition, which has drastically improved my own life. But honestly, all I can say is, Fuck those guys.

Benjamin and Belt's approach to trans medicine exemplifies and distills the construction and use of incoherent sex that this book has explored all the way through. They insisted that sex is mutable and fixed, simultaneously. It was changeable, but only by experts. It persisted past those changes in a nebulous way that could not be measured but surely existed. I opened this book with the first lines of Benjamin's *Transsexual Phenomenon*, in which he named sex as vague, emotionally loaded, definite and indefinite, increasingly studied and increasingly devoid of meaning. His assessment offers an unerring summary of the history of sex science. The question that remains is how to do something different.

Epilogue

Chaotic Good

Between the mid-nineteenth and mid-twentieth centuries, across disciplines and decades, throughout the specialization of fields and the transformation of practices, as the stated social goals of research became increasingly less explicit, two scaffolding features of knowledge production about sex persisted: its incoherence and its importance. With considerable conflict about what it was and what bodies fit into which categories, sex remained paramount to understanding life. As I've argued throughout this book, sex maintained its grip on science and medicine—as a topic of study, as a research variable, as a material to be molded—in large part because it could respond adroitly to any challenge. Inconsistencies and contrary evidence could be smoothed away by a small tweak of criteria, a blatant disregard for the gap between theory and practice, and a turn to simplicity in the face of complexity. A desire for sex to naturalize social hierarchies, whether race, scientific expertise, or gender, superseded a need to address the anomalies that suffused sex research.

The five sites in this book—animal research, eugenics, gynecology, statistical sexology, and trans medicine—have each demonstrated a particular means by which sex science produced and deployed incoherence. Scientists and agriculturalists who focused on animal bodies created conflicting accounts of the nature of sex: one in which evolutionary racial hierarchy depended on a scale from common hermaphroditism to rare sexual dimorphism, with white bodies as the most successfully differentiated, and another in which scientific practices could identify male or female sex even when bodies presented with ambiguous traits. Eugenic researchers called on differing meanings of what sex itself was—malleable or static—depending on whether they were attempting to gain control over bodies in the lab or contend with masses of data about heredity. In Robert Latou Dickinson's gynecological practice, sex as a matter of degree rather than kind stood companionably beside and in stark opposition to a commitment to women remaining women, regardless of what their bodies looked like or did. The authors of the Kinsey Reports, though they acknowledged their great difficulty finding physical differences between men and women that would explain differences in their sexual behavior, nonetheless reinscribed binary sex with the use of new statistical methods. Finally, decades of disagreement and willingness to look past a tremendous diversity of sexed forms culminated in the arrival of trans medicine in the mid-twentieth century. Here, the diagnostic category "transsexual" proved less important than quotidian scientific practices that foregrounded risk and uncertainty over ontological stipulations, and stringent gatekeeping based on those practices narrowed the possibility of accessing surgical transition care. By the end of this arc, questions about the nature of sex, how many categories there might be, and what kinds of bodies might belong in each one gave way to the sequestering of a new kind of incoherence—a mismatch of "sex" and "gender"—in transsexuality, a pathologized minority state of being in contrast to the assumption that, for everyone else, sex was simple and uncomplicated.

In ensuing years, a new way of thinking about sex and gender classification has emerged. The term *cisgender* first appeared in 1994, coined by Dana Leland Defosse in reference to the scientific prefix *cis-*, which in its original use refers to molecules staying put. The category further took off around 2008.[1] I've already problematized the assumption that cisness is ahistorical—that there have always been people whose gender identity matched their birth-assigned sex—in the introduction to this book. Having traced its prehistory, I'll add that cisness, as a category rooted in coherence between sex and gender, only works if this history of science is ignored. Throughout the period investigated here, sex was not coherent. It did not stay put. Coherence has not been the source of a definition of sex, or maleness and femaleness; gender itself only developed as an axis of being in the mid-twentieth century, in the context of trans medicine that made decisions more on the basis of professional anxieties than a robust theory of anything in particular. The very idea that a cisgender type of person would emerge from the congruence—the coherence—of the two always-already incoherent classification systems "sex" and "gender" is, in light of this history, suspect.

Cisness, like the enactments of sex discussed in this book, plasters over the anomalous. Like Harry Benjamin's transsexual classification system, it anchors transness in a specific desire for change and declares everything else a normal relationship between sex and gender, regardless of departure from idealized, binary forms (cisness, after all, is always binary; in this logic, only an intersex person could be nonbinary and cis). We already know from Black transfeminist thought that cisness only works as an analytic if the racialization of gender is not taken seriously.[2] Attending to the history of sex science further demonstrates that the body—whether conceived as birth-assigned or any definition of biologically sexed—with which one's gender might or might not align is not a stable foundation to build on.

There is another peril of ignoring this history: a misplaced appeal to science when countering anti-trans logics. This book has come into existence against the backdrop of an increasing assault on transness in the United States and around the globe, with science consistently trotted out as justification for anti-trans positions. The specifics, at this point a near-constant onslaught of more bad news, matter less than the definitions of sex that anti-trans forces deploy.[i] Legislators, government officials, and malfeasants on the internet have declared, repeatedly, that they know what sex is: It's binary, it's self-evident, and it certainly can't be changed. They've said things like, "Sex means a person's status as male or female based on immutable biological traits identifiable by or before birth."[3] And "Sex is a stable, binary, biological phenomenon."[4] Apparently there are "inherent differences between men and women" that "range from chromosomal and hormonal differences to physiological differences."[5] Or, as a particularly on-the-nose case of this line of argumentation, put it, "There are TWO genders: Male & Female. Trust the Science!"[6] Meanwhile, we've heard that medical transition for trans youth is "dangerous and uncontrolled human medical experimentation" and that no "rigorous studies" have shown that medical transition has any long-term beneficial effects—here, a supposed disregard for the scientific method by trans people and our doctors is the reason that access to medical transition should be limited.[7] Suffice it to say that appeals to "biological sex" and "science" do a lot to stand in for "trans people are gross and we wish they would all die."

i   The following examples come from the first draft of this epilogue because I simply could not bring myself to update them repeatedly over several drafts.

As a result, pro-trans rhetoric often turns to science to push back. There is plenty one could say—plenty that people *have* said—about what contemporary scientists think about sex. Publications from *Nature* to *Teen Vogue* have, in response to anti-trans legislation, laid out the ways that science has found sex to not be binary at all and to be far more complicated than anyone arguing sex that can be easily defined by the state would have it.[8] Social media outrage deriding both trans-exclusionary radical feminist (TERF) and legislative rhetoric as rooted in middle-school-level biology appears just about every time some new story drops involving trans-exclusive claims about biological sex. The scientific retort has become so commonplace that there are now critiques of the tactic—including one that I wrote in 2018 after a leaked Trump administration Health and Human Services memo revealed that the federal government wanted to establish a definition of sex "on a biological basis that is clear, grounded in science, objective and administrable."[9]

The problem, though, is not that state representatives or whoever else have gotten their scientific facts wrong. To frame anti-trans positions as a question of correct or incorrect, or real or fake science is to miss what this disaster is truly about: enabling the state to eradicate some kinds of sex and gender deviance marked as trans, while allowing other kinds to exist unmarked under the safe insistence that sex is always fully knowable. Renderings of stable, biological, binary sex don't make sense under close scrutiny, whether in the context of access to transition-related healthcare, bathroom use, sports participation, or anything else that Republicans/internet trolls/children's authors get mad about. But like nineteenth- and twentieth-century scientists who could ignore the contradictions in their work in favor of using nature and science in the abstract as sources of authority to produce and maintain social hierarchies, those who seek to curtail trans existence today see their knowledge claims as legitimate, no matter how much effort goes into ignoring why those claims make no sense. Sex remains infinitely mobile, its meaning far more related to the needs of the classifier than the classified.

Science that shows that sex really isn't binary is not going to save us, just as science that showed that sex was not binary did not undercut eugenics or lead to widespread queer liberation in the twentieth century. On the contrary, it aided in the development of violent racial theories and policies and solidified the grip of science on sex at the expense of other possibilities. That current efforts to annihilate transness are couched in references to biology and science should come as no surprise.

Sex science, throughout much of the nineteenth and twentieth centuries, was a close ally to those invested in supporting white-supremacist governance, ejecting women from political life, and otherwise making life harder for many of the people still most subject to state violence today. In many ways, despite appeals to what science "really" says about sex, science continues to be an anti-trans ally. Bills and lawsuits are often supported by scientists and medical doctors who have aligned themselves against transness; scientists themselves still have to actively work against a pervasive assumption that sex is a simple research variable and queerness shouldn't evolutionarily have happened.[10] Queer and trans scientists are still fighting for inclusion in their own fields and for more just science education.[11] Though understandable and well intentioned, the center-left turn toward biological truth positions sex science as a benevolent solver of problems and renders invisible its history as a source of violence.[12]

It also concedes knowledge about sex to science, instead of demanding that other ways of knowing be taken seriously. In large part, this aligns with the split between sex and gender over the past several decades within left-ish political discourse. Even though gender as a concept is, itself, a product of scientific and clinical invention, it has come to occupy a space as the "social" or "cultural" opposite of physical and/or biological sex.[13] Consider, for example, the Genderbread Person, a staple of Trans 101s in the early 2010s that I continue to see circulate as a primer on different axes of sexual and gendered identity.[14] The friendly diagram seeks to disentangle what it refers to as "anatomical sex" from "gender identity," "gender expression," and "attraction." Anatomical sex refers to "genitals, chromosomes, hormones, body hair, and more," legitimizing the idea that while trans people might have a gender distinct from their body, their body's physical reality is undeniable. Structurally, sex and gender have also been partitioned into hard science and squishy stuff. Within the walls of academic institutions, "sex" is the domain of biologists; things like "gender and sexuality studies" get shunted over to the humanities and social sciences. This book is, I predict, more likely to be read as being about how scientists *talk* about sex rather than *about* sex. My colleagues and I can write about sex from a place of critique, but opportunities to be taken seriously as experts on sex itself are few and far between.[ii]

ii   There are, of course, always exceptions, and there are two substantial scientific happenings I've recently been part of that do indicate a population of scientists and clinicians excited to learn about sex from humanists and social

There is a circularity here: Sex is the domain of science because science is what we use to understand physical, biological realities; sex is a physical, biological reality; QED, sex is the domain of science. But as I've argued throughout this book, sex was constructed as the domain of science because that was useful for the development of scientific expertise and social control, rather than anything specific to sex. If we take seriously what scientists have said over the past two centuries about sex, it begins to appear that there is no there there. Sex cannot be better understood with better science—and transness cannot be defended with better science—because sex is not a coherent object subject to biased misunderstandings. The multiplicity and incoherence *is* the object. There is no singular *sex* to understand. It has the power to structure our lives because of its incoherence and thus can survive the pointing out of that incoherence. Another tactic is required.

I find myself at a bit of an impasse. On one hand, I want to argue that we (by "we," I mean anyone invested in trans life) need a different way of knowing sex, to wrest it away from science and see what other questions we might be able to ask using different methods. On the other, I'm not certain that sex exists outside of science at all. If sex is just an appeal to the natural, maybe it vanishes in a puff of smoke when we let go of nature as the thing that legitimizes trans existence. I began this book with an argument that trans people don't need to have existed forever in order to exist in the present. The same goes, I think, here: Sex does not naturally have to be more than binary in order to insist on the importance of

---

scientists. One is the 2024 focus issue on sex and gender in *Cell* that centered on trans-inclusive research, interdisciplinary collaboration, and the voices of queer and trans scientists; the other was a multiday National Institutes of Health/National Human Genome Research Institute symposium, again interdisciplinary, on the complexity of sex and the impossibility of resolving it into one binary thing (so much for that being an event that could happen with federal funding). Thank you to Isabel Goldman and Liz Dietz and their respective teams for organizing these events. I will note, though, that I got a bit of hate mail in response to my *Cell* contribution, and a handful of evolutionary biologists trolled the chats of the symposium until a preemptive participant agreement was enforced and they were ejected from the webinar. Which is to say, a lot of work remains to be done among scientists who continue to cling to binary sex defined, in theory, by gamete size. As I noted in my panel's Q&A, I do suspect that none of them go around testing people's gametes before deciding what pronouns to use. Consider also the GenderSci Lab at Harvard, which frames itself as a specifically interdisciplinary laboratory for "scientific research on sex and gender."

trans life. Rather than appeal to the cultural authority of science and nature to make arguments for trans legitimacy, we might instead refuse demands for scientific evidence that we should get to exist—demands that will never be satisfied anyway, because it was never about the evidence. Evidence is a thing that sex seems immune to.

These two positions can coexist. We can demand to know more about sex or, more usefully, about bodies in queer and trans ways that resist ossified categories, that build in questions about why we want to know what we want to know and what the price of knowing might be, and that improve trans (and also cis) healthcare. We can insist that sex is not only the purview of the biological sciences and resist the urge to limit nonscientific inquiries to critiques about gendered influence over how we understand some kind of real thing. At the same time, we can deny the primacy of nature as justification to exist. We do not need to apologize for being here. Rather than fight over definitions of sex, I would prefer to think about how to make bodies feel good to inhabit in new and exciting ways, how to house and feed everyone, how to get beyond the provision of basic life necessities and into the time and space to imagine a world of being trans that doesn't revolve around putting out fires. These approaches—demanding that there are more ways to know sex than science, and denying the utility of sex as a category at all—conflict with each other, but I suspect that may be a source of strength. After all, incoherence is a powerful thing.

I hope this is one of those conclusions that will feel immediately dated. If, next year, in five years, ten years from now, anyone reading this has a chuckle at my expense for making the move of concluding with a current political situation that emphasizes the broad relevance of my argument, then I applaud you for making it beyond our current tribulations to a world where anti-trans legislation seems like a quaint relic from the early to mid-2020s. I, for one, eagerly await a government takeover by the communist nonbinary ants.[15]

Postscript

Hi there, it's me again.[i] Just as this book was heading into production, Donald Trump spent his second first day as president—January 20, 2025—signing executive orders as though each time he scrawled his name he'd get a new golf course. One of these orders was titled "Defending Women from Gender Ideology Extremism and Restoring Biological Truth to the Federal Government."[ii] This snappily monikered text reads, and I quote, "Beans was right."

Ah, one must laugh to keep from screaming.

What it actually says is that the United States federal government will henceforth recognize only "two sexes, male and female," which "are not changeable and are grounded in fundamental and incontrovertible reality." Following a short glossary of its terms, it details a litany of policy ramifications: fortifying "single-sex" spaces, removing "gender identity" from federal paperwork, banning incarcerated people from accessing transition care, blah blah blah, ad nauseam. I won't rehash the whole thing—by the

---

i   Thank you to Salonee Bhaman and Monique Flores Ulysses for last-minute draft help.
ii   Exec. Order No. 14168, 90 Fed. Reg. 8615 (January 30, 2025). All quotes here are from "Defending Women from Gender Ideology Extremism and Restoring Biological Truth to the Federal Government," The White House, updated January 20, 2025, https://www.whitehouse.gov/presidential-actions/2025/01/defending-women-from-gender-ideology-extremism-and-restoring-biological-truth-to-the-federal-government/.

time this book is in your hands or on your screen we'll have been living for an eternity with the consequences of this and Trump's other "glorified press releases designed to create confusion and chaos," as Chase Strangio so aptly called them.[iii] At base, the order is a bullshit appeal to science of the sort I've just described in this book's epilogue, and as such, I'm more interested in its deployment of incoherence than its claims about biology. The important thing to know is that DWGIERBTFG[iv] exemplifies the power of incoherence in producing, maintaining, and enforcing the categories "sex," "female," and "male" as it both establishes a definition of sex so impractical it invites multiple enactments and turns accusations of incoherence against the villainous "gender ideology." In so doing, it provides a text that I can gesticulate wildly at to illustrate why efforts to counter anti-trans violence need to take the ongoing weaponization of incoherence seriously and ensure that our work doesn't stop at fighting over what sex means.

After informing the nation that sex is "an individual's immutable biological classification as either male or female," the order asserts that the defining criterion of that classification is gamete size. "'Female' means a person belonging, at conception, to the sex that produces the large reproductive cell," it avers, while "'male' means a person belonging, at conception, to the sex that produces the small reproductive cell." The fallacy of this construction becomes clear when one considers for the merest moment how fetal development works, or the circularity of sex being male or female and then male and female being which sex you are. Its incoherence, though, isn't limited to word choice.

The ontological claim of sex determined by gamete size, putting aside all of its other problems, concretizes the ability to, in practice, have sex be whatever an empowered party needs it to be in a given moment. Sex, according to the Trump administration, is the current or anticipated future production of eggs or sperm—the large and small "reproductive cell," respectively. Even if this were readily testable in utero or at birth (let alone at conception), though, it's not realistically something that would end up being documented on a broad scale as evidence of individual "biological sex." The presence of XX, XY, or indeed any combination of related

---

iii   Chase Strangio (@chasestrangio), "In advance of inauguration," Instagram, January 19, 2025, https://www.instagram.com/p/DFA-A-DRaQ3/.
iv   In the immortal words of Dar Williams, "and besides, a name like that doesn't make a good acronym." "The Pointless, Yet Poignant, Crisis of a Co-ed," track 8 on *Mortal City*, Razor & Tie, 1996.

chromosomes *is* possible to ascertain at a fetal or neonatal stage, but people customarily choose gender-reveal artillery fire colors based on the results of ultrasounds and the appearance of a fetus' proto-genitals, not expensive and invasive chromosomal tests. Doctors aren't routinely karyotyping infants before declaring "it's a girl!" and sticking an F on the birth certificate. For that matter, bathroom patrons who apparently feel threatened by the trans woman washing her hands three sinks away aren't demanding to measure the size of whatever, if any, gametes she's producing. Sex, now federally mandated as binary, remains a series of proxies stacked on top of each other without even the decency to wear a trench coat; Trump's decree is the predictable legacy of the history documented in this book.

Meanwhile, the order explicitly constructs gender as the absurd foil to reasonable, commonsense "biological truth," deflecting its own incoherence onto transness with a stunning redirection of categorical pandemonium. Gender identity, says the order, is an "ever-shifting concept" that is "self-assessed" and "subjective" and based on the idea of a "vast spectrum" and "infinite continuum" of possibilities "disconnected from biological reality." Gender ideology is "internally inconsistent" and "an inchoate social concept" that can and should be dislodged from American society with the power of straightforward, objective science. In light of what I've argued throughout this book, these phrases barely need analyzing: Sex is objective and scientifically proven to exist, and anyone who thinks otherwise is lost in some namby-pamby dreamland. It is another iteration of incoherence being created so that some other idea can be positioned as comparatively rational, and therefore superior. This transposition also reflects how Trump and his surrogates have countered all manner of accusations by throwing whatever Trump is charged with back at his accusers. As with tax fraud, it seems, so with sex.[v]

This is all to say that regardless of the order's fantasy and decree of truthon gamete size, sex assignments and assumptions will continue to be based on whatever best supports the federal government's mission of trans exclusion: genital morphology, hormone levels, facial and bodily appearance, voice, whatever. Ultimately, the specifics of the order's definitions are less important than the morass they maintain under the guise of certainty. The gamete thing is just an appeal to science in the abstract that

v   See, e.g., Sidney Blumenthal, "Deny, Attack, Reverse—Trump Has Perfected the Art of Inverted Victimhood," *Guardian*, February 1, 2024, among many other news reports describing the same phenomenon.

justifies making life shittier for trans people, as well as for intersex people and anyone else who doesn't fit white cis- and heteronormative gender standards. That definitions of "male" and "female" based on gamete size are unenforceable invigorates a classification system that has never been beholden to internal consistency, only to the power to classify. The wielding of sex as a cudgel to execute binary hegemony becomes easier as the incoherence of the order's stated definitions creates a void that can be filled by multiple enactments of sex.

There's no good way to end this postscript, which I've written in between efforts to help loved ones navigate the throwback bureaucratic regime and cope with the fear that this order has unleashed.[vi] I'll sign off

vi   Applications for gender marker changes on passports were among the first things to go sideways in the aftermath of this executive order. Within a week of the decree of sex as binary and immutable, Secretary of State Marco Rubio issued a suspension of applications for X gender markers, which, because there was initially no guidance from the State Department about how to process the applications already in a passport office's possession (and thus what to do with the existing passports that many applicants had sent in), resulted in applicants' passports being held indefinitely until guidance to renew those passports according to sex assigned at birth was issued on February 10. I witnessed firsthand just such an instance of a passport held in bureaucratic limbo during this period; eventually, a congressperson had to get involved to rescind the X marker application. This ultimately resulted in the issuance of an unnecessarily renewed passport with a binary gender marker, accompanied in the mail by someone else's (based on the photo, probably another person stuck in the same situation) now-voided old passport instead of the correct one. Just one example of the time-, money-, and energy-wasting indignity and chaos of being trans under the second Trump administration. Allegedly, existing passports that reflect a past gender marker change and passports with X gender markers will continue to be valid until their printed expiration date, at which point they'll have to be renewed to reflect one's sex assigned at birth, but this hasn't prevented the spread of intense anxiety and rumors on social media that trans people will not be able to reenter the country if they leave it or will have passports taken from them at the border. Jaclyn Diaz, "Trump's Passport Policy Leaves Trans, Intersex Americans in the Lurch," NPR, February 21, 2025; "Sex Marker in Passports," US Department of State—Bureau of Consular Affairs, last updated May 16, 2025, https://travel.state.gov/content/travel/en/passports/passport-help/sex-marker .html. For an example of the fear that the new passport policies have generated, see @Loose-Kiwi-7876, "Majority of Trans Americans' Passports Now One-Way Tickets," thread on r/MTF, January 21, 2025, https://www.reddit.com/r/MtF /comments/1i6fxvi/majority_of_trans_americans_passports_now_oneway/.

with a revision of how I closed the conclusion: We can't wait for anyone to save us, so be the communist nonbinary ant you want to see in the world.

January 24, 2025[vii]

---

vii   With a couple of tweaks in May.

Notes

Archival Collections

AFB: Albert Francis Blakeslee Papers
AKC: Alfred Kinsey Correspondence
CDP: Charles Davenport Papers
EROR: Eugenics Record Office Records
FWF: Field Worker Files
HBP: Harry Benjamin Papers
JWT: John W. Tukey Papers
LLC: Lawrence Collection
ORP: Oscar Riddle Papers
RMY: Robert Mearns Yerkes Papers

Introduction: A Trans History of Classification

1 Benjamin, *Transsexual Phenomenon*, 3.
2 See Kahan, *Book of Minor Perverts*, for a complementary argument about "multiple and conflicting . . . explanations of sexuality [that] came to exist simultaneously," 4. Our shared word choice was entirely incidental but validating when I first read *Minor Perverts*!
3 On the late consolidation of reproductive science compared to other life sciences, see Clarke, *Disciplining Reproduction*.
4 Foucault, *History of Sexuality*.

5 On the shifting developments of these categories as precursors to the contemporary "transgender," see Meyerowitz, *How Sex Changed*; Stryker, *Transgender History*; and Valentine, *Imagining Transgender*.

6 À la Foucault, *History of Sexuality*. On the invention of homosexuality, see also Halperin, "How to Do the History of Male Homosexuality"; Terry, *American Obsession*; D'Emilio, *Sexual Politics*; and Chauncey, *Gay New York*, among many others. While not all of these scholars agree with Foucault wholesale— D'Emilio and Chauncey put more stock in communities' understandings of themselves than in medical discourse—they all take as axiomatic the idea that sexual categories born of the late nineteenth century fundamentally differ from earlier understandings of sexual behavior. This raises the question of whether homosexuality and transness can be seen as historically comparable categories. On one hand, it seems suspect that they should be conflated and thus require the same methodology. On the other, the *splitting* of transness and homosexuality into separate categories of "sexual orientation" and "gender identity" masks their shared origins and the ways in which trans and other queer people who violate gender norms have historically been thrown under the bus to make gender-normative queer people seem more respectable. See Stryker, *Transgender History*, 151–52, as well as Stryker, "Transgender History, Homonormativity, and Disciplinarity," and Valentine, *Imagining Transgender*.

7 See Gordon's critique of the use of *transvestism* by scholars writing about periods before the twentieth century in *Glorious Bodies*. Stryker, *Transgender History*, 16, 18, 123.

8 Meyerowitz, *How Sex Changed*, 5; Boag, *Re-Dressing*, 52; Sears, *Arresting Dress*, 9.

9 Skidmore, *True Sex*, 10.

10 Manion, *Female Husbands*, 10. Stryker's definition comes from *Transgender History*, 1.

11 For cross-dressing, see Boag, *Re-Dressing*, and Sears, *Arresting Dress*. For female husbands, see Manion, *Female Husbands*. For passing or masquerading, see Manion, "Queer History"; Skidmore, *True Sex*; and LaFleur, "Precipitous Sensations." On physical changes, see Meyerowitz, *How Sex Changed*, and, more speculatively, LaFleur, "Trans Feminine Histories."

12 See, e.g., the story of Joseph Lobdell's encounters with state institutions in Manion, "Queer History," or the mysteriously named Mrs. Nash's outing when a friend changed the deceased Mrs. Nash's clothing for her burial in Boag, *Re-Dressing*, 130–38. On the ways that a heterosexual/homosexual binary relies on stable sex and gender categories, see Stryker, "Transgender History, Homonormativity, and Disciplinarity." See also Scott Larson's absolutely crucial piece on the ethical dimensions of working with these sources of violent outing to determine if historical figures were "really" trans, in "Laid Open," in *Trans Historical*.

13 Bowker and Star, *Sorting Things Out*, 5.

14 Stryker, *Transgender History*, 7. This is, however, Stryker writing a Trans 101; her other work questions the idea that transness is the unnatural move away from natural cisness. Nonetheless, this turn of phrase captures the minoritizing tendency of contemporary mainstream understandings of transness.

15 Stryker, *Transgender History*, 7.

16 This book is what Bowker and Star call an "infrastructural inversion": an analysis that "look[s] closely at technologies and arrangements that, by design and by habit, tend to fade into the woodwork." Bowker and Star, *Sorting Things Out*, 34.

17 Enke, "Education of Little Cis," 60.

18 Sedgwick, *Epistemology of the Closet*, 13.

19 For examples of this work, see the contents of Chess, Gordon, and Fisher, "Early Modern Trans Studies." Other recent examples of this expansive framing include many of the essays in LaFleur, Raskolnikov, and Kłosowska, *Trans Historical*, and Gordon, *Glorious Bodies*. I suspect the pre- and early modernists with literary inclinations are on to something!

20 E.g., Heaney, *New Woman*; Larson, "'Indescribable Being.'"

21 On race in trans studies, see Schuller, *Biopolitics of Feeling*; Snorton, *Black on Both Sides*; and Bey, *Cistem Failure*. This work follows a powerful scholarly repertoire, itself indebted to Black feminist thought; see, e.g., Spillers, "Mama's Baby"; Gilman, *Difference and Pathology*; Somerville, "Scientific Racism"; and Rosen, *Terror in the Heart of Freedom*. On species, see Luciano and Chen, "Has the Queer Ever Been Human?"; Hayward and Weinstein, "Tranimalities"; and Amin, "Trans* Plasticity." On age, see Stockton, *Queer Child*, and Gill-Peterson, *Histories of the Transgender Child*. See also Larson, "'Indescribable Being,'" and the essays in Chess, Gordon, and Fisher, "Early Modern Trans Studies," which stick closer to recognizably trans/trans-adjacent figures but emphasize that the question of whether they're trans or not is far less important than widening the range of texts to which we might apply trans analytics.

22 Stone, "*Empire* Strikes Back," 227.

23 Stryker, "My Words to Victor Frankenstein," 240–41.

24 E.g., the work of Marquis Bey, Emma Heaney, Finn Enke, C. Riley Snorton, and many others cited in this book.

25 Najmabadi, "Beyond the Americas," 18.

26 Stryker, "Transgender History, Homonormativity, and Disciplinarity," 153.

27 See Velocci, "Wrenching Torque," for an account of my own navigation of disciplinarity.

28 Haraway, "Situated Knowledges." See also Moore, Cowles, and Ramalingam, "Dilemmas of Archival Objectivity."

29 See, e.g., Schiebinger, "Skeletons in the Closet"; Traweek, *Beamtimes and Lifetimes*; and Oreskes, "Objectivity or Heroism?" for several discussions of how

investments in masculinity influence knowledge production practices and outcomes.

30 See Vertesi, "Seeing Like a Rover"; Myers, "Molecular Embodiments"; and Prentice, *Bodies in Formation*.

31 See Daston and Galison, *Objectivity*; Gieryn, "Boundary-Work"; and Daston, "Objectivity."

32 See Stone, "*Empire* Strikes Back"; Velocci, "Standards of Care"; Gill-Peterson, *Histories of the Transgender Child*; Meyerowitz, *How Sex Changed*; and Latham, "Making and Treating Trans Problems."

33 J. R. Latham's work is a stellar example of what bridging this gap portends.

34 E.g., Bulmer, "Why Is the Cassowary Not a Bird?"; Hacking, "Making Up People"; Ritvo, *Platypus and the Mermaid*; Winsor, *Reading the Shape of Nature*; Bowker and Star, *Sorting Things Out*; Mol, *Body Multiple*; TallBear, *Native American DNA*; McOuat, "From Cutting Nature at Its Joints"; Burnett, *Trying Leviathan*; and Robertson, "Granular Certainty."

35 See, among others, Foucault, *History of Sexuality*; Halperin, "How to Do the History of Male Homosexuality"; D'Emilio, *Sexual Politics*; D'Emilio, "Capitalism and Gay Identity"; Terry, *American Obsession*; Duggan, *Sapphic Slashers*; Chauncey, "From Sexual Inversion to Homosexuality"; Chauncey, *Gay New York*; Boag, *Same-Sex Affairs*; Kunzel, *Criminal Intimacy*; and Canaday, *Straight State*. Scholars have also shown that some people refused to be labeled with what they saw as pathologizing categories, e.g., Freedman, "'Burning of Letters,'" and Carter, "On Mother-Love."

36 For examples of this approach, see Somerville, "Scientific Racism," and Terry, *American Obsession*, which have as their source base touchstones of the sexology canon like works by Havelock Ellis and Richard von Krafft-Ebing. This is by no means a criticism of analyses of those texts, which were crucial to the formation of sexual categories at the turn of the century. Rather, attention to how science and classification themselves work offers a way to build on those foundations. On *networks*, which I use throughout as a shorthand for the relations between the many kinds of actors (human, animal, and institutional, as well as scientist, administrator, funder, and research subject—the list goes on) upon whose collaborative efforts science depends, see Callon, "Some Elements," and Latour, *Science in Action*.

37 Somerville, "Scientific Racism," 246. On queer reading against the grain, see Wrathall, "Provenance as Text"; Potter, "Queer Hoover"; and Freedman, "'Burning of Letters.'"

38 I look for failure because, as Leigh Star has pointed out, "the normally invisible quality of working infrastructure becomes visible when it breaks"—sex, in this case, being the infrastructure. Star, "Ethnography of Infrastructure," 382. My turn to thinking on failure is particularly indebted to Campos, *Radium and the Secret of Life*.

39 There is a substantial amount of work on hermaphroditism and intersex, some of which has been written by historians of science and STS scholars like Lorraine Daston, Katherine Park, and Katrina Karkazis. Historiographically, though, that work has come to be cited and imagined as *about intersex*, rather than *about sex*. History of sexology and sexual deviance, on one hand, and history of sex science, on the other, constitute largely separate historiographies. See Park and Daston, *Wonders and the Order of Nature*, and Karkazis, *Fixing Sex*.

40 See Schiebinger, *Nature's Body*; Russett, *Sexual Science*; Richardson, *Sex Itself*; and Moscucci, *Science of Woman*.

41 Laqueur, *Making Sex*; Schuller, *Biopolitics of Feeling*.

42 See Karkazis, "Misuses of 'Biological Sex.'" For contemporary accounts of this, see, e.g., McLaughlin et al., "Multivariate Models of Animal Sex"; Garcia-Sifuentes and Maney, "Reporting and Misreporting of Sex Differences"; DuBois and Shattuck-Heidorn, "Challenging the Binary"; Patsopoulos, Tatsioni, and Ioannidis, "Claims of Sex Differences"; Zemenick et al., "Six Principles"; and Ah-King and Ahnesjö, "'Sex Role' Concept."

43 Schiebinger, "Skeletons in the Closet," 46.

44 Lillie, "General Biological Introduction," 3.

45 Mak, *Doubting Sex*, 2.

46 See, for just one example, Whooley, *On the Heels of Ignorance*. I hear physics has a whole principle about it.

47 See Shapin and Schaffer, *Leviathan*; Gilbert and Mulkay, *Opening Pandora's Box*, chap. 4; and Star and Griesemer, "Institutional Ecology."

48 Murphy, *Sick Building Syndrome*, 7. On Indigenous approaches to these questions, see Todd, "Indigenous Feminist's Take."

49 Shapin and Schaffer, *Leviathan*; Latour, *Science in Action*.

50 Barad, *Meeting the Universe Halfway*; Murphy, *Sick Building Syndrome*; Mol, *Body Multiple*; Livingston, *Debility and the Moral Imagination*; and Bowker and Star, *Sorting Things Out*.

51 Mol, *Body Multiple*, 35.

52 Mol, *Body Multiple*, 5.

53 Mol's "Who Knows What a Woman Is" gets delightfully (or, when I first read it, unnervingly) close to making this argument! That piece focuses on competing enactments of sex in different branches of science, however, while I'm more concerned with the enactments of sex themselves.

54 Mol, *Body Multiple*, 178–81.

55 Mol, *Body Multiple*, 178.

56 Star and Griesemer, "Institutional Ecology," 393.

57 Star and Griesemer, "Institutional Ecology," 391.

58 Star, "This Is Not a Boundary Object," 615.

59 On knowledge not being made, see Proctor and Schiebinger, *Agnotology*; Schiebinger, *Plants and Empire*; and Murphy, *Sick Building Syndrome*. See also

Sedgwick's remarks on "ignorance effects" in *Epistemology of the Closet* (4–8): Not knowing and incitements not to know dictate the range of discursive possibility just as much as the creation of knowledge.

60 Gill-Peterson, *Histories of the Transgender Child*, 97.

61 Gay genes and trans brain scans are quintessential examples. See Clare, Grzanka, and Wuest, "Gay Genes in the Postgenomic Era."

62 On politics of refusal, see Simpson, *Mohawk Interruptus*. On state demands for gender legibility, see Beauchamp, *Going Stealth*, and Currah and Moore, "We Won't Know Who You Are."

63 Bowker and Star, *Sorting Things Out*.

64 For an overview, see Love, "Queer."

65 Murphy, *Sick Building Syndrome*; Velocci, "These Uncertain Times."

66 See Bouk, *How Our Days Became Numbered*; Aronowitz, *Risky Medicine*; Adams, Murphy, and Clarke, "Anticipation"; and Radin, "Alternative Facts."

67 Schuller, *Biopolitics of Feeling*; Gill-Peterson, *Histories of the Transgender Child*; Schuller and Gill-Peterson, "Biopolitics of Plasticity," especially Amin, "Trans* Plasticity."

68 See the conclusion.

69 See, e.g., Richardson, *Sex Itself*, and Karkazis and Jordan-Young, *Testosterone*.

70 For a deeper history of the institutions that collectively established American sex science, see Clarke, *Disciplining Reproduction*.

71 Canaday, *Straight State*; see also Currah, *Sex Is as Sex Does*, for a beautiful takedown of the concept of "legal sex."

72 Chauncey, *Gay New York*.

73 Bouk, *How Our Days Became Numbered*, 40n29, 227.

74 For discussions on the study of sex in the European tradition from antiquity to the early modern, see Laqueur, *Making Sex*, esp. chap. 2; Park, "Myth of the 'One-Sex Body'"; Cadden, *Meanings of Sex Difference in the Middle Ages*; and Schiebinger, *Nature's Body*.

75 See, e.g., Beccalossi, *Female Sexual Inversion*; Bauer, *English Literary Sexology*; Kahan, *Book of Minor Perverts*; and Sutton, *Sex Between Body and Mind*.

76 Johnson, *Just Queer Folks*, 30.

77 At this point in time, agricultural experiment stations were among the most important and well-funded spaces of American sex research. These institutions are beyond the scope of this book; for more, see G. Rosenberg, *4-H Harvest*; G. Rosenberg, "No Scrubs"; and Johnson, *Just Queer Folks*, chap. 1.

78 On American leadership in an international eugenics movement, see Klautke, "'Germans Are Beating Us,'" and Allen, "Misuse of Biological Hierarchies."

79 Kline, *Building a Better Race*, among others.

80 *Oxford English Dictionary* (OED), s.v. "binary" (n. & adj.), June 2024, https://doi.org/10.1093/OED/3895868093.

81 See *An Anglo-Saxon Dictionary Online*, s.v. 'Twi-,' accessed August 17, 2024, https://bosworthtoller.com/31232.

82 Capgrave, "Dedication to Edward IV," 91.

83 This and previous examples from OED, s.v. "binary."

Chapter 1. Constructing Sexual Multiplicity in Animal Research

1 Ritvo, *Platypus and the Mermaid*, xii. On the racialization of the human, see Z. Jackson, *Becoming Human*, and Kim, *Dangerous Crossings*.

2 See Bagemihl, *Biological Exuberance*, and Roughgarden, *Evolution's Rainbow*, for a far more thorough accounting of these exceptions than I can provide here.

3 See, respectively, Terry, *American Obsession*; Schuller, *Biopolitics of Feeling*; G. Rosenberg, *4-H Harvest*; and Willey, *Undoing Monogamy*.

4 See Oudshoorn, *Beyond the Natural Body*, and Meyerowitz, *How Sex Changed*.

5 For more on the impact of Kinsey's entomological training on his human sex research, see Drucker, *Classification of Sex*.

6 For the political usages of claims of natural sex difference, see Russett, *Sexual Science*; Bederman, *Manliness and Civilization*; Schiebinger, *Nature's Body*; Rosen, *Terror in the Heart of Freedom*; and Laqueur, *Making Sex*.

7 For an ur-text, see Haraway, *Primate Visions*, as well as Terry, "'Unnatural Acts.'" More recently, scholars have looked to the human/animal divide itself as a site for the regulation of sexuality. See, e.g., Giffney and Hird, *Queering the Non/Human*; Mortimer-Sandilands and Erickson, *Queer Ecologies*; Hayward and Weinstein, "Tranimalities"; Luciano and Chen, "Queer Inhumanisms"; Chen, *Animacies*; and Amin, "Trans* Plasticity."

8 See Foucault, *Herculine Barbin*; Dreger, *Hermaphrodites*; Richardson, *Sex Itself*; Schiebinger, "Skeletons in the Closet"; Sears, *Arresting Dress*; and Manion, "Queer History," among others. The main exception to this is *How Sex Changed*, which suggests that sex became more malleable during the twentieth century. Regina Kunzel, in *Criminal Intimacy*, also highlights that the consolidation of sexual categories where sexuality is concerned, too, was fragmentary and uneven.

9 Alice Dreger wrote about the "Age of the Gonad." See Ha, "Riddle of Sex," for an overview of some of these models.

10 Laqueur, *Making Sex*, 6. For two retorts, see Park and Nye, "Destiny Is Anatomy," and Park, "Myth of the One-Sex Body."

11 Richardson, *Sex Itself*, 36.

12 Many of the historical sources hyphenate the word as *free-martin*.

13 Clarke, *Disciplining Reproduction*.

14 Russett, *Sexual Science*; Schiebinger, *Nature's Body*.

15 Countless instances exist, with many texts quoting each other (almost word for word) on the "absurd" belief of the ancients that hyenas were hermaphroditic or changed sex. See, for just a few examples, Bingley, *Animal Biography*, 225; Calmet, *Taylor's Edition*; Society for the Diffusion of Useful Knowledge, *Penny Cyclopaedia*, 368; Lieber, *Encyclopaedia Americana*, 500; Smith, *Natural History of Dogs*, 272; Buffon, *Natural History*; 368; Reid, *Bush-Boys*, 179; and Littman, *Publications*, 32. This framing of hyena sex was not new to the nineteenth century—John Johnston reported in the seventeenth century that "the Hyaena of old is said (as by Ovid and Pliny) yearly to change sex, and to gender with a male; which though it be false, and disavowed. By Aristotles yet there is a vessel in the Hyaenam, that makes the heedles [*sic*] think it Epicene or double-sexed" (*Description of the Nature of Four-footed Beasts*, 114). Anna Wilson suggests medieval bestiaries like the *Physiologus* might be the post-antiquity origins of the recurring set of myths—that hyenas can imitate human voices, that they lure humans out of their homes at night, and that its laughter sounds like a human vomiting—that often accompany the assertions of sex change ("Sexing the Hyena," 760). See also DeVun, *The Shape of Sex*, especially chapter 3, for a discussion of hyenas' supposedly "unclean" nonbinary sexual morphology as a referent for medieval Jewish deviance. Because many of the nineteenth-century texts do not cite their sources, it is unclear exactly how the "absurd story of the ancients" construction made its way into those texts, though Wilson and DeVun make compelling arguments for the role of the European medieval corpus in their transmission. What is clear is that these narratives were not new in the nineteenth century; the novelty appears rooted in their proliferation through multiple kinds of print media.

16 See Funk, "R. J. Gordon's Discovery," for a critique of the view that Watson was the first to discover the so-called peculiarities of female hyena anatomy as well as a fascinating reproduction of old tropes in a contemporary article.

17 Watson, "On the Female," 369.

18 Watson, "On the Female," 369.

19 Watson, "On the Female," 370.

20 Watson, "On the Male," 425.

21 Watson, "On the Female," 369.

22 Watson, "On the Male," 425.

23 Watson, "On the Male," 425.

24 See Burnett, *Trying Leviathan*.

25 See citations in J. Simpson, "On the Alleged Infecundity," 4.

26 Hunter, "Account of the Free Martin," 291.

27 J. Simpson, "Hermaphroditism," 701.

28 Vasey, *Delineation of the Ox Tribe*, 156.

29 Galton, "Short Notes," 328. Neither Vasey nor Galton defines precisely which male or female characteristics the freemartin combines.

30 Savory, "Description of the Organs," 64.
31 Savory, "Description of the Organs," 66–67.
32 Savory, "Description of the Organs," 66.
33 Savory, "Description of the Organs," 66.
34 Hart, "Structure of the Reproductive Organs," 230.
35 Hart, "Structure of the Reproductive Organs," 234.
36 Hart, "Structure of the Reproductive Organs," 234, 238.
37 Hart, "Structure of the Reproductive Organs," 234.
38 Lillie, "Theory of the Free-Martin," 611.
39 On universal bisexuality, see Meyerowitz, *How Sex Changed*, chap. 1; for an overview of sexual inversion, see Clement and Velocci, "Modern Sexuality," among many others.
40 Lillie, "Theory of the Free-Martin," 612.
41 Lillie, "Theory of the Free-Martin," 612.
42 Lillie, "Theory of the Free-Martin," 612–13.
43 Lillie, "Theory of the Free-Martin," 612.
44 G. Rosenberg, *4-H Harvest*; Johnson, *Just Queer Folks*, esp. chap. 1. On the role of agriculture in American reproductive sciences more broadly, see also Clarke, *Disciplining Reproduction*, especially chapters 2 and 3—as noted earlier, Clarke locates the development of research on the "problems of sex" in the combined forces of biology, medicine, and agriculture.
45 Clarke, *Disciplining Reproduction*, 41–42.
46 G. Rosenberg, *4-H Harvest*, 143.
47 See, e.g., Vasey, *Delineation of the Ox Tribe*.
48 Hunter, "Account of the Free Martin," 280.
49 Goubaux and Barrier, *Exterior of the Horse*, 814–15.
50 Debraw, "Discoveries on the Sex of Bees," 19; Huber, *Natural History of Ants*, 372; F. Butler, "Farmer's Manual"; and Phillips, "Review of Parthenogenesis."
51 The following describe workers as neuter: Kidder, *Kidder's Guide to Apiarian Science*, 32; "Lecture on Bees," 179; "Instinct in Bees," 7; Cook, *Manual of the Apiary*; Case, "Voices from Among the Hives," 193; Munn and Wildman, *Description of the Bar-and-Frame Hive*, 49, 51; Heddon, *Success in Bee-Culture*, 11. The following describe workers as some variation on imperfect or underdeveloped females: "Bee-Culture," 22; Shaw, "Is Parthenogenesis Proven?," 42; "Nomenclature of Bee-Keeping," 67; Hastings, "Bee Culture," 21; Cook, "Peculiarities of Abnormal Bees"; Kebler, "Beeswax and Its Adulterants," 405; Castle, "Sex Determination in Bees and Ants," 390. A few used neuter and underdeveloped female interchangeably: Oliver, "'Ox-Cow' Queen Bees," 258; Abbot, "Scientific Ignorance About Bees," 435. Notably, some overlap of years exists, suggesting that not all were convinced, though interestingly it appears that Cook changed his mind between the 1876 *Manual of the Apiary* and his 1891 article "Peculiarities of Abnormal Bees." Some articles, in this note and throughout the chapter, have no named author.

52 It's worth noting, also, that this was a period in which agriculture was becoming increasingly scientized. See C. Rosenberg, "Rationalization and Reality," 401.

53 Johnson, *Just Queer Folks*, 39.

54 For examples of this process, see Foucault, *History of Sexuality*; Terry, *American Obsession*; and Canaday, *Straight State*, among others.

55 King, "Permanence of Bee-Keeping Industry," 526.

56 King, "Permanence of Bee-Keeping Industry," 526.

57 "Ignorance of the Past Ages," 770.

58 Lubbock, "On the Habits of Ants."

59 "Workers Not Monsters," 28. Later, the same article appeared as "Agricultural News Items: Are Workers Abortions?," 563.

60 Greiner, "He or She," 213; "The Worker Bee—He, She, or It?," 216.

61 Miller, "He or She," 80.

62 Leland, "Anent Ants."

63 Leland, "Anent Ants."

64 "Our Six-Footed Rivals."

65 "Our Six-Footed Rivals."

66 Kim, *Dangerous Crossings*, 18. Mel Chen has expanded a similar line of thinking beyond closeness to the animal or the human to consider how race aligns with animacy or inanimacy—in other words, closeness to life itself—in *Animacies*.

67 Kim, *Dangerous Crossings*, 36. See also Rifkin, *When Did Indians Become Straight?*; Morgensen, "Settler Homonationalism"; and TallBear, "Making Love and Relations." Some of this earlier racial science of sex paid more attention to the impact of the environment on a body; the nineteenth-century scientists I discuss here had largely moved toward a more individuated model in which race and sex were inherent and inherited qualities. On environmental theories of race and sex, see LaFleur, *Natural History of Sexuality*.

68 Starkweather, *Law of Sex*, 51.

69 T. Morgan, *Experimental Zoology*, 368. See Richardson, *Sex Itself*, esp. chap. 3, "How X and Y Became the Sex Chromosomes," for a discussion of Morgan's central role in enshrining sex in the chromosomes.

70 T. Morgan, *Experimental Zoology*, 365.

71 T. Morgan, *Experimental Zoology*, 372–73.

72 For a discussion of the relationship between older hierarchies of nature and nineteenth-century conceptualizations influenced by Lamarck and Darwin, see Bowler, *Evolution*, esp. chaps. 3 and 6.

73 Starkweather, *Law of Sex*, 52–53.

74 T. Morgan, *Experimental Zoology*, 365.

75 For numerous examples of hermaphroditism in medical journals, see Reis, *Bodies in Doubt*.

76 Hunter, "Account of the Free Martin." Francis Galton would eventually lift this phrasing almost word for word for his article "Short Notes on Heredity, &c., in Twins," 327.

77 Home, "On the Propagation of the Species in the Oyster."

78 See Duggan, *Sapphic Slashers*.

79 See Schuller, *Biopolitics of Feeling*, and Somerville, "Scientific Racism."

80 On overcivilization anxieties, see Schuller, *Biopolitics of Feeling*; Boag, *Re-Dressing*; and Bederman, *Manliness and Civilization*.

81 Ellis, *Man and Woman*, esp. chap. 4; Qureshi, *Peoples on Parade*.

82 Harlan, "Description of an Hermaphrodite Orang Outang," 232.

83 Harlan, "Description of an Hermaphrodite Orang Outang," 233.

84 J. Simpson, "Hermaphroditism," 702.

85 Geddes and Thomson, *Evolution of Sex*, 3.

86 Geddes and Thomson, *Evolution of Sex*, 21.

87 Geddes and Thomson, *Evolution of Sex*, 19.

88 Russett in particular uses a "defense" conception (natural difference has to be articulated in order to protect the public sphere from women's intrusion), but that defense framing is often present in works regarding ambiguous sex as a problem needing to be solved in order to preserve social order, as well as histories of transness and cross-dressing that envision more restrictive legislation as an effort to prevent movement between gender categories. Russett, *Sexual Science*. For ambiguous sex, see Reis, *Bodies in Doubt*; and Brown, "'Changed into the Fashion of Man'"; for transness, see Manion, "Queer History"; Sears, *Arresting Dress*; and Boag, *Re-Dressing*.

89 Chauncey, *Gay New York*, esp. chaps. 2 and 3.

90 See, e.g., Martin, "Gender and Sexuality"; Lehring, *Officially Gay*; Clement and Velocci, "Modern Sexuality in Modern Times"; and Heaney, "Introduction." The chapter featuring Loop-the-Loop from *Gay New York* has also been reprinted in Katz, *Gay/Lesbian Almanac*, and Phillips and Reay, *Sexualities in History*.

91 Shufeldt, "Mortuary Customs of the Navajo Indians," 304–5.

92 Biographical detail from Lambrecht, "In Memoriam."

93 Shufeldt, "Mammalogy."

94 Shufeldt, "Distinction Between Anatomy and Comparative Anatomy," 328.

95 Shufeldt, "Observations on the Classification of Birds," 490.

96 For a discussion of nineteenth- and early twentieth-century debates over the definition of species and whether they existed in nature, see McOuat, "From Cutting Nature at Its Joints."

97 Shufeldt, *Chapters on the Natural History*, 31.

98 Shufeldt, *Chapters on the Natural History*, 32.

99 Shufeldt, "Waning of the Interest."

100 Shufeldt, "On the Study of the Question of Sex," 109.

101 Shufeldt, "On the Study of the Question of Sex," 109.

102 Shufeldt, "Need of Parental Enlightenment," 726.

103 Shufeldt, "On the Study of the Question of Sex," 109.

104 Shufeldt, *The Negro*, 77.

105 Shufeldt, *The Negro*, 78. See also N'yongo, *The Amalgamation Waltz*, for a discussion of racial hybridity as sexual discourse.

106 Shufeldt, *The Negro*, 79–80.

107 Shufeldt, *The Negro*, unpaginated dedication.

108 Shufeldt, *The Negro*, 90–91.

109 Shufeldt, *The Negro*, 119.

110 Shufeldt, *The Negro*, 24.

111 Shufeldt, *The Negro*, 88.

112 Shufeldt, *The Negro*, 88. Shufeldt's argument for the modeling of eugenic principles on animal breeding corresponds to Gabriel Rosenberg's and Colin Johnson's respective arguments, in *4-H Harvest* and *Just Queer Folks*, about the agricultural origins of American eugenics and sexual understandings. It also presages the following chapter in this book.

113 Shufeldt, *The Negro*, 145, 156.

114 Shufeldt, *The Negro*, 162. In one of history's great assertions of "I'm not racist, but. . . ," Shufeldt also noted "it is not that I have anything against the black race in the United States."

115 Shufeldt, *The Negro*, 70.

116 Shufeldt, *The Negro*, 74.

Chapter 2. Conflicting Sexes at Two Eugenics Laboratories

1 Davenport, "Biological Experiment Station," 280.

2 There is a robust literature at the nexus of histories of race and histories of sex, especially when factoring in work on miscegenation and reproduction that is less explicitly about the sexed body. In addition to the texts I discuss here, see Schiebinger, *Nature's Body*; J. Morgan, *Laboring Women*; Gilman, *Difference and Pathology*; Mitra, *Indian Sex Life*; and LaFleur, *Natural History of Sexuality*.

3 See, e.g., Somerville, "Scientific Racism"; Boag, *Re-Dressing*; and Duggan, *Sapphic Slashers*, among many others.

4 Somerville, "Scientific Racism," 245–46.

5 Duggan, *Sapphic Slashers*; Terry, *American Obsession*; Boag, *Re-Dressing*.

6 I am also indebted to scholarship that examines the role of methods and theories of agricultural breeding in the formation of American eugenic regimes of sexuality, especially Johnson, *Just Queer Folks*, and G. Rosenberg, *4-H Harvest*.

7 For more on Davenport and Yerkes's relationship, see their correspondence in the Charles Benedict Davenport Papers at the American

Philosophical Society (hereafter CDP) and the Robert Mearns Yerkes Papers at Yale University Manuscripts and Archives (hereafter RMY).

8 Davenport to Yerkes, March 24, 1925, Box 97, Folder Yerkes #7, CDP; Davenport to Lillie, December 24, 1923, Box 64, Folder Lillie, Frank Rattaray #4, CDP; Davenport to Yerkes, February 26, 1930, Box 97, Folder Yerkes #7, CDP.

9 Yerkes to Davenport, July 17, 1911, Box 97, Folder Yerkes, Robert, CDP.

10 Davenport to Yerkes, July 22, 1911, Group 569, Series I, Box 13, Folder 230, RMY; Davenport to Yerkes, August 24, 1911, Group 569, Series I, Box 13, Folder 230, RMY; "Mental Traits," Group 569, Series I, Box 13, Folder 230, RMY; Laughlin, *Eugenics Record Office, Report No. 1*, 9; "Second International Eugenics Conference," 36.

11 "Letchworth Village, Thiells, N.Y.," Group 569, Series II, Box 77, Folder 1470, RMY; Davenport to Yerkes, March 24, 1924, Box 97, Folder Yerkes #7, CDP.

12 See Richardson, *Sex Itself*; Gill-Peterson, *Histories of the Transgender Child*.

13 I'm looking forward to Rana Hogarth's work in progress on the race-crossing studies.

14 For more on the history of sterilization and other methods of limiting the reproduction of Black and brown people, see Roberts, *Killing the Black Body*; Stern, *Eugenic Nation*; and Briggs, *Reproducing Empire*.

15 Rafter, *White Trash*, 17.

16 See Baynton, *Defectives in the Land*; Whatcott, *Menace to the Future*.

17 Of course, nothing is ever really that simple. On Black eugenic arguments for racial equality, see Nuriddin, "Liberation Eugenics," and Nuriddin, "Engineering Uplift."

18 See Stern, *Eugenic Nation*, and G. Rosenberg, *4-H Harvest*, for accounts of eugenic activity in the American West and Midwest; see Kevles, *In the Name of Eugenics*, for an articulation of various eugenic approaches, namely what Kevles terms "mainline" and "reform" eugenics that differed in their acceptance of a Mendelian, single-gene trait model—though Kevles's clear distinction between the two has since become rather more nuanced. On that, see Comfort, *Science of Human Perfection*.

19 Allen, "Eugenics Record Office," 226.

20 See Meyerowitz, *How Sex Changed*; Amin, "Glands, Eugenics, and Rejuvenation"; Sengoopta, *Most Secret Quintessence*; Chiang, *After Eunuchs*.

21 See Chauncey, *Gay New York*; Canaday, *Straight State*; Stryker, *Transgender History*; and Skidmore, *True Sex*.

22 Allen's article is perhaps the most frequently cited. See also Kevles, *In the Name of Eugenics*, esp. chap. 2; Largent, *Breeding Contempt*, esp. chap. 2; and Bix, "Experiences and Voices."

23 Largent, *Breeding Contempt*, 40.

24 See Birn, "Philanthrocapitalism, Past and Present," and Kay, *Molecular Vision of Life*, esp. chap. 1.

25 Allen, "Eugenics Record Office," 229.

26 Allen, "Eugenics Record Office," 230.

27 Largent, *Breeding Contempt*, 43; Largent, "Zoology of the Twentieth Century."

28 Kimmelman, "American Breeders' Association."

29 Kimmelman, "American Breeders' Association," 184–85.

30 Kevles, *In the Name of Eugenics*, 55.

31 Allen, "Eugenics Record Office," 235.

32 Davenport, *Report on the Work of the Station*, 92.

33 Davenport, "Relation of the Association to Pure Research," 66. Another perk of working with agriculturalists: Davenport noted that allying with livestock breeders instilled a sense of "dignity and safety" in eugenic research. Davenport, "Eugenics, a Subject for Investigation," 68. See Kimmelman, "American Breeders' Association," on the relationship between eugenics and the scientization of agriculture, as well as G. Rosenberg, "No Scrubs."

34 Davenport, "Relation of the Association to Pure Research," 66.

35 Davenport to Dickinson, June 10, 1925, Box 34, Folder Robert Latou Dickinson, CDP.

36 Chauncey, "From Sexual Inversion to Homosexuality."

37 Meyerowitz, *How Sex Changed*.

38 Davenport, "Department of Experimental Evolution," *No. 11*, 87.

39 Goodale, *Gonadectomy*, 48.

40 Goodale, *Gonadectomy*, 48.

41 Goodale, *Gonadectomy*, 48.

42 Davenport, "Department of Experimental Evolution," *No. 12*, 98.

43 Davenport, "Department of Experimental Evolution," *No. 15*, 127.

44 For further detail on Riddle's metabolic theory of sex, see Ha, "Riddle of Sex." Riddle's approach was a self-conscious descendant of Geddes and Thomson's work discussed in chapter 1; see Riddle, "Complete Case," 177.

45 Davenport, "Department of Experimental Evolution," *No. 13*, 119.

46 Davenport, "Department of Experimental Evolution," *No. 15*, 129.

47 Riddle, "Complete Sex Transformation," n.d., Box 4, Oscar Riddle Papers (hereafter ORP).

48 Riddle, "Studies on Sex," 1924, Box 129, Folder 6, CDP.

49 Joanne Meyerowitz makes a similar argument regarding the power of post–World War II medicine to transform sexed bodies in *How Sex Changed*.

50 Riddle, "Control of the Sex Ratio," 355.

51 Riddle to Davenport, May 14, 1921, Box 129, Folder Riddle #4, CDP.

52 Blakeslee to Davenport, September 9, 1919, Box 109, Folder Blakeslee #4, Series IIB, CDP.

53 On the queerness of plants and its discursive disappearance, see Subramaniam and Bartlett, "Re-Imagining Reproduction."

54 Blakeslee to Davenport, September 5, 1912, Box 109, Folder Blakeslee #1, Series IIB, CDP; Blakeslee to Davenport, August 30, 1913, Box 109, Folder Blakeslee #1, Series IIB, CDP.

55 Davenport, "Department of Experimental Evolution," *No. 15*, 98–99. Vegetative luxuriance referred, essentially, to size, density, and complexity of growth—in the often erotically tinged language of botany, lushness.

56 Davenport, "Department of Experimental Evolution," *No. 12*, 99.

57 Davenport, "Department of Experimental Evolution," *No. 12*, 99.

58 Blakeslee to Davenport, May 1, 1925, Box 109, Folder Blakeslee #8, Series IIB, CDP.

59 Blakeslee, "Major Lines of Investigation," Box 109, Folder Blakeslee #10, Series IIB, CDP.

60 Blakeslee, "Chemical and Physical Bases for the Difference Between the Sexes: A Study of the Fundamental Difference Between Male and Female in Plants," Box 109, Folder Blakeslee #11, Series IIB, CDP.

61 Satina and Blakeslee, "Studies on Biochemical Differences," 192.

62 Satina and Demerec, "Manoilov's Reaction," 225.

63 Satina and Blakeslee, "Studies of Biochemical Differences," 196.

64 Davenport to Lillie, January 12, 1925, Box 64, Folder Lillie, Frank Rattaray #4, CDP.

65 Davenport, "Eugenics Record Office,"in "Official Records in the History of the Eugenics Record Office," 24–25, RG1: Eugenics Records Office, H. H. Laughlin 1907-1970, Series 3: H. H. Laughlin—Publications, 1912-1940, Box 6, Eugenics Record Office Collection (hereafter EROC), Cold Spring Harbor Laboratory (hereafter CSHL).

66 Bix, "Experiences and Voices," 633.

67 Kidder, "Report of the Advisory Committee on the Eugenics Record Office," June 28, 1935, RG2: Eugenics Record Office, 1908-1973, Series 1: Eugenics Record Office, Administrative 1908-1973, Box 8, EROC, CSHL.

68 Aero Mayflower Transit Company estimate, #1 and #3, March 30, 1948, Folder Deposition of Files to Dight Institute, Box 8, RG2: Eugenics Record Office, 1908-1973, Series 1: Eugenics Record Office, Administrative, EROC, CSHL.

69 C. F. Rogers to E. B. Biesecker, July 3, 1948, Folder Deposition of Files to Dight Institute, Box 8, RG2: Eugenics Record Office, 1908-1973, Series 1: Eugenics Record Office, Administrative, EROC, CSHL.

70 Gitelman, *Paper Knowledge*, 23, 31. See also Bowker and Star, *Sorting Things Out*, for the influence of standardization more broadly and the ways lives are torqued to fit the needs of paperwork. See also conversations in trans studies on the violence of sexed identification paperwork, such as Currah and Moore, "We Won't Know Who You Are," and Beauchamp, *Going Stealth*.

71 Robertson, "Granular Certainty." See also Krajewski, *Paper Machines*. In an earlier period, see Blair, *Too Much to Know*. On the role of computerization in

thinking differently about sex and sexuality, see Drucker, *The Classification of Sex*, esp. chap. 4, and Hicks, "Hacking the Cis-Tem."

72 Eugenics Record Office, "Abridged Record of Family Traits," n.d., Folder Forms—Abridge Record Family Traits, RG2: Eugenics Record Office, 1908–1973, Series 2: Eugenics Record Office—Forms, Pedigrees, 1908–1940, Box 9, EROC, CSHL.

73 Resta, "Crane's Foot," 236.

74 "News and Notes," 306; Resta, "Historical Origin of Pedigree Charts," 167. Davenport et al., "The Study of Human Heredity," names a different meeting as the place where the square and circle model was decided upon; it instead references a committee of the American Association for the Study of the Feebleminded that met in Illinois the same year. While Davenport himself didn't attend that gathering, Henry Herbert Goddard, who headed the ERO Committee on the Heredity of the Feebleminded, did.

75 "Method for Studying the Hereditary History of Patients as Used at the New Jersey State Village for Epileptics," n.d., Box Series I-1, Folder A:0 #1, "New Jersey State Village for Epileptics Schedules and Forms," Eugenics Record Office Records, American Philosophical Society (hereafter EROR).

76 Schott, "Sex Symbols Ancient and Modern."

77 Pearl et al., "Studies on Constitution," 20. "I am sorry," Davenport responded to Pearl in a letter after reading the article, "if [this] system gives you the horrors." Pearl wrote back that he simply had a "fondness for the traditional," which he joked was a "sign of oncoming senescence." Davenport to Pearl, February 21, 1929, Box 79, Folder Pearl, Raymond #5, CDP; Pearl to Davenport, February 23, 1929, Box 79, Folder Pearl, Raymond #5, CDP.

78 Davenport to Pearl, February 21, 1929, Box 79, Folder Pearl, Raymond #5, CDP.

79 See Richardson, *Sex Itself*, on the history of sex linkage in genetics.

80 Eugenics Record Office, "Family Distribution of Personal Traits," n.d., Group 569, Series II, Box 59, Folder 1128, RMY; Davenport and Laughlin, *Eugenical Family Study*, 7.

81 "Method for Studying . . . New Jersey State Village for Epileptics."

82 Davenport and Laughlin, *Eugenical Family Study*, 7; for an example of the use of a diamond to denote the limit of knowledge, see Fieldworker #56 Clara P. Pond, "The Smith Family" pedigree, Series VII, Subseries II, EROR.

83 Davenport and Laughlin, *Eugenical Family Study*, 9.

84 All quotes in this paragraph are from "Research Department Suggestions" notebook, 1915 and 1917, Folder "Meeting of the Field Workers," Box 1, Series VII, EROR.

85 See Amin, "Glands, Eugenics, and Rejuvenation," and Amin, "Trans* Plasticity."

86 For a mid-twentieth-century example of why the "our equipment can only handle binary sex" argument falls short as an explanation for the limits of the government filing systems, see Hicks, "Hacking the Cis-Tem."

87 Kevles, *In the Name of Eugenics*, 52–53.

88 Joseph Gould to Davenport, December 9, 1915, Box 42, Folder Gould, Joseph F., CDP.

89 Davenport to Dickinson, June 10, 1925, Box 34, Folder Robert Latou Dickinson, CDP.

90 Fieldworker #94 Whittier School, report on Louis Rubenstein, October 29, 1921; Fieldworker #88 Esther Bingham, report on Morris Rozell, March 13, 1920; Fieldworker #40 Elizabeth Greene, Janey Garvey report, January 5, 1914; Fieldworker #19 Jaime de Angulo, "Pedigree of the Duncan Family of Bedford, Ind.," February 24, 1912; Fieldworker #91 George A. Brammer, report on Cecil Brennan, March 10, 1921; Fieldworker #94 Whittier School, report on Louis Rubenstein, October 29, 1921; all located in Series VII, Subseries II, EROR.

91 See, e.g., Mumford, "'Lost Manhood' Found"; Bederman, *Manliness and Civilization*; and Boag, *Re-Dressing*.

92 Laughlin, *Report of the Committee*, 63.

93 Laughlin, *Report of the Committee*, 63.

94 Davenport to Earle, November 11, 1914, Box 34, Folder Earle #1, CDP.

95 Davenport to Earle, January 21, 1915, Box 34, Folder Earle #2, CDP.

96 Earle notes on Robert Battey, 1873, "Normal Ovariotomy," filed February 5, 1915, Box 60, Folder A:94348, EROR.

97 Earle notes on Battey, February 5, 1915.

98 Earle notes on F. E. Walker, "The Induced Climacteric," filed March 22, 1915, Box 60, Folder A:94348, EROR.

99 Earle notes on Battey, February 5, 1915, quoting a Dr. Atlee.

100 See Battey, "Normal Ovariotomy."

101 Earle notes on Alfred Gordon, 1914, "Nervous and Mental Disturbances Following Castration in Women," filed March 22, 1915, Box 60, Folder A:94348, EROR.

102 No name or date stamp but Earle's handwriting, notes on F. C. Cave, "Sterilization in Kansas State Home for Feeble-Minded," n.d., Box 60, Folder A:94348, EROR.

103 Earle notes on Gordon, March 22, 1915, quoting Jackson and Atlee.

104 Earle notes on Gordon, March 22, 1915.

105 Earle notes on Battey, February 5, 1915. Earle's notes on Munde, 1899, also note an "increased sexual desire after removal of the ovaries."

106 Earle notes on Gordon, March 22, 1915, quoting Starkey.

107 Earle notes on B. Sherwood-Dunn et al., 1897, "Conservation of the Ovary. Discussion following paper by Sherwood-Dunn," filed March 22, 1915, Box 60, Folder A:94348, EROR.

108 Earle notes on Kurt Boas, 1911, "Ein weiterer Fall von Suicidismus Men-struale," filed March 22, 1915, Box 60, Folder A:94348, EROR.

109 Earle notes on Gordon, filed March 22, 1915.

110 Earle notes on F. O. Polak, 1909, "Final Results in Conservative Surgery on the Ovaries," filed March 22, 1915, Box 60, Folder A:94348, EROR.

111 Moscucci, *Science of Woman*.

112 Record of an Individual Case of Sterilization of Any Type, H. H. LeSeur, Fieldworker #20, EROR.

113 Kline, *Building a Better Race*, esp. chap. 3.

114 LeSeur's notes on C. V. Carringon, "Sterilization of Habitual Criminals," H. H. LeSeur, Fieldworker #20, EROR.

115 Meyerowitz, *How Sex Changed*.

116 Mol, "Who Knows What a Woman Is."

117 "Report of the Advisory Committee on the Eugenics Record Office," June 28, 1935, R2, Series I, Box 8, Folder Report . . . , EROR, CSHL.

118 See Allen, "Eugenics Record Office," for a thorough account of the ERO's closure.

119 In a truly full-circle example, I invite you to peruse McCormick et al., "Sex Differences in Spotted Hyenas," recently published in *Cold Spring Harbor Perspectives in Biology*.

Chapter 3. Maintaining Womanhood in Gynecological Practice

1 See Dreger, *Hermaphrodites*; Reis, *Bodies in Doubt*; Mak, *Doubting Sex*. See also Foucault, *Herculine Barbin*; Brown, "'Changed into the Fashion of Man'"; and Karkazis, *Fixing Sex*.

2 See, e.g., Richardson, *Sex Itself*; Foucault, *Herculine Barbin*; and Reis, *Bodies in Doubt*.

3 Meyerowitz, *How Sex Changed*.

4 See Meyerowitz, *How Sex Changed*; Dreger, *Hermaphrodites*; and Richardson, *Sex Itself*.

5 On trans social worlds and lives before science caught up, see Heaney, *New Woman*; Skidmore, *True Sex*; and Manion, *Female Husbands*.

6 See, e.g., Terry, *American Obsession*, and the entire Foucauldian tradition for the former, and Meyerowitz, *How Sex Changed*; and Amin, "Glands, Eugenics, and Rejuvenation," for the latter.

7 Owens, *Medical Bondage*, 28, 111.

8 Spillers, "Mama's Baby," 68.

9 Snorton, *Black on Both Sides*, 53.

10 Briggs, "Race of Hysteria."

11 Longo, "Rise and Fall."

12 Ehrenreich and English, *For Her Own Good*, 131.

13 Moscucci, *Science of Woman*, 134. Thomas Laqueur has made a similar argument, referring to the relationship between ovary and woman as a "synecdochic leap." Laqueur, *Making Sex*, 176. As demonstrated in chapter 1 of this book, the location of femaleness in nonhumans was not so obviously located in the ovaries as most historians would have it. It will become clearer in this chapter that the ovaries themselves were similarly uncertain definers of sex in humans.

14 Longo, "Rise and Fall," 266.

15 See chapters 1 and 2. For more on the history of animal gonad transplants, see Meyerowitz, *How Sex Changed*; Amin, "Glands, Eugenics, and Rejuvenation"; Amin, "Trans* Plasticity"; Gill-Peterson, *Histories of the Transgender Child*; and Pettit, "Becoming Glandular."

16 Carter, "On Mother-Love," 117.

17 "Dr. R. L. Dickinson, Gynecologist, 89," *New York Times*, November 30, 1950; Kline, *Building a Better Race*, 66.

18 Soloway, "'Perfect Contraceptive,'" 642.

19 Yikes. Dickinson to Margaretta Keller Bowers, December 4, 1944, Box 1, Folder 63, Robert Latou Dickinson Papers, Center for the History of Medicine, Francis A. Countway Library of Medicine, Harvard University (hereafter RLD).

20 See the contents of Box 6, Folder 54, RLD. For Dickinson's card file of sexological citations, see Box 5, Folders 6 and 7, RLD.

21 Handwritten note, "311 Marchand see drawings," Box 6, Folder 54, RLD.

22 Handwritten note, "311 Marchand see drawings," Box 6, Folder 54, RLD.

23 Handwritten note, "Neugebauer," Box 6, Folder 54, RLD.

24 Typed note, "Intersex. By Robert L. Dickinson," Box 6, Folder 54, RLD.

25 Both can be found in Box 6, Folder 54, RLD. One is a pencil version on its own sheet of paper; the ink version is on a typed note titled "Intersex. By Robert L. Dickinson."

26 Handwritten notes, "Neugebauer," Box 6, Folder 54. RLD. On Neugebauer, see Mak, "'So We Must Go Behind Even What the Microscope Can Reveal,'" and Mak, *Doubting Sex*.

27 Dickinson, *Human Sex Anatomy*, fig. 116.

28 Handwritten note, "On the dimensions of the copulator . . . ," Box 6, Folder 54, RLD.

29 Typed note, "Intersex. By Robert Latou Dickinson," Box 6, folder 54, RLD.

30 Typed note, "Intersex. By Robert Latou Dickinson," Box 6, folder 54, RLD.

31 Typed note, "Intersex. By Robert Latou Dickinson," Box 6, folder 54, RLD.

32 Handwritten note, "On the dimensions of the copulator . . . ," Box 6, Folder 54, RLD.

33 Handwritten note, "311 Marchand see drawings," Box 6, Folder 54, RLD.

34 Dickinson, *Human Sex Anatomy*, vii.

35 Handwritten note, "Dimensions of penis," Box 6, Folder 54, RLD.

36 Handwritten note, "Constant use," Box 6, Folder 54, RLD.

37 Handwritten note, "Our only practical interest . . . ," Box 6, Folder 54, RLD.

38 Handwritten note, "311 Marchard see drawings," Box 6, Folder 54, RLD.

39 Handwritten note, "Menstruation is not always available . . . ," Box 6, Folder 54, RLD.

40 Handwritten note, "Menstruation is not always available . . . ," Box 6, Folder 54, RLD.

41 "Brides S 3008 H 433," 1932, Box 3, Folder 57, RLD. Sources with this title structure are case history summaries, with the letters and numbers referring to specific anonymized patients. Though they are all dated 1932 or 1933, they are based on patient encounters that stretched much further back in Dickinson's clinical career.

42 "Youth S 870 H 4238," 1932, Box 4, Folder 1, RLD.

43 "S 64 H 88A," 1932, Box 4, Folder 2, RLD.

44 "S64A H 141," 1933, Box 4, Folder 2, RLD.

45 "S 462 H 462," 1933, Box 6, Folder 3, RLD.

46 "S 177," 1933, Box 6, Folder 3, RLD.

47 "S 72 H 028A," 1933, Box 4, Folder 2, RLD.

48 "S 177," 1933, Box 6, Folder 3, RLD.

49 "Aberration S 476 H 3336," 1932, Box 3, Folder 55, RLD.

50 See especially the "Special Cases" histories, Box 6, Folder 3, RLD.

51 "Vulvar Signs S 773 H 4221," 1932, Box 4, Folder 1, RLD.

52 "S 810," 1933, Box 4, Folder 3, RLD.

53 "S 939 (H 4045)," Box 4, Folder 15, RLD.

54 "S 161 (H 1997 A)," Box 4, Folder 15, RLD.

55 Especially a white woman. See Rich, "Curse of Civilised Woman."

56 For case studies involving irregular menstruation, see esp. Box 4, Folders 1 and 2; on dislike of sex and painful sex, Box 4, Folders 15 and 18; and on sterility, Box 6, Folder 4.

57 Dickinson and Beam, *Thousand Marriages*, 42.

58 Strange, "Menstrual Fictions."

59 Dickinson and Beam, *Thousand Marriages*, ix, xv.

60 Dickinson and Beam, *Thousand Marriages*, 435, 446.

61 Carter, "On Mother-Love," 118.

62 Carter, "On Mother-Love," 121.

63 Carter, "On Mother-Love," 120–23. For more on women rejecting "lesbian" identity at the turn of the twentieth century, see Freedman, "'Burning of Letters Continues.'"

64 Dickinson and Beam, *Thousand Marriages*, 128.

65 Dickinson and Beam, *Thousand Marriages*, 128–29.

66 Dickinson and Beam, *Thousand Marriages*, 366–67.

67 Carter, "On Mother-Love," 122.

68 Dickinson and Beam, *Thousand Marriages*, 73.

69 Dickinson and Beam, *Thousand Marriages*, 82.

70 Dickinson and Beam, *Thousand Marriages*, 3.

71 Dickinson and Beam, *Single Woman*, 214.

72 Traub, "Psychomorphology of the Clitoris"; Somerville, "Scientific Racism"; Groneman, "Nymphomania"; Terry, "Lesbians Under the Medical Gaze." Terry briefly mentions Dickinson's work on vulvar changes in "normal" women in *American Obsession* but primarily focuses her analysis on Dickinson's contributions to George Henry's *Sex Variants* rather than the 1902 study (198–212).

73 See, e.g., Gilman, *Difference and Pathology*; Sharpley-Whiting, *Black Venus*; and Qureshi's account of Sarah Baartman (also known as Sara or Saartjie Bartman), whose genitals were preserved by Cuvier and displayed at the Paris Musée de l'Homme until 1974, in *Peoples on Parade*. See also J. Morgan, *Laboring Women*, for a discussion of "Hottentot" breasts as a marker of sexualized racial difference.

74 Flower and Murie, "Account of the Dissection of a Bushwoman," 207.

75 Dickinson, "Hypertrophies," 226.

76 Dickinson, "Hypertrophies," 235.

77 Dickinson, "Hypertrophies," 236.

78 Dickinson, "Hypertrophies," 237.

79 Dickinson, "Hypertrophies," 236.

80 Dickinson, "Gynecology of Homosexuality," 1077. Moench conducted the gynecological examinations for Henry's study. According to Jennifer Terry's account, Dickinson encouraged Moench to take measurements of each subject's vulva and vagina and to take tracings of vulvar anatomy using a glass plate and crayon. Dickinson later added to the sketches. Terry, *American Obsession*, 203.

81 Dickinson, "Gynecology of Homosexuality," 1082.

82 Dickinson, "Gynecology of Homosexuality," 1094.

83 Dickinson, "Gynecology of Homosexuality," fig. 11, 1110, 1094.

84 Dickinson, "Gynecology of Homosexuality," 1078–79.

85 Dickinson, "Gynecology of Homosexuality," 1082.

86 Terry, *American Obsession*, 208.

87 Dickinson, *Human Sex Anatomy*, 57.

88 Ellis, *Studies in the Psychology of Sex*, 135.

89 Ellis, *Studies in the Psychology of Sex*, 136.

90 Schuller, *Biopolitics of Feeling*, 109. See Musser, "Race and the Integrity of the Line," for another use of Schuller to interpret Dickinson, specifically two drawings of Black women in *Human Sex Anatomy*.

91 Dickinson, *Human Sex Anatomy*, 57.

92 Dickinson, "Hypertrophies," 232.

93 Dickinson, "Hypertrophies," 244. The original text says "bruskin" rather than "brunette." In theory, this could apparently be a corruption of "buskin," an open-toed, laced boot. However, that seems unlikely. I suspect that this may be a printing issue; the word is hyphenated across a line break, "bru-skins," and the following line also begins "skins," suggesting an unintentional doubling where the second half of the word should have been "nette," as in "brunette." In any case, the sentence that follows makes it clear that Dickinson is comparing light-skinned blondes to women with "darker skins."

94 Dickinson, *Human Sex Anatomy*, figs. 77a, 77c.

95 Dickinson, *Human Sex Anatomy*, fig. 84.

96 Dickinson, "Gynecology of Homosexuality," 1097.

97 Dickinson, "Gynecology of Homosexuality," 1096.

98 Dickinson, "Gynecology of Homosexuality," 1125.

99 Dickinson, "Gynecology of Homosexuality," 1109.

100 Dickinson, "Memorandum of Data Needed," December 1946, Box 4, Folder 26, RLD.

101 Dickinson, "Masturbation, Physical Signs in Males," December 1946, Box 4, Folder 26, RLD.

102 Dickinson, "Masturbation: Physical Signs in Males," January 12, 1947, Box 4, Folder 26, RLD.

103 Dickinson, "Masturbation: Physical Signs in Males," January 12, 1947, Box 4, Folder 26, RLD.

104 Dickinson, "Masturbation: Physical Signs in Males," January 12, 1947, Box 4, Folder 26, RLD.

105 Dickinson, *Human Sex Anatomy*, i.

106 Dickinson, *Human Sex Anatomy*, vii.

107 Dickinson, *Human Sex Anatomy*, vii.

108 Daston and Galison, *Objectivity*, 22.

109 Moore and Clarke, "Clitoral Conventions," 256.

110 Dickinson, *Human Sex Anatomy*, 3.

111 Dickinson, *Human Sex Anatomy*, 4–5.

112 Dickinson, *Human Sex Anatomy*, 6.

113 Dickinson, *Human Sex Anatomy*, 3.

114 Dickinson, *Human Sex Anatomy*, 6.

115 Dickinson, *Human Sex Anatomy*, 50.

116 Creadick, *Perfectly Average*, 19.

117 Carter, *Heart of Whiteness*; Creadick, *Perfectly Average*.

118 Dickinson and Beam, *Thousand Marriages*, 26–27.

119 Kline, *Building a Better Race*, 66.

120 Briggs, "Race of Hysteria."

121 Search for Norma Scrapbook, 1945, Oversize Box 20, Folder 2, RLD.

122 Cleveland Health Museum Norma and Normman Order Form, Search for Norma Scrapbook, 1945, Oversize box 20, Folder 2, RLD.

123 Josephine Robertson, "3,700 Sent measurements in Ohio Search for Norma," Search for Norma Scrapbook, 1945, Oversize Box 20, Folder 2, RLD.

124 Creadick, *Perfectly Average*, 34.

125 Mak, "Conflicting Heterosexualities," 416.

126 Mak, "Conflicting Heterosexualities," 418.

127 Mak, "Conflicting Heterosexualities," 419.

128 Handwritten note, "Medical programs," Box 6, Folder 54, RLD.

129 Handwritten notes from examination of B. K., Box 4, Folder 25, RLD.

130 Dickinson to Robert T. Frank, April 30, 1927, Box 4, Folder 25, RLD.

131 Dickinson to Robert J. Wilkin, April 30, 1927, Box 4, Folder 25, RLD.

132 Dickinson to Wilkin, April 30, 1927.

133 Dickinson to Herbert Chase, April 30, 1927, Box 4, Folder 25, RLD.

Chapter 4. Variable Sex in Statistical Research

1 Kinsey, Pomeroy, and Martin, *Sexual Behavior in the Human Male* (hereafter *HM*), 639. How to refer to the research team is a bit of an unsettled question. The reports are often written about as though they were the product of a solitary effort: "Kinsey argued . . ." et cetera. Some scholars have written about the team by their initials: KPM. The first renders the collective nature of the work invisible; the latter leaves out Paul Gebhard, coauthor of *Sexual Behavior in the Human Female* (hereafter *HF*). Neither of these options credits the rest of the staff. In this chapter, I've resorted to a somewhat clumsy use of group nouns like "the authors" or "the Kinsey team" when referring to *Human Male* and *Human Female*, and have attempted to use "Kinsey" only when referring to Alfred specifically.

2 Terry, *American Obsession*.

3 *HM*, 639.

4 See Drucker, *Classification of Sex*, and several articles; Hegarty, *Gentleman's Disagreement*; Reumann, *American Sexual Character*; Igo, *Averaged American*; Terry, *American Obsession*; Chiang, "Effecting Science"; and Chiang, "Liberating Sex."

5 *HM*, 3.

6 Brewer, *Kinsey Interview Kit*.

7 Reumann, *American Sexual Character*, 1.

8 Reumann, *American Sexual Character*; Drucker, "'A Most Interesting Chapter.'"

9 "Sexual Misbehavior in the Human Male" cocktail napkins, *Monogram of California*, 1954, in author's possession.

10 Suresha, "Properly Placed."

11 See Drucker, *Classification of Sex*; Terry, *American Obsession*; and Igo, *Averaged American*.

12 Igo, *Averaged American*, 256. See also Reumann, *American Sexual Character*, for public reactions.

13 Geddes and Curie, *About the Kinsey Report*, 6.

14 Clarke, in American Social Hygiene Association, *Problems of Sexual Behavior*, iii, v.

15 Montagu, in Geddes and Curie, *About the Kinsey Report*, 63, 69.

16 Mead, in American Social Hygiene Association, *Problems of Sexual Behavior*, 58, 60.

17 Igo, *Averaged American*, 234; Reumann, *American Sexual Character*, 1; Cochran, Mosteller, and Tukey (hereafter CMT), *Statistical Problems*, 74.

18 Irvine, *Disorders of Desire*, 17–18.

19 Irvine, *Disorders of Desire*, 42.

20 Chiang, "Liberating Sex"; Chiang, "Effecting Science"; Terry, *American Obsession* (though Terry differentiates between a decline in the idea of a homosexual body and depathologization, and notes backlash to the Reports by psychiatrists).

21 Reumann, *American Sexual Character*.

22 D'Emilio, *Sexual Politics*.

23 Chiang, "Liberating Sex," 53.

24 Davis, *Factors in the Sex Life*; Terman, *Psychological Factors*; Henry, *Sex Variants*.

25 This embrace of large data sets was not limited to sex research. See Igo, *Averaged American*; Bouk, *How Our Days Became Numbered*; and Matthews, *Quantification*, for discussions of the adoption of new mathematical methods in other fields, as well as Desrosières, *Politics of Large Numbers*, and Porter, *Genetics in the Madhouse*, for pre-twentieth-century examples.

26 MacKenzie, *Statistics in Britain*.

27 Fitzpatrick, "Statistical Works"; Davenport, *Statistical Methods*, preface.

28 Little and Garruto, "Raymond Pearl"; Jennings, "Biographical Memoir of Raymond Pearl." Much of Pearl's early work involved research on sex ratios, determination, and development; see, e.g., Pearl and Pearl, "On the Relation of Race Crossing," and Pearl and Surface, "Assumption of Male Secondary Characters."

29 See Stigler, *History of Statistics*; MacKenzie, *Statistics in Britain*; Desrosières, *Politics of Large Numbers*; and Hacking, *Taming of Chance*. Science not being monolithic, the uptake of statistical methods was uneven. For example, many clinicians long doubted that quantitative and statistical approaches could do better than "medical judgment." See Matthews, *Quantification*.

30 Didier, *America by the Numbers*.

31 Didier, *America by the Numbers*, 3.

32 Didier, *America by the Numbers*.

33 Didier, *America by the Numbers*, 149. See also Odell, *Interpretation of the Probable Error*.

34 Didier, *America by the Numbers*, 161.

35 *HF*, 5.

36 *HF*, 5.

37 Drucker, *Classification of Sex*.

38 *HM*, 17.

39 *HM*, 19.

40 *HM*, 34.

41 *HM*, 34.

42 *HM*, 23.

43 *HM*, 24.

44 *HM*, 27.

45 Police being categorized as "semi-skilled" labor (*HM*, 78)—Kinsey said ACAB (All Cops Are Bastards)?

46 *HM*, 75.

47 *HM*, 80.

48 *HF*, 28.

49 Gebhard and Johnson, *Kinsey Data*, 26.

50 CMT, "Statistical Problems," 677, 713; CMT, *Statistical Problems*, 2. See Cornel, "Contested Numbers," on the tensions that emerged throughout the review.

51 CMT, *Statistical Problems*, 192.

52 CMT, "Statistical Problems," 705.

53 CMT, "Statistical Problems," 706.

54 Special Report No. 1 of the ASA committee to NRC committee, 4, n.d., series XI, Kinsey materials, Folder Kinsey Report: American Statistical Association/Special Report #1, John W. Tukey Papers, American Philosophical Society (hereafter JWT).

55 Compare to Shapin and Schaffer's discussion of calibration errors in *Leviathan*.

56 CMT, *Statistical Problems*, 23.

57 *HF*, 26–27.

58 CMT, *Statistical Problems*, 23.

59 CMT, *Statistical Problems*, 23.

60 CMT *Statistical Problems*, 2.

61 CMT, *Statistical Problems*, 3.

62 See chapters 1 and 3.

63 Womack, *Matter of Black Living*.

64 Didier, *America by the Numbers*, chap. 8.

65 Didier, *America by the Numbers*, chap. 9.

66 *HM*, 623.

67 *HM*, 623.

68 *HM*, 623.

69 *HM*, 650.

70 *HM*, 637.

71 *HM*, 637.

72 *HM*, 647.

73 Meyerowitz, *How Sex Changed*, 29; Valentine, *Imagining Transgender*.

74 Meyerowitz, *How Sex Changed*, 29.

75 See Kunzel, *Criminal Intimacy*.

76 Corner, in American Social Hygiene Association, *Problems of Sexual Behavior*, 17.

77 Meyerowitz, *How Sex Changed*, 309n13.

78 Ulrichs, *Riddle of "Man-Manly" Love*, 36. "Urning," as defined by Ulrichs, re-fers to those "whose body is built like a male, and at the same time, whose sexual drive is directed toward men, who are sexually not aroused by women, i.e., are horrified by any sexual contact with women" (34).

79 See Kahan, *Book of Minor Perverts*, on sexological etiology.

80 See Chauncey, "From Sexual Inversion"; Terry, *American Obsession*; and Somerville, "Scientific Racism" for detailed explanations of the varied, often contradictory approaches to sexual inversion (by that and other names).

81 *HM*, 327; Brewer, *Kinsey Interview Kit*, 21.

82 *HM*, 327, 329.

83 For further elaboration of this concept, see Stryker, "Transgender History, Homonormativity"; and Valentine, *Imagining Transgender*.

84 *HM*, 639–35.

85 Godbeer, "Cry of Sodom."

86 LaFleur, *Natural History of Sexuality*.

87 Foucault, *History of Sexuality*, 43.

88 *HF*, 4.

89 *HF*, 57.

90 *HF*, 171, 330, 392.

91 *HF*, 571; Gebhard and Johnson, *Kinsey Data*.

92 *HF*, 572.

93 *HF*, 574, 578.

94 *HF*, 592.

95 On Kinsey's speciation work, see Drucker, *Classification of Sex*, 51–56. On the recapitulation of evolutionary processes in nineteenth-century embry-ology, see Rich, "Monstrosity in Medical Science."

96 *HF*, 641.

97 *HF*, 690.

98 *HF*, 712–13.

99 *HF*, 760.

100 *HF*, 761.

101 Drucker, *Classification of Sex*, esp. chap. 6.

102 Drucker, *Classification of Sex*, 144.

103 Reumann, *American Sexual Character*, 99–102.

104 Reumann, *American Sexual Character*, 108–9.

105 Courdileone, *Manhood and American Political Culture*; Johnson, *Lavender Scare*; Canaday, *Straight State*; May, *Homeward Bound*; and Bishop, *Every Home a Fortress*.

106 See, e.g., Drucker, *Classification of Sex*; Igo, *Averaged American*; and Reumann, *American Sexual Character*, 133.

107 Drucker, *Classification of Sex*, 99–102. Pearl eventually renounced eugenics, but his subsequent work on population control was effectively eugenics by another name, with slightly different tactics and biological reasonings. Allen, "Old Wine"; Murphy, *Economization of Life*.

108 *HM*, vii.

109 *HM*, viii.

110 Dickinson to Gregg, January 13, 1943, Box 2, Folder 4, RLD.

111 Kinsey, *New Introduction*, 387.

112 Kinsey, *New Introduction*, 401.

113 Kinsey, *Methods in Biology*, 222.

114 Kinsey, *Methods in Biology*, 223–24.

115 Kinsey, *Methods in Biology*, 227.

116 Drucker, *Classification of Sex*, 50.

117 Kinsey, *Methods in Biology*, 223–24.

118 *HM*, 6.

119 *HM*, 81.

120 *HM*, 235.

121 *HM*, 76.

122 The 1950 statement would be revised in 1951, 1964, and 1967. See UNESCO, *Four Statements on the Race Question*; Gil-Riaño, "Relocating Anti-Racist Science"; Selcer, "Beyond the Cephalic Index"; and Thakkar, "Reeducation of Race." This was, to an extent, perhaps more theory than practice, given the continued use of race as a biological difference in medicine, intelligence testing, and many other areas of science.

123 Brewer, *Kinsey Interview Kit*, 21.

124 *HM*, 68–69.

125 *HM*, 66–70.

126 *HM*, 332.

127 *HM*, 332.

128 *HM*, 332–33.

129 *HM*, 329.

130 *HM*, 52.

131 *HM*, 387.

132 *HM*, 386.

133 *HM*, 332.

134 *HF*, 325.

135 *HM*, 390.

136 *HM*, 558.

137 *HF*, 79.

138 See Comfort, *Science of Human Perfection*, esp. chap. 2.

139 *HM*, 393.

140 *HM*, 69–70.

141 *HM*, 68, 70.

142 Kline, *Building a Better Race*.

143 The Kinsey studies, of course, were not the only instance of this continued linkage. See, e.g., Moynihan, *Negro Family*, and Benoit Denizet-Lewis,
"Double Lives on the Down Low," *New York Times*, August 3, 2003.

144 *HM*, 89.

145 *HM*, 85.

146 *HM*, 81.

147 *HM*, 34.

148 Lawrence to Kinsey, April 8, 1954, Alfred Kinsey Correspondence, Kinsey
Institute (hereafter AKC). See Meyerowitz, "Sex Research at the Bor-
ders," for a detailed treatment of the relationship between Kinsey and
Lawrence.

149 Lawrence to Kinsey, November 16, 1950, AKC; Lawrence to Benjamin, February 12, 1954, AKC.

150 Drucker, *Classification of Sex*, 163.

151 Meyerowitz, "Sex Research at the Borders," 80.

152 *HF*, 681.

153 Meyerowitz, "Sex Research at the Borders," 89.

154 Meyerowitz, "Sex Research at the Borders," 89. See also correspondence
among Kinsey, Benjamin, and Val Barry, in which Kinsey advised against
Barry's potential surgery, AKC and Harry Benjamin Collection, Kinsey Institute (hereafter HBC).

155 Terry, *American Obsession*, 350.

156 Gebhard and Johnson, *Kinsey Data*, 2.

157 Hegarty, *Gentlemen's Disagreement*, 157.

158 Irvine, *Disorders of Desire*, esp. part II.

Chapter 5. Immaterial Categories in Trans Medicine

1 Belt to Benjamin, February 22, 1960, series IIC, Box 3, Folder Belt,
Dr. Elmer (1959–1962), HBC.

2 See, e.g., Stone, *"Empire* Strikes Back"; Meyerowitz, *How Sex Changed*;
Stryker, *Transgender History*; Gill-Peterson, *Histories of the Transgender Child*.
While not the topic of this chapter, it should be noted that university-
based gender clinics founded in the late 1960s at Johns Hopkins, the University of Minnesota, and Stanford privileged adherence to binary gender
norms in their patient-selection process to what seems like a greater
extent.

3 This remains a part of the trans medicine experience in the present. See
shuster, "Uncertain Expertise," for an elaboration of how clinicians end

up reinforcing a narrow definition of transness and maintain gatekeeping practices as they attempt to deal with their own feelings of uncertainty.

4 On the impact of media coverage of Jorgensen's transition, see Meyerowitz, *How Sex Changed*.

5 Meyerowitz, *How Sex Changed*; Latham, "Axiomatic."

6 Stryker, *Transgender History*; Skidmore, *True Sex*; Manion, *Female Husbands*.

7 Meyerowitz, *How Sex Changed*; Stryker, *Transgender History*; Gill-Peterson, *Histories of the Transgender Child*.

8 Benjamin, *Transsexual Phenomenon*, 22, 14.

9 Benjamin, *Transsexual Phenomenon*, 21.

10 Bowker and Star, *Sorting Things Out*; Mol, *Body Multiple*; Murphy, *Sick Building Syndrome*; Barad, *Meeting the Universe Halfway*.

11 See also Latham, "Axiomatic."

12 Meyerowitz, *How Sex Changed*, 132.

13 Francisco Núñez Chávez to Benjamin [or possibly to Max Thorek, and forwarded to Benjamin], October 27, 1950, series IIC, Box 3, Folder B, VB, HBC.

14 Belt to Benjamin, December 15, 1958, series IIC, Box 3, Folder Belt, Dr. Elmer (1958–1959), HBC.

15 Meyerowitz, *How Sex Changed*, 20. On Hirschfeld and the Institut, see also Marhoefer, *Sex and the Weimar Republic*, and Nunn, "Trans Liminality."

16 Belt to Benjamin, July [n.d.], 1956, Series VIB, Box 23, Folder 34, HBC.

17 See chapter 5 of Gordon, *Glorious Bodies*, for a thorough account of early modern uses of mayhem and their implications for trans bodily integrity.

18 Sherwin, "Legal Problem in Transvestism."

19 Meyerowitz in *How Sex Changed* and Stryker in *Transgender History* both accept Sherwin's perspective in their explanations of mayhem concerns among surgeons in the 1950s.

20 "U'Ren Pressing Gland Charges," *Los Angeles Times*, May 16, 1928.

21 Blue, "Strange Career," 210.

22 "Mother of Kelly Sues Physicians," *Los Angeles Times*, May 23, 1928. See also "Gland Scandal in San Quentin," *Los Angeles Times*, May 14, 1928; "New Gland Quiz to Be Launched," *Los Angeles Times*, May 15, 1928; and "Warden Avers Kelly Willed Body to Science," *Los Angeles Times*, May 25, 1928.

23 "Operation Plot Is Told to Court by Ann Hewitt: Heiress Testifies She Was Tricked by Doctors in Examination," *Washington Post*, August 15, 1936; "Ann Cooper Hewitt Accuses 2 Doctors: Tricked into Sterilization Operation, She Says," *Boston Daily Globe*, August 15, 1936.

24 "Girl's Charge of Operation Investigated: Police Study Ann Cooper Hewitt's Claim That Sterilization Part of Trust Fund Scheme," *Hartford Courant*, January 8, 1936.

25 "Ann Hewitt Is Wed Again: Heiress Takes Radio Artist as Fourth Husband," *New York Times*, September 23, 1947.

26 "Sterilization Case Heiress Would Be Nevada Senator," *Washington Post*, March 18, 1950; "Ann Hewitt, of Sterilization Case, Dies at 40," *New York Herald Tribune*, February 11, 1956.

27 See LaFleur, "Trans Feminine Histories," for further discussion of castration as punishment, as well as an investigation of eighteenth- and nineteenth-century newspaper reports of self-castrations treated as oddities worth reporting but far less common than castration of Black men.

28 Amin, "Glands, Eugenics, and Rejuvenation."

29 Meyerowitz, *How Sex Changed*, 16. See also the first several chapters of Li, *Wondrous Transformations*, for a thorough account of Benjamin's pre–trans medicine career.

30 Schaefer and Wheeler, "Harry Benjamin's First Ten Cases," 77; Wolf-Gould, "History of Transgender Medicine," 509.

31 Schaefer and Wheeler, "Harry Benjamin's First Ten Cases," 78.

32 Benjamin to H. M. Coon, May 31, 1949, series 1, Box 1, Folder B, BV, HBC.

33 Meyerowitz, *How Sex Changed*, 133.

34 Meyerowitz, *How Sex Changed*, 133.

35 Meyerowitz, *How Sex Changed*, 133. See also correspondence between Louise Lawrence and Alfred Kinsey, series 1B, Box 1, Folder 1, Louise Lawrence Collection, Kinsey Institute.

36 Meyerowitz, *How Sex Changed*, 164; Person, "Harry Benjamin," 260; Schaefer and Wheeler, "Harry Benjamin's First Ten Cases," 75. See also Li, *Wondrous Transformations*, for a sustained treatment of Benjamin as caring, sympathetic, and charming, and my review of Li for further discussion of this characterization.

37 H. M. Coon to W. B. Campbell, July 19, 1948, Box 3, Folder B, VB, HBC.

38 Benjamin to Alfred Kinsey, October 4, 1950, File Benjamin, H., AKC.

39 Barry was finally able to access surgery in Sweden in 1953. Benjamin to Kinsey, December 1, 1953, File Benjamin, H., AKC.

40 Benjamin to Barry, December 27, 1949, series IIC, Box 3, Folder B, VB, HBC.

41 Benjamin to Bowman, n.d., series IIC, Box 3, Folder B, VB, HBC.

42 Ecker to Benjamin, October 5, [1953], series IIC, Box 4, Folder E, C, HBC.

43 Ecker to Benjamin, December 3, 1953, series IIC, Box 4, Folder E, C, HBC.

44 Benjamin to Belt, May 12, 1958, series IIC, Box 3, Folder Belt, Dr. Elmer (1958–1959), HBC.

45 Benjamin to Belt, June 11, 1958, series IIC, Box 3, Folder Belt, Dr. Elmer (1958–1959), HBC.

46 Belt to Benjamin, June 12, 1958, series IIC, Box 3, Folder Belt, Dr. Elmer (1958–1959), HBC.

47 Belt to Benjamin, June 23, 1958, series IIC, Box 3, Folder Belt, Dr. Elmer (1958–1959), HBC.

48 Belt to Edie Hutchens, December 1, 1958, series IIC, Box 3, Folder Belt, Dr. Elmer (1958–1959), HBC.

49 Belt to Benjamin, December 15, 1958, series IIC, Box 3, Folder Belt, Dr. Elmer (1958–1959), HBC.

50 Benjamin to Belt, December 30, 1958, series IIC, Box 3, Folder Belt, Dr. Elmer (1958–1959), HBC.

51 Belt never would operate on Hutchens. She was finally able to have surgery in 1963, with Dr. Georges Burou of Casablanca. Benjamin to Belt, December 5, 1963, series IIC, Box 3, Folder Belt, Dr. Elmer (1962–1965), HBC.

52 Benjamin to Belt, March 3, 1958, series VIB, Box 23, Folder 34, HBC.

53 Benjamin to Belt, September 12, 1957, series IIC, Box 3, Folder Belt, Dr. Elmer (1958–1959), HBC.

54 Benjamin to Hartsuiker, November 24, 1954, series IIC, Box 5, Folder Hart-suiker, Dr. F, HBC.

55 Benjamin to Pomeroy, January 8, 1959, File Benjamin, H., AKC.

56 Benjamin to Kinsey, December 3, 1954, File Benjamin, H., AKC. Cf. Li's characterization of Benjamin as a man with a "keen, sympathetic gaze. . . . inclined to recognize his patient as sane and to accept her expressed sense of self," who "listened to the voices of his patients with respect." *Wondrous Transformations*, 143, 146.

57 Benjamin to Morton M. Garfield, March 7, 1967, series IIC, Box 6, Folder N, W, HBC.

58 On delay in trans medicine, see Pitts-Taylor, "'Slow and Unrewarding."

59 Benjamin to B. S., September 29, 1955, series IIC, Box 6, Folder S, B, HBC.

60 Dorta to Benjamin, March 13, 1965, series IIC, Box 4, Folder D, C, HBC.

61 Benjamin to O. S. [Dorta], October 16, 1968, series IIC, Box 6, Folder S, O, HBC. Dorta used several names, hence the discrepancy in initials.

62 Benjamin to M. H., May 15, 1956, series IIC, Box 5, Folder, H, M, HBC.

63 Wagner to Benjamin, January 8, 1967, series VIC, Box 25, Folder 7, HBC.

64 Benjamin to R. W., November 3, 1969, series IIC, Box 8, Folder W, R, HBC. "Safe" here, of course, referring to Benjamin and Belt.

65 Belt to Robert P. McDonald, June 2, 1958, series IIC, Box 3, Folder Belt, Dr. Elmer (1958–1959), HBC.

66 Benjamin to Belt, June 2, 1959, series IIC, Box 3, Folder Belt, Dr. Elmer (1958–1959), HBC.

67 Benjamin to Barry, May 31, 1949, series IIC, Box 3, Folder B, VB, HBC.

68 Benjamin to Belt, June 11, 1958, series IIC, Box 3, Folder Belt, Dr. Elmer (1958–1959), HBC.

69 Benjamin to Pomeroy, June 27, 1957, File Benjamin, H., AKC.

70 Benjamin to Belt, March 3, 1958, series IIC, Box 3, Folder Belt, Dr. Elmer (1958–1959), HBC.

71 Benjamin to W. J. D., August 18, 1955, series IIC, Box 4, Folder D., W.J., HCB.

72 Kinsey to Benjamin, January 5, 1955, File Benjamin, H., AKC.

73 Benjamin to H-J. S., July 22, 1965, series IIC, Box 6, Folder s, Dr. H-J, HBC.

74 Benjamin to Kinsey, June 6, 1955, File Benjamin, H., AKC.

75 Benjamin to editor of *JAMA*, April 20, 1955, File Benjamin, H., AKC.

76 Stone, "*Empire* Strikes Back"; Meyerowitz, *How Sex Changed*; Valentine, *Imagining Transgender*; Gill-Peterson, *Histories of the Transgender Child*.

77 On the construction of reality, see Law, *After Method*.

78 Benjamin to E. W., June 30, 1958, series VIB, Box 24, Folder 22, HBC.

79 Belt to Benjamin, July 29, 1956, series VIB, Box 23, Folder 34, HBC.

80 Belt to B. O., September 5, 1956, series IIC, Box 6, Folder O, B, HBC.

81 Belt to Wolfe, March 24, 1969, series IIC, Box 3, Folder Belt, Dr. Elmer (1965–1971), HBC.

82 Belt to L. W. H, August 29, 1958, series IIC, Box 3, Folder Belt, Dr. Elmer (1958–1959), HBC.

83 Benjamin to D. M., March 15, 1954, series IIC, Box 6, Folder M, D, HBC.

84 Benjamin to Hartsuiker, February 15, 1954, series IIC, Box 6, Folder Hartsuiker, F., HBC.

85 Benjamin to Edmund Brown, November 22, 1949, series IIC, Box 3, Folder B, BV, HBC.

86 Benjamin to Kinsey, June 17, 1952, File Benjamin, H., AKC.

87 Benjamin to Hartsuiker, May 18, 1954, series IIC, Box 5, Folder Hartsuiker, Dr. F., HBC.

88 Benjamin to Belt, April 7, 1958, series VIB, Box 23, Folder 34, HBC.

89 Benjamin to Belt, September 12, 1957, series VIB, Box 23, Folder 34, HBC.

90 Benjamin, introductory speech read before the Section on Obstetrics and Gynecology, New York Academy of Medicine, April 25, 1973, series II E, Box 9, Folder 13, HBC.

91 Ecker, account of Debbie Mayne's life, May 24, 1955, series IIC, Box 4, Folder E, C, HBC.

92 Debbie Mayne to Benjamin, March 15, 1954, series IIC, Box 6, Folder M, D, HBC.

93 Carla Sawyer to Benjamin, November 21, 1954, series IIC, Box 6, Folder S, C, HBC.

94 Sawyer to Benjamin, March 30, 1955, series IIC, Box 6, Folder S, C, HBC.

95 Benjamin to Sawyer, April 7, 1955, series IIC, Box 6, Folder S, C, HBC.

96 Sawyer to Benjamin, July 15, 1955, series IIC, Box 6, Folder S, C, HBC; Benjamin to Sawyer, July 18, 1955, series IIC, Box 6, Folder S, C, HBC.

97 Sawyer to Benjamin, August 14, 1955, series IIC, Box 6, Folder S, C, HBC.

98 Belt to Benjamin, July 1956, series VIB, Box 23, Folder 34, HBC.

99 Belt to Benjamin, December 15, 1958, series IIC, Box 3, Folder Belt, Dr. Elmer (1958–1959), HBC.

100 Hage, Karim, and Laub Sr., "On the Origin of Pedicled Skin Inversion Vaginoplasty," 723.

101 Benjamin to Suzie Watson, April 26, 1956, series IIC, Box 8, Folder W, S, HBC.

102 L. L. to Benjamin, May 19, 1966, series VIB, Box 24, Folder 20, HBC.

103 Benjamin to L. L., May 4, 1966, series VIB, Box 24, Folder 20, HBC.

104 Transcript, "Male TS Meetings, Sunday, January 12, 1969," series III A, Box 14, Folder 16, HBC.

105 Benjamin to P. L., September 17, 1973, series IIC, Box 5, Folder L, P HBC.

106 This narrative has been perpetuated by several historical accounts, most recently Li, *Wondrous Transformations*.

107 Benjamin to Belt, March 8, 1955, series VIB, Box 23, Folder 34, HBC.

108 Doorbar to Benjamin, October 18, 1964, series VIB, Box 24, Folder 43, HBC.

109 Benjamin, *Transsexual Phenomenon*, 22.

110 Benjamin, *Transsexual Phenomenon*, 23.

111 Benjamin, *Transsexual Phenomenon*, 24.

112 Benjamin, *Transsexual Phenomenon*, 22.

113 Benjamin to J. S., November 24, 1961, series IIC, Box 8, Folder S, J, HBC.

114 Patient spreadsheet and intake form, series VIE, Box 28, Folder 20, HBC.

115 See Mol, *Body Multiple*, for a similar process in the relationship between pathological studies and clinical encounters regarding atherosclerosis.

116 On the history of gender clinics, see Drager, "Early Gender Clinics"; Keyes, "Trans Science"; and Roblee et al., "History of Gender-Affirming Surgery."

117 Meyerowitz, *How Sex Changed*, 121. That said, a letter from Belt to Benjamin, dated October 16, 1962, refers to a doctor being sued for "converting a Frolich's syndrome transsexual into a female." Series IIC, Box 3, Folder Belt, Dr. Elmer (1962–1965), HBC.

118 Belt to Burton H. Wolfe, March 24, 1969, series IIC, Box 3, Folder Belt, Dr. Elmer (1965–1971), HBC.

119 There were, of course, various less-official methods of obtaining transition surgeries. Jules Gill-Peterson's current book in progress will detail the history of DIY medical transition. For a fairly recent example of orchiectomies performed by trans women for trans women in a surgical-suite-cum-tractor-barn, see Io Dodds, "'Never Ask Permission': How Two Trans Women Ran a Legendary Underground Surgical Clinic in a Rural Tractor Barn," *Independent*, July 3, 2022.

120 Belt to Burton H. Wolfe, March 24, 1969, series IIC, Box 3, Folder Belt, Dr. Elmer (1965–1971), HBC.

121 Meyerowitz, *How Sex Changed*, esp. chap. 6, "The Liberal Moment." Starting in 1979, however, many of these programs began to shut down, making private surgeons the most viable option for surgical access. Roblee et al., "History of Gender-Affirming Surgery," 55.

122 Paul Walker to Members of the Association, "Re: The Harry Benjamin International Gender Dysphoria Association Standards of Care," August 10, 1981, series VIC, Box 25, Folder 17, HBC.

123 Benjamin to Sawyer, November 12, 1954, series IIC, Box 6, Folder S, C, HBC.

124 Benjamin to B. H., August 5, 1957, series IIC, Box 5, Folder H, B, HBC.

125 Adams, Murphy, and Clarke, "Anticipation," 248.

126 Not cited on principle.

Epilogue: Chaotic Good

1 Enke, "Education of Little Cis," 235. This timing aligns generally with my own experience of when I started seeing it in use in queer spaces; anecdotally, I'd date its broader uptake to the mid-2010s or so.

2 See especially Bey, *Cistem Failure*.

3 Erica L. Green, Katie Benner, and Robert Pear, "'Transgender' Could Be Defined Out of Existence Under Trump Administration," *New York Times*, October 21, 2018. See my reaction in Beans Velocci, "Perspective: The Battle Over Trans Rights Is About Power, Not Science," *Washington Post*, October 29, 2018.

4 Brief of Ryan T. Anderson as Amicus Curiae in Support of Employers, *Harris Funeral Homes, Inc., v. EEOC*, No. 18–107 (August 21, 2019).

5 Mississippi Fairness Act, Mississippi Senate Bill 2536 (2021), https://billstatus.ls.state.ms.us/documents/2021/html/SB/2500-2599/SB2536IN.htm.

6 Katie Shepherd, "Marjorie Taylor Greene Blasted for Attacking Colleague's Transgender Daughter," *Washington Post*, February 25, 2021.

7 Vulnerable Child Compassion and Protection Act, AL HB1 (February 2, 2021), https://www.billtrack50.com/BillDetail/1247206.

8 For an argument that contemporary science favors a complex version of sex beyond a binary in two very different registers, see "Editorial: US Proposal for Defining Gender"; and McNamara, "This Biology Teacher Disproved Transphobia."

9 Green, Benner, and Pear, "'Transgender' Could Be Defined Out of Existence." In addition to the *Washington Post* op-ed cited in note 6, see Anne Fausto-Sterling, "Science Won't Settle Trans Rights," *Boston Review*, February 10, 2020; and Riley Black, "Stop Trying to Out-Science Transphobes," *Slate*, March 3, 2021.

10 For scientists—including, in this case, those who purport themselves to be trans allies—backing the idea that transition care for youth must proceed with caution, see Anacker et al., "Behavioral and Neurobiological Effects." For an example of scientists pushing back against an assumed simplicity

of sex, see Monk et al., "Alternative Hypothesis," especially their notes on terminology.

11 Powell, Terry, and Chen, "How LGBT+ Scientists Would Like to Be Included"; Zemenick et al., "Six Principles."

12 This is not unique to sex science. Scholars in science and technology studies have shown that science and social justice do not necessarily follow the same trajectories. See Nelson, *Social Life of DNA*; Hicks, *Programmed Inequality*; and Benjamin, *Race After Technology*. Eve Kosofsky Sedgwick warned of the relationship between biological claims for the legitimacy of queerness and efforts to intervene and *prevent* queerness based on that knowledge in "How to Bring Your Kids Up Gay."

13 See Gill-Peterson, *Histories of the Transgender Child*, and Eder, *How the Clinic Made Gender*.

14 Killerman, "Genderbread Person."

15 See chapter 1 of this book.

Bibliography

Published Primary Sources

Abbot, E. T. "Scientific Ignorance About Bees." *American Bee Journal* 30, no. 14 (September 29, 1892).

"Agricultural News Items: Are Workers Abortions?" *American Bee Journal* 21, no. 36 (September 9, 1885): 563.

American Social Hygiene Association. *Problems of Sexual Behavior: Proceedings of a Symposium on the First Published Report of a Series of Studies of Sex Phenomena.* 1948.

Anacker, Christoph, et al. "Behavioral and Neurobiological Effects of GnRH Agonist Treatment in Mice—Potential Implications for Puberty Suppression in Transgender Individuals." *Neuropsychopharmacology* 46 (2021): 882–90.

Anderson, Ryan T. Brief as Amicus Curiae in Support of Employers, *Harris Funeral Homes, Inc., v. EEOC*, No. 18-107. August 21, 2019.

"Ann Cooper Hewitt Accuses 2 Doctors: Tricked into Sterilization Operation, She Says." *Boston Daily Globe*, August 15, 1936.

"Ann Hewitt Is Wed Again: Heiress Takes Radio Artist as Fourth Husband." *New York Times*, September 23, 1947.

"Ann Hewitt, of Sterilization Case, Dies at 40." *New York Herald Tribune*, February 11, 1956.

"Bee-Culture." *American Bee Journal* 8, no. 1 (July 1872): 22–24.

Benjamin, Harry. *The Transsexual Phenomenon.* New York: Julian Press, 1966.

Bingley, William. *Animal Biography, or, Anecdotes of the Lives, Manners, and Economy of Animal Creation, Arranged According to the System of Linnaeus.* London: R. Phillips, 1803.

Bostock v. Clayton County, Georgia, 590 U.S. 644 (2020). https://www.supremecourt.gov/opinions/19pdf/17-1618_hfci.pdf.

Brewer, Joan Scherer. *The Kinsey Interview Kit*. Bloomington, IN: Kinsey Institute for Research in Sex, Gender, and Reproduction, 1985.

Buffon, Georges Louis Leclerc, comte de. *Natural History, General and Particular, Vol. VI, The History of Man and Quadrupeds*. Translated by William Smellie. London: T. Cadell and W. Davies, 1812.

Butler, Frederick. "The Farmer's Manual, Including a Treatise on the Management of Bees." *North American Review* (October 1828): 338–59.

Calmet, Augustin. *Taylor's Edition of Calmet's Great Dictionary of the Holy Bible*. Charlestown, NS: Samuel Etheridge, 1812.

Capgrave, John. "Dedication to Edward IV." In *English Prose*, edited by Henry Craik, 91–92. London: Macmillan, 1893.

Case, Norton. Reader's comment in "Voices from Among the Hives." *American Bee Journal* 12, no. 7 (July 1876): 193.

Castle, W. E. "Sex Determination in Bees and Ants." *Science* 19, no. 479 (March 4, 1904): 380–92.

Cochran, William G., Frederick Mosteller, and John W. Tukey. "Statistical Problems of the Kinsey Report." *Journal of the American Statistical Association* 48, no. 264 (1953): 673–716.

Cochran, William G., Frederick Mosteller, and John W. Tukey. *Statistical Problems of the Kinsey Report on Sexual Behavior in the Human Male*. Washington, DC: American Statistical Association, 1954.

Cook, A. J. *Manual of the Apiary*. Lansing: W. S. George, 1876.

Cook, A. J. "Peculiarities of Abnormal Bees." *American Bee Journal* 28, no. 12 (September 17, 1891): 364–66.

Davenport, Charles B. "Biological Experiment Station for Studying Evolution." In *Carnegie Institution of Washington Year Book No. 1, 1902*, 280–82. Washington, DC: Carnegie Institution of Washington, 1903.

Davenport, Charles B. "Department of Experimental Evolution." In *Carnegie Institution of Washington Year Book No. 11*, 83–93. Washington, DC: Carnegie Institution of Washington, 1912.

Davenport, Charles B. "Department of Experimental Evolution." In *Carnegie Institution of Washington Year Book No. 12*, 97–122. Washington, DC: Carnegie Institution of Washington, 1913.

Davenport, Charles B. "Department of Experimental Evolution." In *Carnegie Institution of Washington Year Book No. 13*, 116–33. Washington, DC: Carnegie Institution of Washington, 1914.

Davenport, Charles B. "Department of Experimental Evolution." In *Carnegie Institution of Washington Year Book No. 15*, 121–35. Washington, DC: Carnegie Institution of Washington, 1916.

Davenport, Charles B. "Eugenics, a Subject for Investigation Rather Than Instruction." *American Breeders Magazine* 1, no. 1 (1910): 68–69.

Davenport, Charles B. "The Relation of the Association to Pure Research." *American Breeders Magazine* 1, no. 1 (1910): 66–67.

Davenport, Charles B. *Report on the Work of the Station for Experimental Evolution at Cold Spring Harbor*. Washington, DC: Press of Judd and Detweiler, 1907.

Davenport, Charles B. *Statistical Methods, with Special Reference to Biological Variation.* New York: J. Wiley and Sons, 1899.

Davenport, Charles B. *The Trait Book, Eugenics Record Office Bulletin no. 6.* Cold Spring Harbor, NY: Eugenics Record Office, 1912.

Davenport, Charles B., and Harry H. Laughlin. *How to Make a Eugenical Family Study, Eugenics Record Office Bulletin No. 2.* Cold Spring Harbor, NY: Eugenics Record Office, 1915.

Davenport, Charles B., H. H. Laughlin, David F. Weeks, E. R. Johnstone, and Henry H. Goddard. *The Study of Human Heredity: Methods of Collecting, Charting and Analyzing Data, Eugenics Record Office Bulletin No. 2.* Cold Spring Harbor, NY: Eugenics Record Office, 1911.

Davis, Katherine Bement. *Factors in the Sex Life of Twenty-Two Hundred Women.* New York: Harper and Brothers, 1929.

Debraw, John. "Discoveries on the Sex of Bees, Explaining the Manner in Which Their Species Is Propagated." *Philosophical Transactions of the Royal Society of London* 67 (1777): 15–32.

Denizet-Lewis, Benoit. "Double Lives on the Down Low." *New York Times,* August 3, 2003.

Dickinson, Robert Latou. "The Gynecology of Homosexuality." Appendix to *Sex Variants: A Study of Homosexual Patterns,* by George W. Henry, 1085–146. New York: Paul B. Hoeber, 1959.

Dickinson, Robert Latou. *Human Sex Anatomy.* Baltimore: Williams and Wilkins, 1933.

Dickinson, Robert Latou. "Hypertrophies of the Labia Minora and Their Significance." *American Gynecology* 1 (1902): 225–54.

Dickinson, Robert Latou, and Lura Beam. *The Single Woman: A Medical Study in Sex Education.* New York: Reynal and Hitchcock, 1934.

Dickinson, Robert Latou, and Lura Beam. *A Thousand Marriages: A Medical Study of Sex Adjustment.* Baltimore, MD: Williams and Wilkins, 1931.

Dickinson, Robert Latou, and Abram Belskie. *Framed Model of Forms of Adult Vulvas.* 1945. Terracotta in wooden frame. 25.6 × 33.5 × 11.5 cm. Warren Anatomical Museum, Boston. *OnView,* accessed July 16, 2024. https://collections.countway.harvard.edu/onview/items/show/14568.

Dickinson, Robert Latou, and Abram Belskie. *Norma.* 1939–50. Sculpture (plaster). 24 × 86 cm. Warren Anatomical Museum, Boston. *OnView,* accessed July 16, 2024. https://collections.countway.harvard.edu/onview/items/show/14642.

Dickinson, Robert Latou, and Abram Belskie. *Normman.* 1939–50. Sculpture (plaster). 9 × 94 cm. Warren Anatomical Museum, Boston. *OnView,* accessed July 16, 2024. https://collections.countway.harvard.edu/onview/items/show/14643.

Dickinson, Robert Latou, and Abram Belskie. *Replica of Dickinson-Belskie Model of Infant.* 1945–2007. 25.5 × 50.5 × 24 cm. Warren Anatomical Museum, Boston. *OnView,* accessed July 16, 2024. https://collections.countway.harvard.edu/onview/items/show/14468.

"Dr. R. L. Dickinson, Gynecologist, 89." *New York Times*, November 30, 1950.

"Editorial: US Proposal for Defining Gender Has No Basis in Science." *Nature* 563, no. 5 (2018): n.p.

Ellis, Havelock. *Man and Woman: A Study of Human Secondary Sex Characters*. London: Walter Scott, 1894.

Ellis, Havelock. *Studies in the Psychology of Sex*. Vol. 5. Philadelphia: F. A. Davis, 1912.

Flower, W. H., and James Murie. "Account of the Dissection of a Bushwoman." *Journal of Anatomy and Physiology* 1, no. 2 (1867): 189–208.

Galton, Francis. "Hereditary Character and Talent." *Macmillan's Magazine* 12 (1865): 157–66, 318–27.

Galton, Francis. "Short Notes on Heredity, &c., in Twins." *Journal of the Anthropological Institute of Great Britain and Ireland* 5 (1876): 324–29.

Gebhard, Paul. "Sexuality in the Post-Kinsey Era." In *Changing Patterns of Sexual Behavior: Proceedings of the Fifteenth Annual Symposium of the Eugenics Society, London, 1978*, edited by W. H. G. Armytage, R. Chester, and John Peel, 45–57. London: Academic Press, 1980.

Gebhard, Paul, and Alan Johnson. *The Kinsey Data: Marginal Tabulations of the 1938–1963 Interviews Conducted by the Institute for Sex Research*. Bloomington: Indiana University Press, 1979.

Geddes, Donald Porter, and Enid Curie. *About the Kinsey Report: Observations by 11 Experts on "Sexual Behavior in the Human Male."* New York: Signet Books, 1948.

Geddes, Patrick, and J. Arthur Thomson. *The Evolution of Sex*. New York: Scribner and Welford, 1890.

"Girl's Charge of Operation Investigated: Police Study Ann Cooper Hewitt's Claim That Sterilization Part of Trust Fund Scheme." *Hartford Courant*, January 8, 1936.

"Gland Scandal in San Quentin." *Los Angeles Times*, May 14, 1928.

Goodale, H. D. *Gonadectomy in Relation to the Secondary Sexual Characters of Some Domestic Birds*. Washington, DC: Carnegie Institute of Washington, 1916.

Goubaux, Armand, and Gustave Barrier. *The Exterior of the Horse*, 2nd ed. Translated by Simon J. J. Harger. Philadelphia: J. P. Lippincott, 1892.

Green, Erica L., Katie Benner, and Robert Pear. "'Transgender' Could Be Defined Out of Existence Under Trump Administration." *New York Times*, October 21, 2018.

Greiner, F. "He or She: Its Use in Different Languages." *Gleanings in Bee Culture* 27 (March 15, 1899): 213.

Harlan, Richard. "Description of an Hermaphrodite Orang Outang Lately Living in Philadelphia." *Journal of the Academy of Natural Sciences of Philadelphia* 5, no. 2 (1827): 229–36.

Hart, D. Berry. "The Structure of the Reproductive Organs in the Free-Martin, with a Theory of the Significance of the Abnormality." *Proceedings of the Royal Society of Edinburgh* (1909–10): 230–41.

Hastings, George H. "Bee Culture." *Scientific American* 59, no. 2 (July 14, 1888): 21.

Heddon, James. *Success in Bee-Culture as Practiced and Advised.* Dowagiac, MI: Times, 1885.

Henry, George. *Sex Variants: A Study of Homosexual Patterns.* New York: P.B. Hoeber, 1941.

Hinman, Frank, Jr. "Advisability of Surgical Reversal of Sex in Female Pseudohermaphroditism." *Journal of the American Medical Association* 146, no. 5 (June 2, 1951): 423–29.

Hirschfeld, Magnus. *Transvestites: The Erotic Drive to Crossdress.* Translated by Michael A. Lombardi-Nash. Buffalo, NY: Prometheus Books, 1991.

Home, Everard. "On the Propagation of the Species in the Oyster." In *Supplement to the Foregoing Lectures on Comparative Anatomy.* Vol. 5. London: Longman, Rees, Orme, Brown, and Green, 1828.

Huber, M. P. *The Natural History of Ants.* Translated by J. R. Johnson. London: Longman, Hurst, Rees, Orme, and Brown, 1820.

Hunter, John. "Account of the Free Martin." *Philosophical Transactions of the Royal Society of London* 69 (1779): 279–93.

"Ignorance of the Past Ages." *American Bee Journal* 18, no. 49 (December 6, 1882).

"Instinct in Bees." *American Bee Journal* 8, no. 1 (July 1872).

International Eugenics Conference. *Report of the Second International Congress of Eugenics.* Baltimore: Williams and Wilkins, 1923.

Jackson, Thomas. *Stories About Animals.* New York: Cassel, Petter, and Galpin, [1876?].

Jennings, H. S. "Biographical Memoir of Raymond Pearl, 1879–1940." *National Academy of Sciences of the United States of America Biographical Memoirs,* Volume 22, 14th Memoir (1942).

Johnston, John. *A Description of the Nature of Four-footed Beasts.* Translated by J.P. Amsterdam. Printed for the Widow of John Jacobsen Schipper, and Stephen Swart, 1678.

Kebler, Lyman F. "Beeswax and Its Adulterants." *American Bee Journal* 36, no. 26 (June 25, 1896).

Killerman, Sam. "Genderbread Person v4.0." The Genderbread Person, last updated 2018, https://www.itspronouncedmetrosexual.com/2018/10/the-genderbread-person-v4/.

Kidder, K. P. *Kidder's Guide to Apiarian Science.* Chicago: R. Blanchard, 1858.

King, A. J. "Permanence of Bee-Keeping Industry." *American Bee Journal* 16, no. 11 (November 1880).

Kinsey, Alfred. *Methods in Biology.* Philadelphia: J.B. Lippincott, 1937.

Kinsey, Alfred. *New Introduction to Biology.* Philadelphia: J.B. Lippincott, 1938.

Kinsey, Alfred, Wardell Pomeroy, and Clyde Martin. *Sexual Behavior in the Human Male.* Philadelphia: W.B. Saunders, 1948.

Kinsey, Alfred, Wardell Pomeroy, Clyde Martin, and Paul Gebhard. *Sexual Behavior in the Human Female.* Philadelphia: W.B. Saunders, 1953.

Lambrecht, Kalman. "In Memoriam: Robert Wilson Shufeldt, 1850–1934." *Auk: A Quarterly Journal of Ornithology* 52, no. 4 (October 1935): 359–61.

Lang, Andrew. *The Red Book of Animal Stories.* New York: Longman's, Green, 1899.

Laughlin, Harry H. *The Eugenical Aspects of Deportation: Hearings Before the Committee on Immigration and Naturalization*. Washington, DC: US Government Printing Office, 1928.

Laughlin, Harry H. *Eugenics Record Office, Report No. 1*. Cold Spring Harbor, NY: Eugenics Record Office, 1913.

Laughlin, Harry H. *Report of the Committee to Study and Report on the Best Practical Means of Cutting off the Defective Germ-Plasm in the American Population*. Cold Spring Harbor, NY: Eugenics Record Office, 1914.

"Lecture on Bees." *American Bee Journal* 2, no. 9 (March 1867).

Legman, Gershon. "Minority Report on Kinsey." Preface to Norman Lockridge, *The Sexual Conduct of Men and Women*. New York: Bridgehead Books, 1956.

Leland, E. R. "Anent Ants." *Popular Science Monthly* 7 (July 1, 1875).

Lieber, Frances, ed. *Encyclopaedia Americana: A Popular Dictionary of Arts, Sciences, Literature, History Politics, and Biography*. Vol. 4. Philadelphia: Thomas, Copwerthwait, 1840.

Lillie, Frank R. "General Biological Introduction." In *Sex and Internal Secretions: A Survey of Recent Research*, 2nd ed., edited by Edgar Allen. Baltimore: Williams and Wilkins, 1939.

Lillie, Frank R. "The Theory of the Free-Martin." *Science* 43, no. 1113 (April 28, 1916): 611–13.

Littman, Enno. *Publications of the Princeton Expedition of Abyssinia*. Vol. 2. Leyden: Late E. J. Brill, 1910.

Lubbock, John. "On the Habits of Ants." *Popular Science Monthly* 11 (May 1, 1887).

Masters, William H., and Virginia E. Johnson. *Human Sexual Response*. Boston: Little, Brown, 1966.

McNamara, Brittney. "This Biology Teacher Disproved Transphobia with Science." *Teen Vogue*, March 6, 2017.

Mead, Margaret. *Male and Female: A Study of the Sexes in a Changing World*. New York: William Morrow, 1949.

Miller, C. C. "He or She: Confab Between Dr. Miller and E. E. Hasty." *Gleanings in Bee Culture* 27 (February 1, 1899).

Mississippi Fairness Act, Mississippi Senate Bill 2536 (2021). http://billstatus.ls.state.ms.us/documents/2021/html/SB/2500-2599/SB2536IN.htm.

Morgan, Thomas Henry. *Experimental Zoology*. New York: Macmillan, 1907.

"Mother of Kelly Sues Physicians." *Los Angeles Times*, May 23, 1928.

Moynihan, Daniel Patrick. *The Negro Family: The Case for National Action*. Washington, DC: US Government Printing Office, 1965.

Munn, Augustus W., and Daniel Wildman. *A Description of the Bar-and-Frame Hive*. London: John Van Voorst, 1884.

"New Gland Quiz to Be Launched." *Los Angeles Times*, May 15, 1928.

"News and Notes." *American Breeders Magazine* 1, no. 1 (1910).

"Nomenclature of Bee-Keeping." *American Bee Journal* 20, no. 5 (January 30, 1884).

Nott, John Fortune. *Wild Animals Photographed and Described*. London: Sampson Low, Marston, Searle, and Rivington, 1886.

Odell, Charles W. *The Interpretation of the Probable Error and Coefficient of Correlation*. Urbana: University of Illinois Press, 1926.

Oliver, Henry K. "'Ox-Cow' Queen Bees." *American Bee Journal* 12, no. 10 (October 1876).

"Operation Plot Is Told to Court by Ann Hewitt: Heiress Testifies She Was Tricked by Doctors in Examination." *Washington Post*, August 15, 1936.

Otis, Margaret. "A Perversion Not Commonly Noted." *Journal of Abnormal Psychology* 8, no. 2 (1913): 113–16.

"Our Six-Footed Rivals." *Popular Science Monthly* 12 (January 12, 1878): 350–58.

Pearl, Maud DeWitt, and Raymond Pearl. "On the Relation of Race Crossing to the Sex Ratio." *Biological Bulletin* 15, no. 4 (1908): 194–205.

Pearl, Raymond, and Frank M. Surface. "The Assumption of Male Secondary Characters by a Cow with Cystic Degeneration of the Ovaries." *Maine Agricultural Experiment Station Bulletin* 237 (March 1915): 63–85.

Pearl, Raymond, Alan C. Sutton, W. T. Howard Jr., and Margaret G. Rioch. "Studies on Constitution. I. Methods." *Human Biology* 1, no. 1 (January 1929): 10–56.

Phillips, Everett F. "A Review of Parthenogenesis." *Proceedings of the American Philosophical Society* 42, no. 174 (May–December 1903): 278–79.

Reid, Mayne. *The Bush-Boys, or, the History and Adventures of a Cape Farmer and His Family in the Wild Karoos of Southern Africa*. Boston: Ticknor and Fields, 1856.

Riddle, Oscar. "A Case of Complete Sex-Reversal in the Adult Pigeon." *American Naturalist* 58, no. 655 (1924): 167–81.

Riddle, Oscar. "The Control of the Sex Ratio." *Journal of the Washington Academy of Sciences* 7, no. 11 (June 4, 1917): 319–56.

Satina, Sophia, and A. F. Blakeslee. "Studies on Biochemical Differences between (+) and (–) Sexes in Mucors: 2. A Preliminary Report on the Manoilov Reaction and Other Tests." *Proceedings of the National Academy of Sciences* 12, no. 3 (March 1926): 191–96.

Satina, Sophia, and M. Demerec. "Manoilov's Reaction for Identification of the Sexes." *Science* 62, no. 1601 (September 5, 1925): 225–26.

Savory, W. S. "A Description of the Organs of Generation of a Hermaphrodite Sheep." *Medico Chirurgical Transactions* 42 (1859): 63–66.

"Second International Eugenics Conference." *Eugenical News* 5, no. 5 (May 1920).

Shaw, J. W. K. "Is Parthenogenesis Proven?" *American Bee Journal* 18, no. 3 (January 18, 1882).

Shepherd, Katie. "Marjorie Taylor Greene Blasted for Attacking Colleague's Transgender Daughter." *Washington Post*, February 25, 2021.

Sherwin, Robert Veit. "The Legal Problem in Transvestism." *American Journal of Psychotherapy* 8, no. 2 (1954): 243–24.

Shufeldt, R. W. "Biography of a Passive Pederast." *American Journal of Urology and Sexology* 13 (1917): 451–60.

Shufeldt, R. W. *Chapters on the Natural History of the United States*. New York: Studer Brothers, 1897.

Shufeldt, R. W. "The Distinction Between Anatomy and Comparative Anatomy." *Science* 7, no. 166 (April 9, 1886): 328.

Shufeldt, R. W. *Lectures on Biology, reprinted from The American Field*. Chicago, 1892.

Shufeldt, R. W. "Mammalogy—An Anatomical and Taxonomic Consideration of the Group to Which Man Belongs." *Medical Record* 98, no. 17 (October 23, 1920): 673–83.

Shufeldt, R. W. "Modesty Among the North American Indians." *Alienist and Neurologist* 36, no. 4 (November 1915): 341–49.

Shufeldt, R. W. "Mortuary Customs of the Navajo Indians." *American Naturalist* 25, no. 292 (April 1891): 303–6.

Shufeldt, R. W. "The Need of Parental Enlightenment in the Mentality and Physical Organization of Children." *Pacific Medical Journal* 54, no. 12 (December 1, 1911): 722–26.

Shufeldt, R. W. *The Negro, A Menace to American Civilization*. Boston: Richard G. Badger, 1907.

Shufeldt, R. W. "Observations on the Classification of Birds." *Proceedings of the Academy of Natural Sciences of Philadelphia* 50 (1898): 489–99.

Shufeldt, R. W. "On the Study of the Question of Sex." *Alienist and Neurologist* 39, no. 2 (April 1, 1918): 109–18.

Shufeldt, R. W. "The Suppression of the Literature of Human Topographical Anatomy in This Country." *Pacific Medical Journal* 52, no. 3 (March 1, 1909): 146–53.

Shufeldt, R. W. "A Waning of the Interest in Comparative Anatomy in this Country." *Arena* 36, no. 205 (December 1, 1906): 629–31.

Shufeldt, Robert Wilson, Sr. *Liberia: The U.S. Navy in Connection with the Foundation, Growth and Prosperity of the Republic of Liberia*. Washington: John L. Ginck, 1877.

Simpson, James Y. *Cyclopaedia of Anatomy and Physiology*. Vol. 2. London: Sherwood, Gilbert, and Piper, 1847.

Simpson, James Y. "On the Alleged Infecundity of Females Born Co-Twins with Males, with Some Notes on the Average Proportion of Marriages Without Issue in General Society." Pamphlet reprint from *Edinburgh Medical and Surgical Journal* 158 (1844).

Smith, Charles Hamilton. *Natural History of Dogs*. Vol. 2. London: Henry G. Bohn, 1865.

Society for the Diffusion of Useful Knowledge. *Penny Cyclopaedia of the Society for the Diffusion of Useful Knowledge*. Vol. 12. London: Charles Knight, 1838.

Starkweather, George B. *The Law of Sex: Being an Exposition of the Natural Law by Which the Sex of Offspring Is Controlled in Man and the Lower Animals*. London: J. and A. Churchill, 1883.

"Sterilization Case Heiress Would Be Nevada Senator." *Washington Post*, March 18, 1950.

"Tenth Annual Meeting." *Eugenical News* 7, no. 7 (July 1922): 77–83.

Terman, Lewis M. *Psychological Factors in Marital Happiness*. New York: McGraw-Hill, 1938.

Ulrichs, Karl Heinrich. 1870. *The Riddle of "Man-Manly" Love*. Translated by Michael A. Lombardi-Nash. Prometheus, 1994.

UNESCO. *Four Statements on the Race Question*. Paris: UNESCO, 1969.

"U'Ren Pressing Gland Charges." *Los Angeles Times*, May 16, 1928.

Vasey, George. *Delineation of the Ox Tribe; or, The Natural History of Bulls, Bisons, and Buffaloes*. London: G. Biggs, 1851.

Vulnerable Child Compassion and Protection Act. AL HB1 (February 2, 2021). https://www.billtrack50.com/BillDetail/1247206.

"Warden Avers Kelly Willed Body to Science." *Los Angeles Times*, May 25, 1928.

Watson, M. "On the Female Generative Organs of *Hyaena crocuta*." *Proceedings of the Zoological Society of London* 24 (1887): 369–78.

Watson, M. "On the Male Generative Organs of the *Hyaena crocuta*." *Proceedings of the Zoological Society of London* 27 (1878): 416–28.

Woolf, Philip. "Neuter Insects." *Popular Science Monthly* 15 (August 1, 1879).

"The Worker-Bee—He, She, or It." *American Bee Journal* 39, no. 14 (1899): 216.

"Workers Not Monsters." *American Bee Journal* 1, no. 2 (February 1861).

Yerkes, Robert M. "Eugenic Bearing of Measurements of Intelligence." *Eugenics Review* 14, no. 4 (1923): 225–45.

Secondary Sources

Abildgaard, Julie, et al. "Use of Antidepressants in Women After Prophylactic Bilateral Oophorectomy: A Danish National Cohort Study." *Psycho-Oncology* (November 2019): 655–62.

Adams, Vincanne, Michelle Murphy, and Adele Clarke. "Anticipation: Technoscience, Life, Affect, Temporality." *Subjectivity* 28 (2009): 246–65.

Ah-King, Malin, and Ingrid Ahnesjö. "The 'Sex Role' Concept: An Overview and Evaluation." *Evolutionary Biology* 40, no. 4 (2013): 461–70.

Allen, Garland E. "The Eugenics Record Office at Cold Spring Harbor, 1910–1940: An Essay in Institutional History." *Osiris* 2 (1986): 225–64.

Allen, Garland E. "The Misuse of Biological Hierarchies: The American Eugenics Movement, 1900–1940." *History and Philosophy of the Life Sciences* 5, no. 2 (1983): 105–28.

Allen, Garland E. "Old Wine in New Bottles: From Eugenics to Population Control in the Work of Raymond Pearl." In *The Expansion of American Biology*, edited by Keith R. Benson, Jane Maienschein, and Ronald Rainger, 231–61. New Brunswick, NJ: Rutgers University Press, 1991.

Amin, Kadji. "Glands, Eugenics, and Rejuvenation in *Man into Woman*: A Biopolitical Genealogy of Transsexuality." *TSQ: Transgender Studies Quarterly* 5, no. 4 (2018): 589–605.

Amin, Kadji. "Trans* Plasticity and the Ontology of Race and Species." *Social Text* 38, no. 2 (2020): 49–71.

Amin, Kadji. "We Are All Nonbinary: A Brief History of Accidents." *Representations* 158, no. 1 (2022): 106–19.

Aronowitz, Robert. *Risky Medicine: Our Quest to Cure Fear and Uncertainty*. Chicago: University of Chicago Press, 2015.

Bagemihl, Bruce. *Biological Exuberance: Animal Homosexuality and Natural Diversity*. New York: St. Martin's Press, 1999.

Barad, Karen. *Meeting the Universe Halfway: Quantum Physics and the Entanglement of Matter and Meaning*. Durham, NC: Duke University Press, 2007.

Bauer, Heike. *English Literary Sexology: Translations of Inversion, 1860–1930*. New York: Palgrave Macmillan, 2009.

Baynton, Douglas C. *Defectives in the Land: Disability and Immigration in the Age of Eugenics*. Chicago: University of Chicago Press, 2016.

Beauchamp, Toby. *Going Stealth: Transgender Politics and U.S. Surveillance Practices*. Durham, NC: Duke University Press, 2019.

Beccalossi, Chiara. *Female Sexual Inversion: Same-Sex Desires in Italian and British Sexology, c. 1870–1920*. New York: Palgrave Macmillan, 2012.

Bederman, Gail. *Manliness and Civilization: A Cultural History of Gender and Race in the United States, 1880–1917*. Chicago: University of Chicago Press, 2005.

Benjamin, Ruha. *Race After Technology: Abolitionist Tools for the New Jim Code*. Medford, MA: Polity Press, 2019.

Bey, Marquis. *Cistem Failure: Essays on Blackness and Cisgender*. Durham, NC: Duke University Press, 2022.

Birn, Anne-Emanuelle. "Philanthrocapitalism, Past and Present: The Rockefeller Foundation, the Gates Foundation, and the Setting(s) of the International/Global Health Agenda." *Hypothesis* 12, no. 1 (2014): e8.

Bishop, Thomas. *Every Home a Fortress: Cold War Fatherhood and the Family Fallout Shelter*. Amherst: University of Massachusetts Press, 2020.

Bix, Amy Sue. "Experiences and Voices of Eugenics Field-Workers: 'Women's Work' in Biology." *Social Studies of Science* 27, no. 4 (1997): 625–68.

Black, Riley. "Stop Trying to Out-Science Transphobes." *Slate*, March 3, 2021.

Blair, Ann. *Too Much to Know: Managing Scholarly Information Before the Modern Age*. New Haven, CT: Yale University Press, 2011.

Bloom, David A., and Barry A. Kogan. "Conversations with Frank Hinman, Jr." *Urology* 57, no. 4 (2001): 843–46.

Blue, Ethan. "The Strange Career of Leo Stanley: Remaking Manhood and Medicine at San Quentin State Penitentiary, 1913–1951." *Pacific Historical Review* 78, no. 2 (2009): 210–41.

Boag, Peter. *Re-Dressing America's Frontier Past*. Berkeley: University of California Press, 2011.

Boag, Peter. *Same-Sex Affairs: Constructing and Controlling Homosexuality in the Pacific Northwest*. Berkeley: University of California Press, 2003.

Bouk, Dan. *How Our Days Became Numbered: Risk and the Rise of the Statistical Individual*. Chicago: University of Chicago Press, 2015.

Bowker, Geoffrey C., and Susan Leigh Star. *Sorting Things Out: Classification and Its Consequences*. Cambridge, MA: MIT Press, 1999.

Bowler, Peter J. *Evolution: The History of an Idea*, 25th anniversary edition. Berkeley: University of California Press, 2009.

Briggs, Laura. "The Race of Hysteria: 'Overcivilization' and the 'Savage' Woman in Late Nineteenth-Century Obstetrics and Gynecology." *American Quarterly* 52, no. 2 (2000): 246–73.

Briggs, Laura. *Reproducing Empire: Race, Sex, Science, and U.S. Imperialism in Puerto Rico*. Berkeley: University of California Press, 2003.

Brown, Kathleen. "'Changed into the Fashion of Man': The Politics of Sexual Difference in a Seventeenth-Century Anglo-American Settlement." *Journal of the History of Sexuality* 6, no. 2 (October 1995): 171–93.

Bulmer, Ralph. "Why Is the Cassowary Not a Bird? A Problem of Zoological Taxonomy Among the Karam of the New Guinea Highlands." *Man* 2, no. 1 (March 1967): 5–25.

Burnett, D. Graham. *Trying Leviathan: The Nineteenth-Century New York Court Case That Put the Whale on Trial and Challenged the Order of Nature*. Princeton, NJ: Princeton University Press, 2007.

Butler, Frederick. "The Farmer's Manual, Including a Treatise on the Management of Bees." *North American Review* (October 1828): 338–59.

Butler, Judith. *Gender Trouble*. New York: Routledge, 1990.

Cadden, Joan. *The Meanings of Sex Difference in the Middle Ages: Medicine, Science, and Culture*. Cambridge: Cambridge University Press, 1993.

Callon, Michel. "Some Elements of a Sociology of Translation: Domestication of the Scallops and the Fishermen of St. Brieuc Bay." *Sociological Review* 32, no. 1 (1984): 196–233.

Campos, Luis. *Radium and the Secret of Life*. Chicago: University of Chicago Press, 2015.

Canaday, Margot. *The Straight State: Sexuality and Citizenship in Twentieth-Century America*. Princeton, NJ: Princeton University Press, 2009.

Carter, Julian B. *The Heart of Whiteness: Normal Sexuality and Race in America, 1880–1940*. Durham, NC: Duke University Press, 2007.

Carter, Julian B. "On Mother-Love: History, Queer Theory, and Nonlesbian Identity." *Journal of the History of Sexuality* 14, nos. 1–2 (January/April 2005): 107–38.

Chauncey, George. "From Sexual Inversion to Homosexuality: Medicine and the Changing Conceptualization of Female Deviance." *Salmagundi* 58/59 (Fall–Winter 1983): 114–46.

Chauncey, George. *Gay New York: Gender, Urban Culture, and the Making of the Gay Male World, 1890–1940*. New York: Basic Books, 1994.

Chen, Mel. *Animacies: Biopolitics, Racial Mattering, and Queer Affect*. Durham, NC: Duke University Press, 2013.

Chen, Mel. "Animals Without Genitals: Race and Transubstantiation." In *The Transgender Studies Reader 2*, edited by Susan Stryker and Aren Z. Aizura, 168–77. New York: Routledge, 2013.

Chess, Simone, Colby Gordon, and Will Fisher, eds. "Early Modern Trans Studies." Special issue, *Journal for Early Modern Cultural Studies* 19, no. 4 (Fall 2019).

Chiang, Howard. *After Eunuchs: Science, Medicine, and the Transformation of Sex in Modern China*. New York: Columbia University Press, 2018.

Chiang, Howard. "Effecting Science, Affecting Medicine: Homosexuality, the Kinsey Reports, and the Contested Boundaries of Psychopathology in the

United States, 1948–1965." *Journal of the History of the Behavioral Sciences* 44, no. 3 (2008): 300–318.

Chiang, Howard. "Liberating Sex, Knowing Desire: Scientia Sexualis and Epistemic Turning Points in the History of Sexuality." *History of the Human Sciences* 23, no. 5 (2010): 42–69.

Chu, Andrea Long. "My New Vagina Won't Make Me Happy." *New York Times*, November 24, 2018.

Clare, Stephanie D., Patrick R. Grzanka, and Joanna Wuest. "Gay Genes in the Postgenomic Era: A Roundtable." *GLQ: A Journal of Lesbian and Gay Studies* 29, no. 1 (2023): 109–28.

Clarke, Adele E. *Disciplining Reproduction: Modernity, American Life Sciences, and the "Problems of Sex."* Berkeley: University of California Press, 1998.

Clayton, Aubrey. *Bernoulli's Fallacy: Statistical Illogic and the Crisis of Modern Science.* New York: Columbia University Press, 2022.

Clement, Elizabeth, and Beans Velocci. "Modern Sexuality in Modern Times (1880s–1930s)." In *The Routledge History of Queer America*, edited by Don Romesburg, 52–67. New York: Routledge, 2018.

Cleves, Rachel Hope. *Charity and Sylvia: A Same-Sex Marriage in Early America.* New York: Oxford University Press, 2014.

Comfort, Nathaniel. *The Science of Human Perfection: How Genes Became the Heart of American Medicine.* New Haven, CT: Yale University Press, 2014.

Cornel, Tabea. "Contested Numbers: The Failed Negotiation of Objective Statistics in a Methodological Review of Kinsey et al.'s Sex Research." *History and Philosophy of the Life Sciences* 43, no. 1 (2021): 13.

Courdileone, K. A. *Manhood and American Political Culture in the Cold War.* New York: Routledge, 2004.

Creadick, Anna G. *Perfectly Average: The Pursuit of Normality in Postwar America.* Amherst: University of Massachusetts Press, 2010.

Crenshaw, Kimberlé. "Mapping the Margins: Intersectionality, Identity Politics, and Violence Against Women of Color." *Stanford Law Review* 43, no. 6 (July 1991): 1241–99.

Currah, Paisley. *Sex Is as Sex Does: Governing Transgender Identity.* New York: New York University Press, 2022.

Currah, Paisley, and Lisa Jean Moore. "We Won't Know Who You Are: Contesting Sex Designations in New York City Birth Certificates." *Hypatia: A Journal of Feminist Philosophy* 24, no. 3 (Summer 2009): 113–35.

Dame-Griff, Avery. *The Two Revolutions: A History of the Transgender Internet.* New York: New York University Press, 2023.

Daston, Lorraine. "Objectivity and the Escape from Perspective." *Social Studies of Science* 22 (1992): 597–618.

Daston, Lorraine, and Peter Galison. *Objectivity.* New York: Zone Books, 2007.

de Beauvoir, Simone. *The Second Sex.* Translated by H. M. Parshley. New York: Vintage Books, 1989.

D'Emilio, John. "Capitalism and Gay Identity." In *The Lesbian and Gay Studies Reader*, edited by Henry Abelove, Michèle Aina Barale, and David M. Halperin, 467–76. New York: Routledge, 1993.

D'Emilio, John. *Sexual Politics, Sexual Communities: The Making of a Homosexual Minority in the United States, 1940–1970*. Chicago: University of Chicago Press, 1983.

D'Emilio, John, and Estelle Freedman. *Intimate Matters: A History of Sexuality in America*, 3rd ed. Chicago: University of Chicago Press, 2012.

Desrosières, Alain. *The Politics of Large Numbers: A History of Statistical Reasoning*. Cambridge, MA: Harvard University Press, 2002.

DeVun, Leah. *The Shape of Sex: Nonbinary Gender from Genesis to the Renaissance*. New York: Columbia University Press, 2021.

Didier, Emmanuel. *America by the Numbers: Quantification, Democracy, and the Birth of National Statistics*. Cambridge, MA: MIT Press, 2020.

Dodds, Io. "'Never Ask Permission': How Two Trans Women Ran a Legendary Underground Surgical Clinic in a Rural Tractor Barn." *The Independent*, July 3, 2022.

Drager, Emmett Harsin. "Early Gender Clinics, Transsexual Etiology, and the Racialized Family." *GLQ: A Journal of Lesbian and Gay Studies* 29, no. 1 (2023): 13–26.

Dreger, Alice. *Hermaphrodites and the Medical Invention of Sex*. Cambridge, MA: Harvard University Press, 2000.

Drucker, Donna J. *The Classification of Sex: Alfred Kinsey and the Organization of Knowledge*. Pittsburgh: University of Pittsburgh Press, 2014.

Drucker, Donna J. "'A Most Interesting Chapter in the History of Science': Intellectual Responses to Alfred Kinsey's *Sexual Behavior in the Human Male*." *History of the Human Sciences* 25, no. 1 (2012): 75–98.

DuBois, L. Zachary, and Heather Shattuck-Heidorn. "Challenging the Binary: Gender/Sex and the Bio-logics of Normalcy." *American Journal of Human Biology* 33, no. 5 (2021): e23623.

Duggan, Lisa. *Sapphic Slashers: Sex, Violence, and American Modernity*. Durham, NC: Duke University Press, 2000.

Eder, Sandra. *How the Clinic Made Gender: The Medical History of a Transformative Idea*. Chicago: University of Chicago Press, 2022.

Ehrenreich, Barbara, and Deirdre English. *For Her Own Good: Two Centuries of the Experts' Advice to Women*, rev. ed. New York: Anchor Books, 2005.

Ekins, Richard. "Science, Politics and Clinical Intervention: Harry Benjamin, Transsexualism and the Problem of Heteronormativity." *Sexualities* 8, no. 3 (2005): 306–28.

Enke, A. Finn. "The Education of Little Cis: Cisgender and the Discipline of Opposing Bodies." In *The Transgender Studies Reader 2*, edited by Susan Stryker and Aren Z. Aizura, 234–47. New York: Routledge, 2013.

Farquhar, Cynthia M., Sally A. Harvey, Yi Yu, Lynn Sadler, and Alistair W. Stewart. "A Prospective Study of 3 Years of Outcomes After Hysterectomy with and Without Oophorectomy." *American Journal of Obstetrics and Gynecology* 194, no. 3 (2006): 711–17.

Fausto-Sterling, Anne. "Science Won't Settle Trans Rights." *Boston Review*, February 10, 2020.

Fausto-Sterling, Anne. *Sexing the Body: Gender Politics and the Construction of Sexuality*. New York: Basic Books: 2000.

Feimster, Crystal Nicole. *Southern Horrors: Women and the Politics of Rape and Lynching*. Cambridge, MA: Harvard University Press, 2011.

Feinberg, Leslie. *Transgender Warriors: Making History from Joan of Arc to Dennis Rodman*. Boston: Beacon Press, 1996.

Ferguson, Roderick A. *Aberrations in Black: Toward a Queer of Color Critique*. Minneapolis: University of Minnesota Press, 2003.

Fitzpatrick, Paul J. "Statistical Works in Early American Statistics Courses." *American Statistician* 10, no. 5 (1956): 14–19.

Ford, Clellan S., and Frank A. Beach. *Patterns of Sexual Behavior*. New York: Harper and Brothers, 1951.

Foucault, Michel. *Herculine Barbin: Being the Recently Discovered Memoirs of a Nineteenth-Century French Hermaphrodite*. Translated by Richard McDougall. New York: Pantheon, 1980.

Foucault, Michel. *The History of Sexuality, Volume One: An Introduction*. Translated by Robert Hurley. New York: Vintage Books, 1990.

Foucault, Michel. *The Order of Things: An Archaeology of the Human Sciences*. Translated by Alan Sheridan. New York: Vintage Books, 1994.

Frampton, Sally. *Belly-Rippers, Surgical Innovation and the Ovariotomy Controversy*. London: Palgrave Macmillan, 2018.

Freedman, Estelle B. "'The Burning of Letters Continues': Elusive Identities and the Historical Construction of Sexuality." *Journal of Women's History* 9, no. 4 (Winter 1998): 181–200.

Fryer, Roland G., Jr. "Guess Who's Been Coming to Dinner? Trends in Interracial Marriage over the 20th Century." *Journal of Economic Perspectives* 21, no. 2 (Spring 2007): 71–90.

Funk, Holger. "R. J. Gordon's Discovery of the Spotted Hyena's Extraordinary Genitalia in 1777." *Journal of the History of Biology* 45 (2012): 301–28.

Garcia-Sifuentes, Yesenia, and Donna Maney. "Reporting and Misreporting of Sex Differences in the Biological Sciences." *Elife* (2021). https://doi.org/10.7554%2FeLife.70817.

Gieryn, Thomas F. "Boundary-Work and the Demarcation of Science from Non-Science: Strains and Interests in the Professional Ideologies of Science." *American Sociological Review* 48, no. 6 (1983): 781–95.

Giffney, Noreen, and Myra J. Hird. *Queering the Non/Human*. Burlington, VT: Ashgate, 2008.

Gilbert, Nigel, and Michael Mulkay. *Opening Pandora's Box: A Sociological Analysis of Scientists' Discourse*. Cambridge: Cambridge University Press, 1984.

Gill-Peterson, Jules. *Histories of the Transgender Child*. Minneapolis: University of Minnesota Press, 2018.

Gill-Peterson, Jules. *A Short History of Trans Misogyny*. New York: Verso, 2024.

Gilman, Sander L. *Difference and Pathology: Stereotypes of Sexuality, Race, and Madness*. Ithaca, NY: Cornell University Press, 1985.

Gil-Riaño, Sebastián. "Relocating Anti-Racist Science: The 1950 UNESCO Statement on Race and Economic Development in the Global South." *British Journal of the History of Science* 51, no. 2 (2018): 281–303.

Gitelman, Lisa. *Paper Knowledge: Toward a Media History of Documents*. Durham, NC: Duke University Press, 2014.

Glick, Megan H. *Infrahumanisms: Science, Culture, and the Making of Modern Personhood*. Durham, NC: Duke University Press, 2018.

Godbeer, Richard. "'The Cry of Sodom': Discourse, Intercourse, and Desire in Colonial New England." *William and Mary Quarterly* 52, no. 2 (1995): 259–86.

Godbeer, Richard. *Sexual Revolution in Early America*. Baltimore: Johns Hopkins University Press, 2002.

Goodwin, Willard E., and Ralph R. Landes. "A Chat with Elmer Belt." *Urology* 10, no. 4 (1977): 398–402.

Gordon, Colby. *Glorious Bodies: Trans Theology and Renaissance Literature*. Chicago: University of Chicago Press, 2024.

Gould, Stephen Jay. *The Mismeasure of Man*. New York: W. W. Norton, 1981.

Groneman, Carol. "Nymphomania: The Historical Construction of Female Sexuality." *Signs: Journal of Women in Culture and Society* 19, no. 2 (Winter 1994): 337–67.

Ha, Nathan. "The Riddle of Sex: Biological Theories of Sexual Difference in the Early Twentieth Century." *Journal of the History of Biology* 44 (2011): 505–46.

Haager, Julia B. "'Sex Education's Many Sides': Eugenics and Sex Education in New York City's Progressive Reform Organizations." *Journal of the Gilded Age and Progressive Era* 21, no. 2 (2022): 74–92.

Hacking, Ian. "Making Up People." In *Reconstructing Individualism: Autonomy, Individuality, and the Self in Western Thought*, edited by Thomas C. Heller, Morton Sosna, and David E. Wellbrey, 222–36. Stanford, CA: Stanford University Press, 1985.

Hacking, Ian. *The Taming of Chance*. Cambridge: Cambridge University Press, 1990.

Haeberle, Erwin J. "The Transatlantic Commuter: An Interview with Harry Benjamin on the Occasion of His 100th Birthday." *Sexualmedizin* 14 (1985): 1–5.

Hage, J. Joris, Refaat B. Karim, and Donald R. Laub Sr. "On the Origin of Pedicled Skin Inversion Vaginoplasty: Life and Work of Dr. Georges Burou of Casablanca." *Annals of Plastic Surgery* 59, no. 6 (December 2007): 723–39.

Halberstam, Jack. *The Queer Art of Failure*. Durham, NC: Duke University Press, 2011.

Haley, Sarah. *No Mercy Here: Gender, Punishment, and the Making of Jim Crow Modernity*. Chapel Hill: University of North Carolina Press, 2016.

Halperin, David M. "How to Do the History of Male Homosexuality." *GLQ: A Journal of Lesbian and Gay Studies* 6, no. 1 (2000): 87–124.

Haraway, Donna. *Primate Visions: Gender, Race, and Nature in the World of Modern Science*. New York: Routledge, 1989.

Haraway, Donna. "Situated Knowledges: The Science Question in Feminism and the Privilege of Partial Perspective." In *Simians, Cyborgs, and Women: The Reinvention of Nature*, 183–201. New York: Routledge, 1991.

Hayward, Eva, and Jami Weinstein, eds. "Tranimalities." Special issue, *TSQ: Transgender Studies Quarterly* 2, no. 2 (May 2015).

Heaney, Emma, ed. *Feminism Against Cisness*. Durham, NC: Duke University Press, 2024.

Heaney, Emma. "Introduction: Sexual Difference Without Cisness." In *Feminism Against Cisness*, edited by Emma Heaney. Durham, NC: Duke University Press, 2024.

Heaney, Emma. *The New Woman: Literary Modernism, Queer Theory, and the Feminine Allegory*. Evanston, IL: Northwestern University Press, 2017.

Hegarty, Peter. *Gentlemen's Disagreement: Alfred Kinsey, Lewis Terman, and the Sexual Politics of Smart Men*. Chicago: University of Chicago Press, 2013.

Hicks, Mar. "Hacking the Cis-Tem." *IEEE Annals of the History of Computing* 41 (January–March 2019): 20–33.

Hicks, Mar. *Programmed Inequality: How Britain Discarded Women Technologists and Lost Its Edge in Computing*. Cambridge, MA: MIT Press, 2017.

Hird, Myra J. "Animal Transsex." In *The Transgender Studies Reader 2*, edited by Susan Stryker and Aren Z. Aizura, 156–67. New York: Routledge, 2013.

Igo, Sarah. *The Averaged American: Surveys, Citizens, and the Making of a Mass Public*. Cambridge, MA: Harvard University Press, 2008.

Irvine, Janice. *Disorders of Desire: Sexuality and Gender in Modern American Sexology*. Philadelphia: Temple University Press, 2005.

Jackson, Zakiyyah Iman. *Becoming Human: Matter and Meaning in an Antiblack World*. New York: New York University Press, 2020.

Johnson, Colin. *Just Queer Folks: Gender and Sexuality in Rural America*. Philadelphia: Temple University Press, 2013.

Johnson, David K. *The Lavender Scare: The Cold War Persecution of Gays and Lesbians in the Federal Government*. Chicago: University of Chicago Press, 2004.

Jordanova, Ludmilla. *Sexual Visions: Images of Gender in Science and Medicine Between the Eighteenth and Twentieth Centuries*. Madison: University of Wisconsin Press, 1989.

Kahan, Benjamin. *The Book of Minor Perverts: Sexology, Etiology, and the Emergences of Sexuality*. Chicago: University of Chicago Press, 2019.

Karkazis, Katrina. *Fixing Sex: Intersex, Medical Authority, and Lived Experience*. Durham, NC: Duke University Press, 2008.

Karkazis, Katrina. "The Misuses of 'Biological Sex.'" *Lancet* 394, no. 10212 (2019): 1898–99.

Karkazis, Katrina, and Rebecca M. Jordan-Young. *Testosterone: An Unauthorized Biography*. Cambridge, MA: Harvard University Press, 2019.

Katz, Jonathan Ned. *Gay/Lesbian Almanac*. New York: Harper and Row, 1983.

Kay, Lily E. *The Molecular Vision of Life: Caltech, the Rockefeller Foundation, and the Rise of the New Biology*. New York: Oxford University Press, 1993.

Kevles, Daniel J. *In the Name of Eugenics: Genetics and the Uses of Human Heredity*. New York: Alfred A. Knopf, 1985.

Keyes, Os. "Trans Science: Medicine, Happiness, and the Politics of Proof." PhD diss., University of Washington, 2024.

Kim, Claire Jean. *Dangerous Crossings: Race, Species, and Nature in a Multicultural Age*. New York: Cambridge University Press, 2015.

Kimmelman, Barbara A. "The American Breeders' Association: Genetics and Eugenics in an Agricultural Context, 1903–13." *Social Studies of Science* 13, no. 2 (1983): 163–204.

Klautke, Egbert. "'The Germans Are Beating Us at Our Own Game': American Eugenics and the German Sterilization Law of 1933." *History of the Human Sciences* 29, no. 3 (2016): 24–43.

Kline, Wendy. *Building a Better Race: Gender, Sexuality, and Eugenics from the Turn of the Century to the Baby Boom*. Berkeley: University of California Press, 2001.

Komagamine, Tomoko, Norito Kokubun, and Koichi Hirata. "Battey's Operation as a Treatment for Hysteria: A Review of a Series of Cases in the Nineteenth Century." *History of Psychiatry* 31, no. 1 (March 2020): 55–66.

Krajewski, Markus. *Paper Machines: About Cards and Catalogs, 1548–1929*. Cambridge, MA: MIT Press, 2011.

Kuhn, Thomas S. *The Structure of Scientific Revolutions*, 4th ed. Chicago: University of Chicago Press, 2012.

Kunzel, Regina. *Criminal Intimacy: Prison and the Uneven History of Modern American Sexuality*. Chicago: University of Chicago Press, 2007.

LaFleur, Greta. *The Natural History of Sexuality in Early America*. Baltimore: Johns Hopkins University Press, 2018.

LaFleur, Greta. "Precipitous Sensations: Herman Mann's *The Female Review* (1797), Botanical Sexuality, and the Challenge of Queer Historiography." *Early American Literature* 48, no. 1 (2013): 93–123.

LaFleur, Greta. "Trans Feminine Histories, Piece by Piece, or, Vernacular Print and the History of Gender." In *Feminism Against Cisness*, edited by Emma Heaney, 83–107. Durham, NC: Duke University Press, 2024.

LaFleur, Greta, Masha Raskolnikov, and Anna Kłosowska, eds. *Trans Historical: Gender Plurality Before the Modern*. Ithaca, NY: Cornell University Press, 2021.

Laqueur, Thomas. *Making Sex: Body and Gender from the Greeks to Freud*. Cambridge, MA: Harvard University Press, 1990.

Largent, Mark. *Breeding Contempt: The History of Coerced Sterilization in the United States*. New Brunswick, NJ: Rutgers University Press, 2008.

Largent, Mark A. "'Zoology of the Twentieth Century': Charles Davenport's Prediction for Biology." *Bulletin of the American Philosophical Society Library* 2 (2002): n.p.

Larson, Scott. "'Indescribable Being': Theological Performances of Genderlessness in the Society of the Publick Universal Friend, 1776–1819." *Early American Studies: An Interdisciplinary Journal* 12, no. 3 (2014): 576–600.

Larson, Scott. "Laid Open." In *Trans Historical: Gender Plurality Before the Modern*, edited by Greta LaFleur, Masha Raskolnikov, and Anna Kłosowska, 350–65. Ithaca, NY: Cornell University Press, 2021.

Latham, J. R. "Axiomatic: Constituting 'Transexuality' and Trans Sexualities in Medicine." *Sexualities* 22, nos. 1–2 (2019): 13–30.

Latham, J. R. "Making and Treating Trans Problems: The Ontological Politics of Clinical Practices." *Studies in Gender and Sexuality* 18, no. 1 (2017): 40–61.

Latour, Bruno. "Give Me a Laboratory and I Will Raise the World." In *Science Observed: Perspectives on the Social Study of Science*, edited by Karin Knorr-Cetina and Michael Mulkay, 141–70. London: Sage, 1983.

Latour, Bruno. *Science in Action: How to Follow Scientists and Engineers Through Society*. Cambridge, MA: Harvard University Press, 1987.

Latour, Bruno, and Steve Woolgar. *Laboratory Life: The Construction of Scientific Facts*. Beverly Hills, CA: Sage, 1979.

Law, John. *After Method: Mess in Social Science Research*. London: Routledge, 2004.

Lehring, Gary. *Officially Gay: The Political Construction of Sexuality by the US Military*. Philadelphia: Temple University Press, 2003.

Levine, Philippa. "Naked Truths: Bodies, Knowledge, and the Erotics of Colonial Power." *Journal of British Studies* 52, no. 1 (2013): 5–25.

Lewis, Sophie. *Enemy Feminisms: TERFs, Policewomen, and Girlbosses Against Liberation*. Chicago: Haymarket Books, 2025.

Li, Alison. *Wondrous Transformations: A Maverick Physician, the Science of Hormones, and the Birth of the Transgender Revolution*. Chapel Hill: University of North Carolina Press.

Little, Michael A., and Ralph M. Garruto. "Raymond Pearl and the Shaping of Human Biology." *Human Biology* 82, no. 1 (2010): 77–102.

Livingston, Julie. *Debility and the Moral Imagination in Botswana*. Bloomington: Indiana University Press, 2005.

Longino, Helen E. *Science as Social Knowledge: Values and Objectivity in Scientific Inquiry*. Princeton, NJ: Princeton University Press, 1990.

Longo, Lawrence D. "The Rise and Fall of Battey's Operation: A Fashion in Surgery." *Bulletin of the History of Medicine* 53, no. 2 (Summer 1979): 244–67.

Luciano, Dana, and Mel Y. Chen. "Has the Queer Ever Been Human?" *GLQ: A Journal of Lesbian and Gay Studies* 21, nos. 2–3 (2015): 183–207.

Love, Heather. "Queer." *TSQ: Transgender Studies Quarterly* 1, nos. 1–2 (2014): 172–76.

Luciano, Dana, and Mel Y. Chen, eds. "Queer Inhumanisms." Special issue, *GLQ: A Journal of Lesbian and Gay Studies* 21, nos. 2–3 (2015).

Lui, Mary Ting Li. *The Chinatown Trunk Mystery: Murder, Miscegenation, and Other Dangerous Encounters in Turn-of-the-Century New York City*. Princeton, NJ: Princeton University Press, 2005.

MacKenzie, Donald A. *Statistics in Britain, 1865–1930: The Social Construction of Scientific Knowledge*. Edinburgh: Edinburgh University Press, 1981.

Mak, Geertje. "Conflicting Heterosexualities: Hermaphroditism and the Emergence of Surgery Around 1900." *Journal of the History of Sexuality* 24, no. 3 (2015): 402–27.

Mak, Geertje. *Doubting Sex: Inscriptions, Bodies and Selves in Nineteenth-Century Hermaphrodite Case Histories*. Manchester: Manchester University Press, 2012.

Mak, Geertje. "'So We Must Go Behind Even What the Microscope Can Reveal': The Hermaphrodite's 'Self' in Medical Discourse at the Start of the Twentieth Century." *GLQ: A Journal of Lesbian and Gay Studies* 11, no. 1 (2005): 65–94.

Manion, Jen. *Female Husbands: A Trans History*. Cambridge: Cambridge University Press, 2020.

Manion, Jen. "The Queer History of Passing as a Man in Early Pennsylvania." *Pennsylvania Legacies* 16, no. 1 (Spring 2016): 6–11.

Marhoefer, Laurie. *Sex and the Weimar Republic: German Homosexual Emancipation and the Rise of the Nazis*. Toronto: University of Toronto Press, 2015.

Martin, Karin A. "Gender and Sexuality: Medical Opinion on Homosexuality, 1900–1950." *Gender and Society* 7, no. 2 (June 1993): 246–60.

Matthews, J. Rosser. *Quantification and the Quest for Medical Certainty*. Princeton, NJ: Princeton University Press, 1995.

May, Elaine Tyler. *Homeward Bound: American Families in the Cold War Era*. New York: Basic Books, 1988.

McCormick, Kevin S., Kay E. Holekamp, Laura Smale, Mary L. Weldele, Stephen E. Flickman, and Ned J. Place. "Sex Differences in Spotted Hyenas." *Cold Spring Harbor Perspectives in Biology* 14, no. 6 (2022).

McGloughlin, J. F., Kinsey Brock, Isabella Gates, Anisha Pethkar, Marcus Piattoni, Alexis Rossi, and Sara Lipshutz. "Multivariate Models of Animal Sex: Breaking Binaries Leads to a Better Understanding of Ecology and Evolution." *Integrated Comparative Biology* 63, no. 4 (2023): 891–906. https://doi.org/10.1093/icb/icad027.

McOuat, Gordon. "From Cutting Nature at Its Joints to Measuring It: New Kinds and New Kinds of People in Biology." *Studies in the History and Philosophy of Science* 32, no. 4 (December 2001): 613–43.

Meyerowitz, Joanne. *How Sex Changed: A History of Transsexuality in the United States*. Cambridge, MA: Harvard University Press, 2002.

Meyerowitz, Joanne. "Sex Research at the Borders of Gender: Transvestites, Transsexuals, and Alfred C. Kinsey." *Bulletin of the History of Medicine* 75, no. 1 (2001): 72–90.

Milam, Erika Lorraine. *Looking for a Few Good Males: Female Choice in Evolutionary Biology*. Baltimore: Johns Hopkins University Press, 2010.

Mitchinson, Wendy. "A Medical Debate in Nineteenth-Century English Canada: Ovariotomies." *Histoire Sociale / Social History* 17, no. 33 (1984): 133–47.

Mitra, Durba. *Indian Sex Life: Sexuality and the Colonial Origins of Modern Social Thought*. Princeton, NJ: Princeton University Press, 2020.

Mol, Annemarie. *The Body Multiple: Ontology in Medical Practice*. Durham, NC: Duke University Press, 2005.

Mol, Annemarie. "Who Knows What a Woman Is . . . : On the Differences and the Relations Between the Sciences." *Medicine Anthropology Theory* 2, no. 1 (2015).

Monk, Julia, Erin Giglio, Ambika Kamath, Max R. Lambert, and Caitlin E. McDonough. "An Alternative Hypothesis for the Evolution of Same-Sex Sexual Behaviour in Animals." *Nature Ecology and Evolution* 3 (2018): 1622–31.

Moore, Lisa Jean, and Adele E. Clarke. "Clitoral Conventions and Transgressions: Graphic Representations in Anatomy Texts, c1900–1991." *Feminist Studies* 21, no. 2 (1995): 255–301.

Moore, Taylor M., Henry M. Cowles, and Chitra Ramalingam, eds. "Dilemmas of Archival Objectivity." Special section, *Historical Studies in the Natural Sciences* 53, no. 1 (February 2023).

Morgan, Jennifer L. *Laboring Women: Reproduction and Gender in New World Slavery*. Philadelphia: University of Pennsylvania Press, 2004.

Morgensen, Scott Lauria. "Settler Homonationalism: Theorizing Settler Colonialism Within Queer Modernities." *GLQ: A Journal of Lesbian and Gay Studies* 16, nos. 1–2 (2010): 105–13.

Mortimer-Sandilands, Catriona, and Bruce Erickson, eds. *Queer Ecologies: Sex, Nature, Politics, and Desire*. Bloomington: Indiana University Press, 2010.

Moscucci, Ornella. *The Science of Woman: Gynaecology and Gender in England, 1800–1929*. Cambridge: Cambridge University Press, 1990.

Mumford, Kevin J. "'Lost Manhood' Found: Male Sexual Impotence and Victorian Culture in the United States." *Journal of the History of Sexuality* 3, no. 1 (1992): 33–57.

Murphy, Michelle. *The Economization of Life*. Durham, NC: Duke University Press, 2017.

Murphy, Michelle. *Sick Building Syndrome and the Problem of Uncertainty: Environmental Politics, Technoscience, and Women Workers*. Durham, NC: Duke University Press, 2006.

Musser, Amber Jamilla. "Race and the Integrity of the Line: Sexology and Representations of Pleasure." *Social Text* 39, no. 3 (2021): 17–35.

Myers, Natasha. "Molecular Embodiments and the Body-Work of Modeling in Protein Crystallography." *Social Studies of Science* 38, no. 2 (2008): 163–99.

Nelson, Alondra. *The Social Life of DNA: Race, Reparations, and Reconciliation After the Genome*. Boston: Beacon Press, 2016.

Najmabadi, Afsaneh. "Beyond the Americas: Are Gender and Sexuality Useful Categories of Analysis?" *Journal of Women's History* 18, no. 1 (2006): 11–21.

Nunn, Zavier. "Trans Liminality and the Nazi State." *Past and Present* 260, no. 1 (August 2023): 123–57.

Nuriddin, Ayah. "Engineering Uplift: Black Eugenics as Black Liberation." In *Nature Remade: Engineering Life, Envisioning Worlds*, edited by Luis A.

Campos, Michael R. Dietrich, Tiago Saraiva, and Christian C. Young, 186–202. Chicago: University of Chicago Press, 2021.

Nuriddin, Ayah. "Liberation Eugenics: African Americans and the Science of Black Freedom Struggles, 1890–1970." PhD diss., Johns Hopkins University, 2021.

N'yongo, Tavia. *The Amalgamation Waltz: Race, Performance, and the Ruses of Memory*. Minneapolis: University of Minnesota Press, 2009.

Ordover, Nancy. *American Eugenics: Race, Queer Anatomy, and the Science of Nationalism*. Minneapolis: University of Minnesota Press, 2003.

Oreskes, Naomi. "Objectivity or Heroism? On the Invisibility of Women in Science." *Osiris* 11 (1996): 87–113.

Oudshoorn, Nelly. *Beyond the Natural Body: An Archaeology of Sex Hormones*. London: Routledge, 1994.

Owens, Deirdre Cooper. *Medical Bondage: Race, Gender, and the Origins of American Gynecology*. Athens: University of Georgia Press, 2017.

Park, Katherine. "The Myth of the One-Sex Body." *Isis* 14, no. 1 (March 2023): 150–75.

Park, Katharine, and Lorraine Daston. *Wonders and the Order of Nature, 1150–1750*. Cambridge, MA: MIT Press, 1998.

Park, Katharine, and Robert A. Nye. "Destiny Is Anatomy: *Making Sex: Body and Gender from the Greeks to Freud* by Thomas Laqueur." *New Republic* (February 18, 1991): 53–57.

Pascoe, Peggy. *What Comes Naturally; Miscegenation Law and the Making of Race in America*. New York: Oxford University Press, 2009.

Patsopoulos, Nikolaos, Athina Tatsioni, and John Ioannidis. "Claims of Sex Differences: An Empirical Assessment in Genetic Associations." *Journal of the American Medical Association* 298, no. 8 (2007): 880–93.

Pauly, Philip J. *Biologists and the Promise of American Life: From Meriwether Lewis to Alfred Kinsey*. Princeton, NJ: Princeton University Press, 2002.

Person, Ethel. "Harry Benjamin: Creative Maverick." *Journal of Gay and Lesbian Mental Health* 12, no. 3 (2008): 259–75.

Pettit, Michael. "Becoming Glandular: Endocrinology, Mass Culture, and Experimental Lives in the Interwar Age." *American Historical Review* 118, no. 4 (2013): 1052–76.

Phillips, Kim M., and Barry Reay, eds. *Sexualities in History: A Reader*. New York: Routledge, 2002.

Pitts-Taylor, Victoria. "'A Slow and Unrewarding and Miserable Pause in Your Life': Waiting in Medicalized Gender Transition." *Health* 24, no. 6 (2020): 646–64.

Plowes, Nicola. "An Introduction to Eusociality." *Nature Education Knowledge* 3, no. 10 (2010).

Porter, Theodore. *Genetics in the Madhouse: The Unknown History of Human Heredity*. Princeton, NJ: Princeton University Press, 2018.

Potter, Claire Bond. "Queer Hoover: Sex, Lies, and Political History." *Journal of the History of Sexuality* 15, no. 3 (September 2006): 355–81.

Powell, Kendall, Ruth Terry, and Sophia Chen. "How LGBT+ Scientists Would Like to Be Included and Welcomed in STEM Workplaces." *Nature Career Feature* (October 19, 2020).

Pratt, Mary Louise. *Imperial Eyes: Travel Writing and Transculturation*. New York: Routledge, 1992.

Prentice, Rachel. *Bodies in Formation: An Ethnography of Anatomy and Surgery Education*. Durham, NC: Duke University Press, 2012.

Proctor, Robert N., and Londa Schiebinger, eds. *Agnotology: The Making and Unmaking of Ignorance*. Stanford, CA: Stanford University Press, 2008.

Qureshi, Sadiah. *Peoples on Parade: Exhibitions, Empire, and Anthropology in Nineteenth-Century Britain*. Chicago: University of Chicago Press, 2011.

Radin, Joanna. "Alternative Facts and States of Fear: Reality and STS in an Age of Climate Fictions." *Minerva* 57, no. 4 (2019): 411–31.

Radin, Joanna. *Life on Ice: A History of New Uses for Cold Blood*. Chicago: University of Chicago Press, 2017.

Rafter, Nicole Hahn. *White Trash: The Eugenic Family Studies, 1877-1919*. Boston: Northeastern University Press, 1988.

Reilly, Zachary P., Timothee F. Fruhauf, and Stephen J. Martin. "Barriers to Evidence-Based Transgender Care: Knowledge Gaps in Gender-Affirming Hysterectomy and Oophorectomy." *Obstetrics and Gynecology* 134, no. 4 (September 2019): 714–17.

Reis, Elizabeth. *Bodies in Doubt: An American History of Intersex*. Baltimore: Johns Hopkins University Press, 2009.

Resta, Robert G. "The Crane's Foot: The Rise of the Pedigree in Human Genetics." *Journal of Genetic Counseling* 2, no. 4 (1993): 235–60.

Resta, Robert G. "Historical Origin of Pedigree Charts." *Journal of Genetic Counseling* 3, no. 2 (November 1994): 165–67.

Reumann, Miriam J. *American Sexual Character: Sex, Gender and National Identity in the Kinsey Reports*. Berkeley: University of California Press, 2005.

Rich, Miriam. "The Curse of Civilised Woman: Race, Gender, and the Pain of Childbirth in Nineteenth-Century American Medicine." *Gender and History* 28, no. 1 (2016): 57–76.

Rich, Miriam. "Monstrosity in Medical Science: Race-Making and Teratology in the Nineteenth-Century United States." *Isis* 114, no. 3 (2023): 513–36.

Richardson, Sarah S. *Sex Itself: The Search for Male and Female in the Human Genome*. Chicago: University of Chicago Press, 2013.

Rifkin, Mark. *When Did Indians Become Straight? Kinship, the History of Sexuality, and Native Sovereignty*. New York: Oxford University Press, 2011.

Ritvo, Harriet. *The Platypus and the Mermaid and Other Figments of the Classifying Imagination*. Cambridge, MA: Harvard University Press, 1998.

Roberts, Dorothy. "Crossing Two Color Lines: Interracial Marriage and Residential Segregation in Chicago." *Capital University Law Review* 45, no. 1 (Winter 2017): 1–32.

Roberts, Dorothy. *Killing the Black Body: Race, Reproduction and the Meaning of Liberty*. New York: Vintage, 1997.

Robertson, Craig. "Granular Certainty, the Vertical Filing Cabinet, and the Transformation of Files." *Administory* 4, no. 1 (2019): 71–86.

Robin, Libby. "The Platypus Frontier: Eggs, Aborigines, and Empire in 19th Century Queensland." In *Dislocating the Frontier: Essaying the Mystique of the Outback*, edited by Deborah Bird Rose and Richard Davis, 99–120. Canberra: Australian National University Press, 2005.

Roblee, Cole, Os Keyes, Gaines Blasdel, Caleb Haley, Megan Lane, Lauren Marquette, Jessica Hsu, and William M. Kuzon Jr. "A History of Gender-Affirming Surgery at the University of Michigan: Lessons for Today." *Seminars in Plastic Surgery* 38, no. 1 (2024): 53–60.

Rocca, W. A., et al. "Increased Risk of Cognitive Impairment or Dementia in Women Who Underwent Oophorectomy Before Menopause." *Neurology* 69, no. 11 (2007): 1074–83.

Rosen, Hannah. *Terror in the Heart of Freedom: Citizenship, Sexual Violence, and the Meaning of Race in the Postemancipation South*. Chapel Hill: University of North Carolina Press, 2009.

Rosenberg, Charles. "Rationalization and Reality in the Shaping of American Agricultural Research, 1875–1914." *Social Studies of Science* 7, no. 4 (1977): 401–22.

Rosenberg, Gabriel N. *The 4-H Harvest: Sexuality and the State in Rural America*. Philadelphia: University of Pennsylvania Press, 2015.

Rosenberg, Gabriel N. "No Scrubs: Livestock Breeding, Eugenics, and the State in the Early Twentieth-Century United States." *Journal of American History* 107, no. 2 (2020): 362–87.

Roughgarden, Joan. *Evolution's Rainbow: Diversity, Gender, and Sexuality in Nature and People*. Berkeley: University of California Press, 2004.

Rubis, June Mary. "The Orang Utan Is Not An Indigenous Name: Knowing and Naming the Maias as a Decolonizing Epistemology." *Cultural Studies* 34, no. 5 (2020): 811–34.

Russett, Cynthia Eagle. *Sexual Science: The Victorian Construction of Womanhood*. Cambridge, MA: Harvard University Press, 1989.

Schaefer, Leah Cahan, and Connie Christine Wheeler. "Harry Benjamin's First Ten Cases (1938–1953): A Clinical Historical Note." *Archives of Sexual Behavior* 24, no. 1 (1995): 73–93.

Schiebinger, Londa. *Nature's Body: Gender in the Making of Modern Science*. Boston: Beacon Press, 1993.

Schiebinger, Londa. *Plants and Empire: Colonial Bioprospecting in the Atlantic World*. Cambridge, MA: Harvard University Press, 2004.

Schiebinger, Londa. "Skeletons in the Closet: The First Illustrations of the Female Skeleton in Eighteenth-Century Anatomy." *Representations* 14 (1986): 42–82.

Schoen, Johanna. *Choice and Coercion: Birth Control, Sterilization, and Abortion in Public Health and Welfare*. Chapel Hill: University of North Carolina Press, 2005.

Schott, G. D. "Sex Symbols Ancient and Modern: Their Origins and Iconography on the Pedigree." *British Medical Journal* 331 (December 24–31, 2005): 1509–10.

Schuller, Kyla. *The Biopolitics of Feeling: Race, Sex, and Science in the Nineteenth Century*. Durham, NC: Duke University Press, 2017.

Schuller, Kyla, and Jules Gill-Peterson, eds. "The Biopolitics of Plasticity." Special issue, *Social Text* 38, no. 2 (2020).

Sears, Clare. *Arresting Dress: Cross-Dressing, Law, and Fascination in Nineteenth-Century San Francisco*. Durham, NC: Duke University Press, 2015.

Sedgwick, Eve Kosofsky. *Epistemology of the Closet*. Berkeley: University of California Press, 1990.

Sedgwick, Eve Kosofsky. "How to Bring Your Kids Up Gay." *Social Text* 29 (1991): 18–27.

Selcer, Perrin. "Beyond the Cephalic Index: Negotiating Politics to Produce UNESCO's Scientific Statements on Race." *Current Anthropology* 53, no. S5 (2012): S173–84.

Sengoopta, Chandak. *The Most Secret Quintessence of Life: Glands, Sex, and Bodies, 1850–1950*. Chicago: University of Chicago Press, 2006.

Shapin, Steven, and Simon Schaffer. *Leviathan and the Air-Pump: Hobbes, Boyle, and the Experimental Life*. Princeton, NJ: Princeton University Press, 1985.

Sharpley-Whiting, T. Denean. *Black Venus: Sexualized Savages, Primal Fears, and Primitive Narratives in French*. Durham, NC: Duke University Press, 1999.

Shepard, Nikita. "To Fight for an End to Intrusions into the Sex Lives of Americans: Gay and Lesbian Resistance to Sexual Surveillance and Data Collection, 1945–1972." In *Queer Data Studies*, edited by Patrick Keilty, 49–82. Seattle: University of Washington Press, 2023.

shuster, stef. "Uncertain Expertise and the Limitations of Clinical Guidelines in Transgender Healthcare." *Journal of Health and Social Behavior* 57, no. 3 (September 2016): 319–32.

Simpson, Audra. *Mohawk Interruptus: Political Life Across the Borders of Settler States*. Durham, NC: Duke University Press, 2014.

Skidmore, Emily. *True Sex: The Lives of Trans Men at the Turn of the 20th Century*. New York: New York University Press, 2017.

Snorton, C. Riley. *Black on Both Sides: A Racial History of Trans Identity*. Minneapolis: University of Minnesota Press, 2017.

Soloway, Richard. "The 'Perfect Contraceptive': Eugenics and Birth Control Research in Britain and America in the Interwar Years." *Journal of Contemporary History* 30, no. 4 (1995): 637–64.

Somerville, Siobhan. "Scientific Racism and the Emergence of the Homosexual Body." *Journal of the History of Sexuality* 5, no. 2 (October 1994): 243–66.

Spillers, Hortense J. "Mama's Baby, Papa's Maybe: An American Grammar Book." *Diacritics* 17, no. 2 (Summer 1987): 64–81.

Star, Susan Leigh. "The Ethnography of Infrastructure." *American Behavioral Scientist* 43, no. 3 (1999): 377–91.

Star, Susan Leigh. "This Is Not a Boundary Object: Reflections on the Origin of a Concept." *Science, Technology, and Human Values* 35, no. 5 (September 2010): 601–17.

Star, Susan Leigh, and James R. Griesemer. "Institutional Ecology, 'Translations' and Boundary Objects: Amateurs and Professionals in Berkeley's Museum of Vertebrate Zoology, 1907–39." *Social Studies of Science* 19, no. 3 (August 1989): 387–420.

Stern, Alexandra Minna. *Eugenic Nation: Faults and Frontiers of Better Breeding in Modern America*. Berkeley: University of California Press, 2005.

Stigler, Stephen M. *The History of Statistics: The Measurement of Uncertainty Before 1900*. Cambridge, MA: Harvard University Press, 1990.

Stockton, Kathryn Bond. *The Queer Child, or Growing Sideways in the Twentieth Century*. Durham, NC: Duke University Press, 2009.

Stone, Sandy. "The *Empire* Strikes Back: A Posttranssexual Manifesto." In *The Transgender Studies Reader*, edited by Susan Stryker and Stephen Whittle, 221–35. New York: Routledge, 2006.

Strange, Julie-Marie. "Menstrual Fictions: Languages of Medicine and Menstruation, c. 1850–1930." *Women's History Review* 9, no. 3 (2000): 607–28.

Stryker, Susan. "My Words to Victor Frankenstein Above the Village of Chamounix: Performing Transgender Rage." *GLQ: A Journal of Lesbian and Gay Studies* 1 (1994): 237–54.

Stryker, Susan. *Transgender History*. Berkeley, CA: Seal Press, 2008.

Stryker, Susan. "Transgender History, Homonormativity, and Disciplinarity." *Radical History Review* 100 (Winter 2008): 145–57.

Studd, John. "Ovariotomy for Menstrual Madness and Premenstrual Syndrome—19th Century History and Lessons for Current Practice." *Gynecological Endocrinology* 22, no. 8 (June 2006): 411–15.

Subramaniam, Banu, and Madelaine Bartlett. "Re-Imagining Reproduction: The Queer Possibilities of Plants." *Integrative and Comparative Biology* 63, no. 4 (2023): 946–59.

Suresha, Ron Jackson. "'Properly Placed Before the Public': Publication and Translation of the Kinsey Reports." *Journal of Bisexuality* 8, nos. 3–4 (2008): 203–28.

Sutton, Katie. *Sex Between Body and Mind: Psychoanalysis and Sexology in the German-Speaking World, 1890s–1930s*. Ann Arbor: University of Michigan Press, 2019.

TallBear, Kim. "Making Love and Relations Beyond Settler Sex and Family." In *Making Kin Not Population*, edited by Adele E. Clarke and Donna Haraway, 145–64. Chicago: University of Chicago Press, 2018.

TallBear, Kim. *Native American DNA: Tribal Belonging and the False Promise of Genetic Science*. Minneapolis: University of Minnesota Press, 2013.

Terry, Jennifer. *An American Obsession: Science, Medicine, and Homosexuality in Modern Society*. Chicago: University of Chicago Press, 2002.

Terry, Jennifer. "Lesbians Under the Medical Gaze: Scientists Search for Remarkable Differences." *Journal of Sex Research* 27, no. 3 (August 1990): 317–39.

Terry, Jennifer. "'Unnatural Acts' in Nature: The Scientific Fascination with Queer Animals." *GLQ: A Journal of Lesbian and Gay Studies* 6, no. 2 (2000): 151–93.

Thakkar, Sonali. "The Reeducation of Race: From UNESCO's 1950 Statement on Race to the Postcolonial Critique of Plasticity." *Social Text* 38, no. 2 (2020): 73–96.

Todd, Zoë. "An Indigenous Feminist's Take on the Ontological Turn: 'Ontology' Is Just Another Word for Colonialism." *Journal of Historical Sociology* 29, no. 1 (2016): 4–22.

Traub, Valerie. "Psychomorphology of the Clitoris." *GLQ: A Journal of Lesbian and Gay Studies* 2 (1995): 81–113.

Traweek, Sharon. *Beamtimes and Lifetimes: The World of High Energy Physicists.* Cambridge, MA: Harvard University Press, 1988.

Valentine, David. *Imagining Transgender: An Ethnography of a Category.* Durham, NC: Duke University Press, 2007.

Velocci, Beans. "Denaturing Cisness: Trans History as Method." In *Feminism Against Cisness,* edited by Emma Heaney. Durham, NC: Duke University Press, 2024.

Velocci, Beans. "Perspective: The Battle over Trans Rights Is About Power, Not Science." *Washington Post,* October 29, 2018.

Velocci, Beans. Review of *Wondrous Transformations: A Maverick Physician, the Science of Hormones, and the Birth of the Transgender Revolution,* by Alison Li. *Journal of the History of the Behavioral Sciences* 60, no. 3 (Summer 2024).

Velocci, Beans. "Standards of Care: Uncertainty and Risk in Harry Benjamin's Transsexual Classifications." *TSQ: Transgender Studies Quarterly* 8, no. 4 (2021): 462–80.

Velocci, Beans. "These Uncertain Times." *Avidly: A Channel of the Los Angeles Review of Books,* June 4, 2020.

Velocci, Beans. "Wrenching Torque: On Being Professionally Nonbinary." *Historical Studies in the Natural Sciences* 52, no. 3 (2022): 476–84.

Vertesi, Janet. "Seeing Like a Rover: Visualization, Embodiment, and Interaction on the Mars Exploration Rover Mission." *Social Studies of Science* 42 (2012): 393–414.

Warner, John Harley. *The Therapeutic Perspective: Medical Practice, Knowledge, and Identity in America, 1820–1885.* Cambridge, MA: Harvard University Press, 1986.

Whooley, Owen. *On the Heels of Ignorance: Psychiatry and the Politics of Not Knowing.* Chicago: University of Chicago Press, 2019.

Whatcott, Jess. *Menace to the Future: A Disability and Queer History of Carceral Eugenics.* Durham, NC: Duke University Press, 2024.

Willey, Angela. *Undoing Monogamy: The Politics of Science and the Possibilities of Biology.* Durham, NC: Duke University Press, 2016.

Wilson, Anna. "Sexing the Hyena: Intraspecies Readings of the Female Phallus." *Signs: Journal of Women in Culture and Society* 28, no. 3 (Spring 2003): 755–90.

Winsor, Mary P. *Reading the Shape of Nature: Comparative Zoology at the Agassiz Museum.* Chicago: University of Chicago Press, 1991.

Wolf-Gould, Carolyn. "History of Transgender Medicine in the United States." In *The SAGE Encyclopedia of LGBTQ Studies,* edited by Abbie E. Goldberg, 508–12. Thousand Oaks, CA: SAGE, 2016.

Womack, Autumn. *The Matter of Black Living: The Aesthetic Experiment of Racial Data, 1880–1930*. Chicago: University of Chicago Press, 2022.

Wrathall, John D. "Provenance as Text: Reading the Silences Around Sexuality in Manuscript Collections." *Journal of American History* 79, no. 1 (June 1992): 165–78.

Zemenick, Ash, Shaun Turney, Alex Webster, Sarah Jones, and Marjorie Weber. "Six Principles for Embracing Gender and Sexual Diversity in Postsecondary Biology Classrooms." *BioScience* 72, no. 5 (2022): 481–92.

Index

Note: Italicized page numbers refer to figures

ants, 25, 35, 38, 45, 47–49, 227, 232
Army Medical Museum, 59
atavism, 64
Atlantic magazine, 219

Baartman, Sarah, 53, 253n.73
Banta, Arthur, 78
Barbosa, Jose Jesus, 211
Barry, Val, 191–97, 202, 207–8, 260n.154
Bateson, William, 43
Battey, Robert, 95, 109–10
Beam, Lura: books with Dickinson, 111–12, 120, 134–35, 137; and the Kinsey Reports, 147, 150, 153, 155; on womanhood, 106, 122–25. *See also Single Woman, The* (Dickinson and Beam); *Thousand Marriages, A* (Dickinson and Beam)
Beauvoir, Simone de, 10n.xiii, 124n.xviii
bees, 25, 35, 38, 45–49, 51, 56–57, 98, 160, 169, 206
Belskie, Abram, 135, 137, *139*, 140
Belt, Elmer: and fear of repercussions, 200–204, 206–10; and the HBIGDA Standards of Care, 215–19; and psychiatric evaluations, 197–200; and trans medicine, 28, 186–87, 192, 210–14, 219–20
Benjamin, Harry, *202*; early patients of, 191–96; and fear of repercussions, 200–205; and the HBIGDA Standards of Care, 214–19; and psychiatric evaluations, 197–200; relationship with Kinsey, 181, 195, 207; and the Sex Orientation Scale, 212–14; and trans medicine, 28, 186–91, 209–11, 219–20; *Transsexual Phenomenon, The*, 1–2, 4n.vi, 220; and womanhood, 205–10
Betsey (Sims patient), 108–9
binary (male and female): and animal research, 25, 34–35, 37, 40, 45, 49, 51–53, 57, 62–65; and anti-trans rhetoric, 232n.vii; and eugenics, 26, 68, 71, 73, 85, 88–92, 98–101; and gynecological research, 26, 104–7, 113, 125, 133–34, 137, 140–44; and the Kinsey Reports, 27, 145–46, 149–50, *154*, 155, 160–64, 168, 170, 180, 183; and

sex science, 1–3, 24, 222–24, 226; and trans history, 4, 6, 11–13, 17, 21; and trans medicine, 28, 187–88, 205, 207, 213; use of term, 2n.ii, 29–30
biological sex (use of term), 17–18
Biometrika, 150
biometry, 26, 74, 151
birds, 60–61, 63–64, 78, 82, 91, 127n.xxi. *See also* ornithology; pigeons
birth control, 26, 98n.xxii, 111–12, 138, 148. *See also* contraception
bisexuality. *See* universal bisexuality
B. K. (Dickinson patient), 142–44
Black men, 50, 57, 64, 178, 181, 262n27
Blackness, 50, 109, 131–33, 149, 180
Blackwell, Elizabeth, 131
Black women: and the Kinsey Reports, 178, 180n.xxxiv; medical experimentation on, 36n.v, 108–9; and sexual dimorphism, 40n.ix, 50, 63; and sexual malleability, 70n.viii, 126–27, 130, 132
Blakeslee, Albert, 26, 80–83, 89
Bloomington Asylum for the Insane (New York), 87
Bostock v. Clayton County, 5n.vii
Bowker, Geoffrey, 22, 89, 235n.16
Bowman, Karl, 197
Boyd, Samuel, 193
breasts, 2, 96, 104n.ii, 117, 121, 127, 253n.73
British Agriculture, 47
Brooklyn Institute of Arts and Sciences, 74
Bryant, Louise Stevens, 135n.xxix
bulls, 35, 41
Burnett, Graham, 45n.xi
Burou, Georges, 210–11
Butler, Judith, 10n.xiii

California, 112, 207
Canaday, Margot, 93n.xx
capitalism, 74
Carlson, Carroll, 198, 208
Carnegie Institution of Washington (CIW), 67n.iii, 73–75, 78, 81, 83, 85, 101

Carrington, C. V., 99

Carter, Julian, 112n.ix, 123–25

castration: and Dickinson's research, 121; and eugenics, 64, 94, 97; and sex science, 49, 77; and trans medicine, 191–93, 196–99, 210, 262n.27

cattle. *See* bulls; freemartins

Cell (journal), 225–26n.ii

Chase, Herbert, 142–43

Chauncey, George, 58, 234n.6

Chen, Mel, 16n.xx, 242n.66

Chiang, Howard, 148, 154n.vii; *After Eunuchs*, 5n.vi

chickens, 75, 77–78

chimpanzees, 75

chromosomes, 2, 33, 71, 76, 78–79, 205n. xvi, 223, 225, 230

cisness: and anti-trans rhetoric, 231; history of, 5–6, 9–11, 14, 19, 28, 187, 188n.v, 222–23, 235n.14; and the Kinsey Reports, 180; and knowledge production, 15–16; use of term, 5–6n.viii, 266n.i; and women, 97n.xii, 105, 124

Civil War, 59

Clarke, Adele, 36, 136, 241n.44

Clarke, Charles Walter, 147

classification: and animal research, 37, 40–41, 43–47, 49, 56–57, 60–65; and anti-trans rhetoric, 230–31; and cisgender, 6n.viii, 222–23; and eugenics, 26, 82, 84–85, 90–91, 100; and gynecological research, 106–7, 119n.xv, 123–27, 129, 133, 135n.xxix, 141; and the Kinsey Reports, 154, 168, 171; and race, 34, 50, 58, 72; and trans history, 2, 8–9, 11–12, 15–16, 19, 22, 30, 236n.36; and trans medicine, 186n.i, 188–91, 200, 203–5, 211. *See also* taxonomies

clitorises: and animal research, 39–40; and Dickinson's research, 117–19, *118*, 121–22, 126–33, 137, 140; and the Kinsey Reports, 162–63, 166

CMT (Kinsey Reports review). *See* Cochran, William; Mosteller, Frederick; Tukey, John

Cochran, William, 157n.xii, 158–59

Cold Spring Harbor, NY. *See* Eugenics Record Office (ERO); Station for Experimental Evolution (SEE)

Cold Spring Harbor Laboratory (CSHL), 66n.i, 101. *See also* Station for Experimental Evolution (SEE)

Cole, Leon J., 43

colonialism, 11, 24, 26, 32, 36n.v, 59, 62

Committee for Research in Problems of Sex (CRPS), 27, 70–71, 75n.x, 76, 83, 157, 162, 171

communism, 48, 227, 232

contraception, 36n.v, 112. *See also* birth control

Cook, Albert John, 46, 241n.51

Cooper, John M., 155–54n.ix

Cope, Edward Drinker, 63

Corner, George Washington, 162

Crenshaw, Kimberlé, 50n.xv, 69n.vii

cross-dressing, 242n.67

cross-gender, 7

Cuvier, Georges, 40, 126, 253n.73

Cyclopaedia of Anatomy and Physiology, 41

Darwin, Charles, 49–50n.xiv, 74, 172

Daston, Lorraine, 135–36, 237n.39

Davenport, Charles: and eugenics, 26, 72–75, 102, 246n.33, 248n.74; and founding of SEE and ERO, 66–72; and the Kinsey Reports, 151, 171; relationship with Benjamin, 194n.viii; relationship with Dickinson, 26, 76, 92, 112; and research at ERO, 84–93; and research at SEE, 75–83; *Statistical Methods with Special Reference to Biological Variation*, 151; and sterilization, 93–96, 110n. vi; *Trait Book*, 85–86

Davenport, Gertrude Crotty, 74

Davis, Katherine Bement, 150

de Angulo, Jaime, 99n.xxiii

"Defending Women from Gender Ideology Extremism . . ." Executive Order, 227–32

Defosse, Dana Leland, 222

D'Emilio, John, 234n.6

Denmark, 191, 195

Feinberg, Leslie: *Transgender Warriors*, 4n.vi, 6n.ix
female (use of term), 30, 104n.ii
female husbands, 7–8, 213n.xxi
femaleness: and animal research, 25, 31–35, 37–38, 47–49, 65, 251n.13; and anti-trans rhetoric, 223, 228–31; and the binary, 1–3, 6n.viii, 11–14, 23–24; and cisness, 222; and eugenics, 77–78, 80–83, 93–94, 98n.xxii, 101; and gynecological research, 27, 104–6, 108–9, 111, 113–20, 116, 122, 125–29, 134–35, 137–38; in sex science, 17–18; and statistical sexology, 145–46, 149–50, 154–55, 161–70, 182–85; and trans medicine, 28, 186–89, 204–5. *See also* binary (male and female)
femininity: and Dickinson and Beam's research, 106, 109–10, 115, 120, 122; and eugenics, 94–97; and the Kinsey Reports, 148, 160, 162; and race, 12; and trans people, 12, 196n.xii, 199, 202–6, 214. *See also* womanhood
feminism: on animal research, 32; Black trans, 69n.viii, 223; on sex and gender, 10n.xiii, 11, 18, 124, 184; STS, 15; and TERFs, 104n.iii, 224
Feminism Against Cisness (Heaney), 11
Femme Forum, The, 4n.vi
fetishism, 2, 149n.ii
Fishbein, Morris, 194
Fisher, R. A., 150
Foucault, Michel, 21–23n.xxv, 36, 164n.xviii, 234n.6
Frampton, Sally, 109n.v
France, 87
Frank, Robert T., 143
freemartins, 25, 35, 38, 41–45, 49, 52–53, 56n.xviii, 77n.xi, 98, 111
Freud, Sigmund, 153
fungi, 66, 81–82

Galison, Peter, 136
Galton, Francis, 41–42, 68n.iv, 74, 150, 240n.29, 243n.76
gamete size, 81, 100, 226n.ii, 230–31

Gay New York, 58, 243n.90
Gebhard, Paul, 146, 152n.v, 156–57, 174, 181n.xxxvi, 182–83, 255n.1
Geddes, Patrick, 54, 56, 57n.xix, 58, 160
gender (definitions of), 9–14, 30
Genderbread Person, 225
gender clinics, 12, 209n.xviii, 214, 260n.2. *See also individual clinics*
gender dysphoria, 5n.vii, 12, 190, 195, 205, 212–13, 215, 216n.xxvi
gender identity, 5, 9, 33, 191, 203, 222, 225, 228–29, 231, 234n.6
gender studies, 10n.xiv, 13
genealogy. *See* pedigree charts
genetics: in animal research, 43, 51; author's relationship with, 66n.i; and eugenics, 67, 71, 73–75, 77, 86–87, 101; and race, 175; in sex science, 26, 184, 216, 230; and universal bisexuality, 162
Germany, 24, 87, 113
Gill-Peterson, Jules, 12, 16, 21, 71, 265n.119; *Histories of the Transgender Child,* 5n.vi, 7n.x, 11n.xiv
Glick, Megan, 193n.vii
Goddard, Henry Herbert, 248n.74
Goldschmidt, Richard, 163
gonads, 43, 91, 98; transplants, 194
Goodale, Hubert Dana, 77–78, 81–83, 89
Google Ngram, 17n.xxi, 29
Great Depression, 151, 160
Griesemer, James, 20
gynecology: and Dickinson's research, 26–27, 103–13, 138, 141, 143, 222, 253n.80; and eugenics, 179; and freemartins, 42–43, 98; and sex science, 3, 23, 221; and womanhood, 35, 117, 120–23

Hamburger, Christian, 191
Harding, Lorna, 206
Harlan, Richard, 53–54, 55
Harriman, Edward Henry, 74
Harriman, Mary Williamson, 74
Harrison, Margaret, 201
Harry Benjamin International Gender Dysphoria Association (HBIGDA), 190–91; Standards of Care, 214–19

Hart, David Berry, 42–43, 45n.xviii
Hartsuiker, Frederick, 200, 207–8
Harvard University, 73; GenderSci Lab, 226n.ii
Heaney, Emily: *New Woman, The*, 4n.vi, 11n.xiv
Hegarty, Peter, 158n.xv, 184
Henry, George, 128, 150, 253n.80
hermaphroditism: and animal research, 25, 31, 38, 40–44, 46, 65, 90; and Davenport's research, 77; and Dickinson's research, 114–17, 128, 131; *versus* intersex, 34n.iv, 237n.39; and the Kinsey Reports, 162, 166; racialization of, 34, 48–58, *55*, 221; and sex science, 19. *See also* intersex
heteronormativity, 16, 231
heterosexuality: and Dickinson and Beam's research, 122, 124, 126, 128, 130, 135, 137; and eugenics, 92; and the Kinsey Reports, 145–46, 161, 163, 172n.xxv, 176, 179; and sex science, 2
Hewitt, Ann Cooper, 193
Hinman, Frank, Jr., 197
Hinman, Frank, Sr., 197
Hird, Myra, 16n.xx
Hirschfeld, Magnus, 4n.vi, 24, 153, 166, 192
history (of science), 4n.v, 14–22, 32–33, 222, 237n.39
history (of sex science), 18, 33, 65, 100, 183–84, 220, 223, 225, 237n.39
history (of sexology), 59, 237n.39
history (of sexuality), 5–6, 16, 33, 69, 145, 244n.2. *See also* trans history
Hollywood Presbyterian Hospital, 199
homosexuality: and animal research, 32, 43, 52; and Dickinson's research, 27, 103, 123, 125–26, 128–30, 133, 140; and eugenics, 92–93; and the Kinsey Reports, 145–48, 157, 160–65, 170, 176, 178–79, 184; and race, 58, 69; and sex science, 2, 10, 234n.6; and trans medicine, 209. *See also* lesbianism
hormone replacement therapies, 97n.xxii
hormones: and animal science, 43; and the Kinsey Reports, 162, 167–68; and

ovariotomies, 110–11; and sex science, 33, 73, 83, 102, 225, 231; and trans medicine, 12, 106, 143, 187, 189, 191, 195, 205, 209n.xviii, 215–16
horses, 44, 64
Human Sex Anatomy (Dickinson), 113, 117–19, *118*, 132–37
Hunter, John, 41–42, 52
Hutchens, Evie V., 186–87, 198–200, 202–4, 263n.51
Huxley, T. H., 46, 64
hyenas, 25, 35, 38–41, 45–46, 49, 53, 63, 152, 240nn.15–16
Hymenoptera (order). *See* ants; bees; sawflies; wasps
hypersexuality, 50, 180n.xxxiv
hysterectomies, 103, 121–22

immigration, 23, 32, 93n.xx
incarceration, 98, 146, 157, 165, 193, 228–29
incoherence: and animal research, 33, 35–36, 41, 46; and anti-trans rhetoric, 229–31; definition of, 19–22; and Dickinson's research, 107–8, 112, 134, 137, 140, 144; and eugenics, 68, 75, 100–101; and the Kinsey Reports, 164, 168, 170; and racial hierarchies, 57–58, 64–65; of sex, 1–3, 10, 14, 18, 23, 26–28, 221–23, 226–27; and trans medicine, 187, 189–91, 200, 202, 218, 220
Indiana University, 171, 183
Indigenous peoples, 32, 38, 50, 59–60
infantilism, 102
insects. *See* ants; bees; wasps
Institute for Sex Research, 182–83
Institut für Sexualwissenschaft, 192
interracial relationships, 176, 178–79
intersex: and animal research, 38; and cisness, 223; and Dickinson's research, 104–6, 113–21, *116*, *118*, 128–29, 132–33, 137, 141–43; and ERO's research, 92; *versus* hermaphroditism, 34n.iv, 237n.39; and the Kinsey Reports, 148, 162n.xvii, 184; and sex science, 13–14, 27; and transness, 187n.iii, 197, 231. *See also* hermaphroditism

University of Chicago, 73–74

University of Minnesota: Dight Institute for the Promotion of Human Genetics, 85, 99n.xxii, 260n.2

US Army, 59, 62

US Senate, 193

uteruses, 42–43, 104n.ii, 105, 121, 206

vaginas: and animals, 42–43; and Benjamin's research, 186; and Dickinson's research, 131–32, 137, 140, 253n.80; and Kinsey's research, 195; and sex science, 104n.ii, 126; and surgery, 108, 121, 198–99

vaginoplasties, 192, 197, 205–6, 210–11

von Krafft-Ebing, Richard, 23, 69, 153, 236, 236n.36

von Mikulicz-Radecki, Felix, 112

von Neugebauer, Frank, 117

vulvas: and Dickinson's research, 103, 113, 121, 125–34, 140; and hyenas, 39

Wagner, Stephen, 201–2, 202

Walker, F. E., 95

Walker, Mary, 131

Wallace, Rhonda, 201

wasps, 32, 45n.x, 151–52

Watson, Morrison, 38–41, 240n.16

whiteness: and animal research, 25, 31n.i; and eugenics, 112; and the Kinsey Reports, 171; and sexual dimorphism, 35, 50, 56, 131, 138; and taxonomy, 58, 62, 64–65

white supremacy: and animal research, 26, 32, 37, 57–59, 63; and eugenics, 67,

72, 140; and gynecological research, 108–13, 144; and sex science, 225

white womanhood, 26–27, 140, 144

Wilkin, Robert J., 142–43

Williams, Edith, 205

Wilson, Anna, 240n.15

Wolfe, Burton, 206

Womack, Autumn, 160

womanhood: and Dickinson and Beam's research, 104–7, 120–27, 130, 134, 140, 144, 160; and gynecological research, 27, 35, 104n.ii, 108–10, 113; and the Kinsey Reports, 168; and sterilization, 95, 97n.xxii; and transness, 12, 205, 207. *See also* femininity

women (use of term), 30, 104n.ii

Worden, Frederic, 209

worker insects. *See* ants; bees

World Professional Association for Transgender Health (WPATH). *See* Harry Benjamin International Gender Dysphoria Association (HBIGDA), Standards of Care

World War I, 59–60

World War II, 148

Yerkes, Robert Mearns, 70, 171

Young, Hugh Hampton, 197

zoology: and agriculture, 44; and Davenport's research, 73; and hermaphroditism, 52; and Kinsey's research, 166; and sex science, 3, 23–25, 31, 36, 91, 111; and Shufeldt's research, 59; and taxonomies, 39

www.ingramcontent.com/pod-product-compliance
Lightning Source LLC
Chambersburg PA
CBHW032343280326
41935CB00008B/436